GREEN WORLD GREEN ENERGY

GREEN WORLD GREEN ENERGY

A Search for a New Way of Living

M.K. Ghosh Roy

GYAN

Green World Green Energy
A Search for a New Way of Living

ISBN : 978-81-212-1168-0

Published in 2013 in India by
Gyan Publishing House
23, Main Ansari Road, Daryaganj,
New Delhi - 110002
Phones : 23282060, 23261060
Fax : (011) 23285914
E-mail: books@gyanbooks.com
website: gyanbooks.com

Laser Type Setting : Rajender Vashist, Delhi
Printed at : Chawla Offset Printers, Delhi

"To the families of my
daughter and granddaughter
of the new generation
Rina and **Vijay Kumar,**
Anjali and **Janardhan Raj Urs**"

Contents

Preface 9

PART I
CONTOURS OF A GREEN WORLD

1. Green Living – A New Beginning 19
2. Technology Solution Towards Green Sustainability : Some Indian Resource Case Studies 36
3. Food Security, World Hunger and Millennium Development Goals 49
4. Green Agriculture and Organic Food 62
5. Sustainability of Global Fisheries Resources 81
6. Conserving Forests Resources for Ecological Sustainability and Mitigation of Global Warming 93
7. Green Corporate, Issues of Sustainability and Governance 110
8. Water Security for Healthy Living 125
9. Global Warming and Ecological Sustainability 165
10. Green Energy for Security, Sustainability and Prosperity 204

PART II
GLOBAL BASKET OF VIABLE GREEN ENERGY

11. An outline of Proven Clean Fossil Energy: Clean Coal and Gas Power in New Century 229
12. Hydropower: Reliable Non-Fossil Power 256
13. Nuclear Energy and Aspects of Safety and Viability 262

14.	Renewable Green Bio-Energy: Ethanol and Bio-Diesel	289
15.	In Search of Viable Green Energy: Solar Energy	307
16.	Wind: Most Widely Used Green Energy	319
17.	Futuristic Green Energy – Hydrogen Energy	333
	Appendix	344
	Index	369

Preface

Man's Adversarial Tryst with Bounteous Nature

Nature is resplendent with the majestic mountains, boundless oceans, deep forests, endless greeneries, rivers, lakes, waterways, azure sky, glowing sunrise and lingering sunset, birds and animals and no less the humans. As long as the human life evolved around that nature, both the humans and the nature flourished in peace and harmony. But the man's tryst with nature turned soon adversarial. In its civilisational quest, man took to rewriting the natural laws and overexploiting the natural resources. Over millions of years, nature had fossilized bio-remnants into precious fuels – coals (called black diamonds) and petroleum (called liquid gold). Man needed huge energy to drive the new ways of life – to run the cars, buses, trains, ship, airplanes, to drive the industry and to lighten the living and working spaces. Thus, these precious natural fuels got consumed on a gargantuan scale. Resultantly, man polluted the air burning the fossil fuel and contaminated the water with the process effluents and brutalized the earth with wanton mining and unending constructions. Today, the world is suddenly confronted with the unprecedented environment and energy crises. As we know, there is today worldwide activism to resuscitate the ailing earth under UN aegis and otherwise for a new way of living.

Recipe for Sustainability and Survival

When the man has supposedly reached the pinnacle of success and achievement, the world has come to realise the Gandhian dictum that the earth has enough for everybody but not enough for everybody's greed.

There is great affluence and wasteful abundance in the northern America and Europe that coexist with rampant penury and physical affliction in the large areas of South America, Asia and Africa. That exposes the inconvenient truth that the much glorified man's progress is flawed and unsustainable.

There is indeed much technological excellence and industrial development, but at great cost by way of overexploitation of scarce resources and energy. The industrial raw materials and energy resources are not unlimited. Fossil fuel shall exhaust in the not too distant future; the minerals are also exhausting fast to scarcity; water is getting polluted and depleted to scarcity, unless it is used economically, reprocessed and regenerated. Thus, again we see that the present much enamoured human progress model is unsustainable.

Brundtland Report on Sustainable Development

The realisation has descended upon the major part of humanity that the world must search for a sustainable model of living and development. The public acclamation of such a proposition was first found in the UN Report authored by Groh Harlem Brundtland ("Our Common Heritage", WCED 1987) which inter alia enunciated:

> "Sustainable Development is the development that meets the needs of the present without compromising the ability of the future generation to meet their needs."

The above Brundtland Report created a profound effect on the concept of economic development globally. It had for the first time highlighted two pioneering concepts.

Needs of the Poor: "The concept of needs, in particular the essential needs of the world's poor, to which overriding priority should be given"; and

Carrying Capacity of Environment: "The idea of limitations, imposed by the state of technology and social organisation that acts on the environment's ability to meet the present and the future needs". (WCED 1987:43)

Earth Summit at Rio

Influenced by the highly revealing recommendations of the Brundtland Report, the UN General Assembly called for the UN Conference on Environment and Development (UNCED) at Rio de Janeiro in 1992 (popularly known as Earth Summit).

After prolonged deliberations by over a hundred participating nations, the Earth Summit arrived, inter alia, at the following major international agreements:

- Very well-known Agenda 21 – a very comprehensive set of action programmes towards sustainable development.
- Rio Declaration of a series of twenty seven Principles defining the rights and responsibilities of individual Party countries towards sustainability of Environment and Development;

Besides the above Agenda 21 and the Rio Declaration, following two very important Conventions were signed at the Earth Summit:

- Framework of the Convention on Climate Change signed by 153 countries (UNFCCC);
- Convention on Biological Diversity, signed by 153 countries.

Millennium Development Goals

At the advent of the 21st Century, the United Nations declared the Millennium Development Goals (MDGs) setting specific targets of poverty alleviations, health and sanitation, literacy and women's empowerment of the Third World. There are eight main goals, under which there are eighteen numbers of targets to be achieved by 2015. The MDGs have been more elaborately described in Chapter 3.

Concept of Green Living and Green Resources

The latest abiding living philosophy is to restrict the world consumptive capacity to the natural renewable or regenerative capacity of the earth, the so called *Green Limit*; and such way of life may be called the *Green Living*.

That implies that all resources – raw materials, minerals and energy should be used with utmost economy and efficiency and not wasted away. Any waste that is a process by-product should be appraised as to its further use with recycling or direct reuse. This is the first doctrine of Sustainable living. Secondly, for all exhaustible resources like energy fuel, it should be forthwith reviewed if the industry and utilities can switch over to natural energy sources like biomass, solar and wind, etc which are inexhaustible or renewable and well-known today as Green energy for they do not emit polluting gases; for biomass renews itself, solar and wind energies are inexhaustible; this is possibly the next principle of sustainable green thinking. For water, what we use, we must regenerate about the same amount of potable fresh water, for available fresh water is very scarce commodity, comprising less than 1 per cent of the gross global water capacity. For agriculture, the method adopted should not harm the earth ecology or the mother soil; we should go for *'organic farming'* using organic manure in preference to chemical fertilizer and insecticides to avoid soil degradation; to prevent the adverse climatic change, forest cover shall be restored and new forestry be developed. These are all part of the green philosophy, to think green and live green.

Green Technology

Now the challenge we face is how to make the present world order sustainable. Here is the imperative of green technology. Earth grows annually enough agricultural produces, fish and other food products, which can be further augmented by application of appropriate technology. As long as the people of the earth live within the available natural resources, there will not be any shortage; these natural resources are renewable every year. Similarly, the global system can be supplied with enough natural energy resources from sun, wind and other natural energy sources, which are renewable every year; if the global demand for energy is kept within these limits, these energy sources remain inexhaustible. However, the renewable energy could be greatly augmented many fold by application of newly developing Green Technologies.

Further these energies do not produce any harmful emission, the so-called green house gases (GHG), and the earth atmosphere will remain pollution free and the global warming will not occur. Living within the earth's natural carrying capacity is what we termed earlier as *Green Living*.

The technologies that are required to enable us to live with these green resources within the earth's carrying capacities are what we mean as *Green Technologies*. Gradually green technologies have to be developed and expanded to cover the entire gamut of our living on this planet.

Climate Change and Green Energy

The earth is receiving the heat energy from the solar radiation all the time. At the same time the earth is radiating away heat to the open universe. So assuming heat balance, the earth should remain at the same temperature.

Unfortunately, the earth is covered with a blanket of gaseous pollution consisting of carbon dioxide, methane, nitrous oxide and CFC/HFC which allows the short wave solar radiation to pass but entraps the longer wave heat emission from the earth; the entrapped heat has been slowly warming up the earth since the beginning of industrial age. As this entrapment of heat is similar to the Greenhouse phenomenon, this class of gases are also known as Greenhouse Gases.

The UN-sponsored Inter-Governmental Panel on Climate Change (IPCC) has now confirmed that the earth temperature has already risen by 0.7°C since the pre-industrial era. Such rise in temperature is affecting climate very seriously with change in rain pattern, occurrence of more intense drought, storm and flood and last and not the least the rise in the sea-level; consequently the global agriculture and economy in general will be grievously affected.

The worst finding by IPCC and the international scientific community that the carbon dioxide, which is the greatest contributor to global warming, is anthropogenic in character, i.e., the carbon dioxide is being generated by human activities, particularly by the combustion of fossil fuel which are hydrocarbons like high CO_2- producing coal, petroleum and less CO_2-producing petro-gases like LPG, Natural Gas (NG) and Compressed Natural Gas (CNG).

All the giant power plants of the world burn mostly coal and millions of motor-transports burns fossil oil. They are the principal emitters of Greenhouse gases polluting the environment.

The hunts are therefore for non-carbon fuels or so-called Green Energies, which provide energy without generating greenhouse gases that is without causing global warming.

Many Facets of Green World

Now we like to say a few words about the organisation of the book. We have overviewed the various facets of the conceptual model of our Green World in Part I of the book which contains ten Chapters. While in the Chapter 1 we give a general introduction to the Green World, in the later chapters we gradually unfold the contour of the Green World through the green technology, food security, green agriculture, world fisheries, water security; we reflect upon the environment challenges of the green world through forest degradation, global warming and energy security.

It is the overall objective of the benign green movement to prevent human hunger. Food security for all humans inside and across nations over Asia and Africa and perhaps beyond is the absolute prerequisite. This concern is very well reflected in the UN Millennium Development Goals, including its commendable emphasis on child malnutrition. Global warming and adverse climate, already mentioned, affect soil productivity and so is the adverse long-term effect of chemical fertilizers. Thus, the green agriculture is important to enhance the crop production to ensure food security.

The corporates fulfill many revolutionary roles in the modern society creating wide-ranging services and manufactured goods and finally gainful employment. We witness the transformation of business into Green corporates to complement the growth of the Green world in Chapter 7.

There are many other public services like enlightened education system and humanitarian health services which go to nourish the green world, which we have not been able to cover in this compendium; the task has to be hopefully taken up in future.

Global Basket of Green Energy

We have devoted a full Part (Part II of this Treatise) offering a series of energy options for the incoming green world.

The coal is the dominant fuel for large power plants, the fuel that is readily available till alternative non-carbon fuel source is technologically and commercially established. Hence to meet the growing needs of power, particularly in the emerging countries, clean coal technology is being increasingly adopted, which promises to reach the thermal efficiency to about 45 per cent with Supercritical Evaporation technology rising from the 25 per cent efficiency level of old boilers. Hence its carbon emission per unit energy output is reduced quite substantially. Further to contain and control the emission of carbon dioxide, CCS (carbon capture and Sequestration) technology is under development which separates out the CO_2 from the emission gases and compresses and stores it in the various voids and caverns underground.

Secondly, Natural Gas (NG) which, of all the fossil fuels, has the least carbon content at 15.3 kg/GJ compared to 26.8 of coal and 20.0 of oil. Besides, natural gas has wide and efficient applications as transport energy, domestic heating and also as power plant energy. Hence pending the evolution of universal non-carbon energy, both clean coal and natural gas supported by CNG/LNG have yet to play important roles for at least two ensuing decades.

The Hydropower and Nuclear power, the two prominent well-developed non-carbon energy sources, are already fulfilling a large part of the present global power needs. But the hydropower is capital intensive and its availability is likely to fall short of the world's needs. Nuclear energy is also capital intensive and it poses safety hazards as experienced lately by tsunami-hit Fukushima nuclear plant. In spite of the repeated assertion of the majority scientific community of its tremendous progress in nuclear safety, the public fear persists at the national and international levels. Though for large countries like China and India, the nuclear energy can meet a large part of its enormous future power needs.

Next we have the basket of renewable sources like biomass, bio-ethanol and bio-diesel and wind and solar energies which are now under intense scrutiny for large scale power applications. The world is looking forward to them to save the earth of the catastrophe of the intractable problems of carbon emissions and the consequent global warming. There would be really two huge beneficial contributions of the renewable energy sources. Firstly, they are inexhaustible sources of energy as long as total demand is within the earth's capacity to renew them over a crop season or year. Secondly, they do not contribute to the global stock of carbon emission. The bio-fuels do emit carbon dioxide, but they are reabsorbed back into the biomasses that are regenerated continuously. Therefore there is no cumulative carbon accumulation in the atmosphere.

We will now shift our attention to a futuristic non-carbon fuel Hydrogen which has immense potentiality. As and when the hydrogen energy technology is fully developed with proven economic and commercial viability, it will impact the world in such a fundamental way that a whole new Hydrogen Economy will open up.

Hydrogen is enormously energy-rich, one kilogram of hydrogen contains 120 MJ of energy compared to 45 MJ of gasoline or petrol. The problem with hydrogen is that it is the lightest material that exists and hence storage of high volume hydrogen raises yet unresolved logistic problem. Secondly, though an unlimited amount of hydrogen is available in the ocean waters, hydrogen does not exist in free form; extracting hydrogen from water requires energy expenditure which makes it costly; further net energy recovery from hydrogen goes down substantially; thus the economic viability is affected too.

Readership

This book may serve as a sequel to the author's earlier compendium *Sustainable Development - Environment, Energy and Water Resources* (Ane-India CRC Press USA). But the present book is naturally more focused on problems of sustainability concerning securities of food, energy, water and seeks viable solution spaces in the light of the Green philosophy unfolding over the crisis-ridden global milieu. The book will thus offer insights and new ideas to the Graduate students and researchers in wide range of topics of social, political, environmental and energy-related applications.

In undertaking to write this book, I have drawn heavily on the pioneering investigative works of many eminent UN and other academic institutions and scientists, professors and social activists, to all of whom I am greatly indebted. I have tried to acknowledge their very valuable contributions at the appropriate places and I apologise in advance to those whom I might have missed to mention and acknowledge.

M.K. Ghosh Roy

PART I
CONTOURS OF A GREEN WORLD

1

Green Living–A New Beginning

I

Concept of Green Living

At the peak of the technological excellence, the world has now come to realise the Gandhian dictum that the earth has enough for everybody but not enough for everybody's greed. In the present system there is too much overexploitation of energy and other resources, there is too much wastage, albeit a large percentage of the people are surviving in abject poverty and penury in Asia and Africa. But the today's industrial raw materials and energy resources are not unlimited. Fossil fuel petroleum shall exhaust in the not too distant future; the minerals are also exhausting fast to scarcity; water is getting polluted and depleted towards acute scarcity unless water is reprocessed and regenerated.

In the face of engulfing catastrophe, the philosophy of 'Green Thinking' has now descended upon the saner and wiser part of the humanity and its truism is now gradually enlightening the rest very fast. That implies that all resources – raw materials, minerals and energy should be used with utmost economy and efficiency and not wasted away. Any waste that is a process by-product should be appraised as to its further use with recycling or direct reuse. This is the first doctrine of sustainable living. Secondly, for all exhaustible resources like energy fuel, it should be forthwith reviewed if the industry and utilities can switch over to natural energy sources like biomass, solar and wind energy which are inexhaustible or renewable and well-known today as Green energy, for they do not emit polluting gases; for biomass renews itself, solar and wind energies are inexhaustible; this is possibly the next principle of sustainable green thinking. For water, what we use, we must regenerate at least the same amount of potable fresh water, for available fresh water is very scarce commodity, comprising less than 1 per cent of the gross global water capacity. For agriculture, the method adopted should not harm the earth ecology or the mother soil; we should go for *'organic farming'* using organic manure in preference to chemical fertilizer and insecticides; to prevent the adverse climatic change, forest cover shall be restored and new forestry be developed. These are all part of the green philosophy, to think green and live green.

Now following the principles of green thinking, new technologies will truly need to be developed urgently to economically generate green energy, to produce organic food in enough quantity for the growing population and to keep them in good health.

Green Technology

The present way of human civilisation particularly in the western part of the world is said to be unsustainable. For it is consuming energy and material resource at such a fast rate that in course of few decades, the fossil energy like petroleum will become scarce or be exhausted. In the same way, some of the valuable minerals like iron ore, aluminum, copper, etc will gradually become scarce. The world has only a limited amount of these resources and hence theoretically these resources are exhaustible over a period of time.

Furthermore, because of the gargantuan scale of consumption of materials and energy, the global atmosphere is getting highly polluted so much so that the earth is getting heated up. In turn such global warming is leading to change of climate affecting agriculture productivity, likely to cause draught, storm and floods and the apprehension of vast coastal zones and island states getting submerged by rising sea levels.

Now the challenge we face is how to make the present world order sustainable. Here is the imperative of green technology. Earth grows annually enough agricultural produce, fish and other food products, which can be further augmented by application of appropriate technology; as long as the people of the earth live within these natural resources, there will not be any shortage; these natural resources are renewable every year. Similarly, the global system can be supplied with enough natural energy resources from sun, wind and other natural energy sources, which are renewable every year; if the global demand for energy is kept within these limits, these energy sources remain inexhaustible. However, the renewable energy could be greatly augmented many fold by application of newly developing Green Technology. Further, these energies do not produce any harmful emission, the so-called Green House Gases (GHG), and the earth atmosphere will remain pollution free and the global warming will not occur. Living within the earth's natural carrying capacity is what we term as Green Living.

The technologies that are required to enable us to live with these green resources within the earth's carrying capacities are what we mean as Green Technologies. The philosophy of life that encourages us to seek Green living is what we call Green Thinking.

In time to come, green thinking has to become truly a part of life that is environment friendly in all its implications. No part of our living system shall give rise to pollution that can contaminate and harm the earth's ecosystem. Natural water must be kept contamination-free, pure and usable for all purposes. Land-use and agriculture shall not degrade the land.

Gradually green technologies have to be developed to cover the entire gamut of our living on this planet. The extent and range of these technologies may comprise the application lists in Box 1.1.

Box 1.1 Important Areas of Green Technologies

Air, Atmosphere, and Weather	Nature and Ecosystems
Business and Economics	Policy, Law and Government
Consumer Products	Pollution and Toxins
Energy and Fuels	Recreation and Travel
Food and Agriculture	Recycling and Waste
Global Warming and Climate Change	Science and Technology
Homes and Buildings	Sustainability and Development
Human Health and Welfare	Transportation
Humanities and Education	Water, Oceans and Ice

Source: Based on Echo-world

II

Green Home, Green Building

Reference: (Mike Valenti www.EnviroCitizen.org) 18 January 2011

For Green living, our homestead shall also be enriched with green thinking. Here comes the relevance of Green Building which enshrines green technology.

The concept of green building involves and integrates a variety of strategies during the design, construction and operation of building projects. The green building materials and products represent one of the important strategies in the design of a building. Such use of green building materials and products results in dual benefits; firstly it means saving of non-renewable and hence exhaustible materials and resources. It also means the avoidance of environmental impacts arising from the mining, extraction, transport, processing, fabrication and disposal of the conventional exhaustible building materials. Going deeper into the philosophy of what constitutes the basic criteria of green building; we may identify the following hallmarks of a green building:

Resource Efficiency;
Energy Use Efficiency;
Water Consumption Efficiency;
Indoor air quality enhancement;
Enhancement of Aesthetics Value;
Preference for Green materials and energies;
Financial Saving in general.

Resource efficiency is achievable in several ways; by using materials which have some recycled components; such materials will have been manufactured with less expenditure of energy. There are also some classes of materials which are particularly sustainable, such as green wood from sustainable forest and certified as such by an approved agency like Rainforest Alliance. In both the above cases, the exhaustible materials are saved and also secondarily the environmental impacts are lessened. Therefore, products manufactured with resource-efficient processes require reduced energy consumption, minimal wastes and reduced greenhouse gas emission. Locally available materials reduce transportation and energy usage and hence raise the resource efficiency.

Indoor air quality is enhanced by utilising materials that do not produce carcinogens or irritants or volatile organic compounds. Similarly, moisture-resistant materials inhibit growth of biological contaminants and reduce the pollution of indoor air quality. Poorly ventilated spaces in building are likely to have poor air quality.

The US Green Building Council has established a set of comprehensive Rules called LEED (Leadership in Energy Efficient Design). A building, that has been designed complying with LEED and built under their inspection, is duly certified by the USGBC. The highlights of the USGBC Rules LEED 2009 are shown in Box 1.2.

Box 1.2: USGBC
US Green Building Council

LEED 2009 introduces new requirements:

- Improved Energy Performance
- Water Usage Reduction
- Green Building Certification Institute will now oversee the Certification Process
- Building Performance Partnership: will now collect data from certified building to ensure performance with design
- USGBC now proposes updating of codes every 2 or 3 years

The Chicago Center for Green Technology (see Box 1.3) educates public on green technologies.

Box 1.3: Chicago Center for Green Technology

The Chicago Center for Green Technology is a Green Building with Platinum LEED status, which offers educational tours on daily basis. There are also community workshops put on by the center, which give examples of how some environment friendly technologies can be incorporated into residential homes (see contents below):

1. History	3. Sustainable Features	5. References
2. Renovation	4. Community Involvement	

III

Green Energy

The fossil fuels like coal and petroleum when combusted emit greenhouse gases (GHG) consisting of principally carbon dioxide and nitrous oxide, which then reside in the atmosphere for long years. Like the heat entrapment in greenhouse for horticulture in winter, these gases are called greenhouse gases as they prevent the infrared heat radiation from the earth to escape to the outer space and gradually heats up the earth's environment. The resultant global warming has raised the average temperature by about 0.7°C since the industrialisation started and it is apprehended that the same may go up by about 3°C by the turn of the 21st century. Scientists, supported by IPCC (Inter-Governmental Panel on Climate Change) studies believe that such abrupt rise of the global average temperature will very seriously change the climate, shifting the climatic zones, prolonging summer, intensifying drought, increasing rain, floods and storms. Thus, it is necessary to move away from the fossil fuels in the long run, though they may continue to exhaust GHG in the next few decades, hopefully at lesser scale.

The worldwide search is on alternative energy sources which will not emit greenhouse gases. The most common renewable energy sources are solar and wind energies, which are inexhaustible and the technology for their development is being intensively pursued. This energy is also called Green Energy, for they do not emit any GHG.

Two of the commonest forms of green energy are solar energy and wind energy. The technology for their electrical power generation utility is not yet fully developed reaching commercial viability, but worldwide research and development activity has been accelerated. Electrical power generation from solar energy can follow two technological routes – Thermal and Photovoltaic. In thermal process, steam is generated from solar heat radiation and a steam turbine drives an electrical generator. In the photovoltaic process, a solar cell, sensitive to sun radiation, directly generates low voltage current, which is then converted to grid quality power in a utility. In Fig. 1.1, we show a Solar Tower where water is heated up by concentrated solar energy reflected by multiple mirrors. Solar power will be dealt in further details in a later Chapter.

Fig. 1.1. PS10 Solar Tower Plant near Seville, Spain. Mirrors concentrate the sun's power on to a central tower Receiver, generating steam driving a steam turbine Green Energy- Fuel from Solar Radiation

Here in this introductory chapter we present a very novel form of solar energy which is in the embryonic stage of development. This process mimics the plant photosynthesis, that is, the way plant generates energy in the form of sugar from water and carbon dioxide in the presence of sunlight. The scheme is shown in Fig. 1.2.

In the Fig. 1.2, sunlight heats a ceria cylinder which breaks down water or carbon dioxide. The machine uses the Sun's rays and the metal oxide ceria to break down carbon dioxide or water into carbon monoxide and hydrogen, which can be stored and transported. In contrast, the conventional photovoltaic panels typically generate the electricity they generate *in situ*, and cannot deliver power at night. Further details are published in the journal *Science Reporter*.

The prototype, which was devised by researchers in the US and Switzerland, uses a quartz window and cavity to concentrate sunlight into a cylinder lined with cerium oxide, also known as ceria. Ceria has a natural propensity to exhale oxygen as it heats up and inhale it as it cools down. If as in the prototype, water and/or carbon dioxide are pumped into the vessel, the ceria will rapidly strip the oxygen from them as it cools, creating hydrogen and/or carbon monoxide.

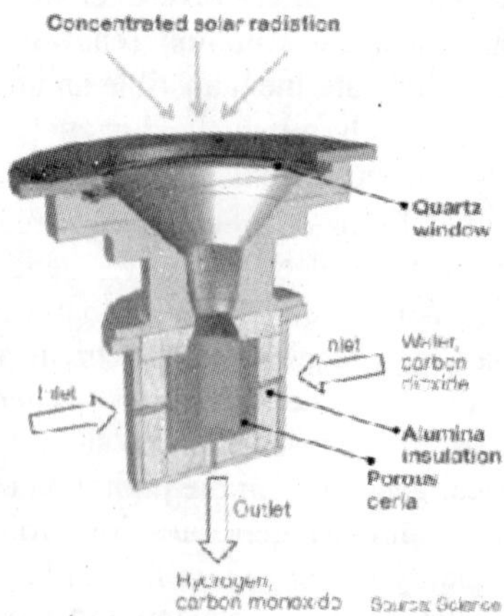

Fig. 1.2. Solar Energy: Direct Photosynthesis of Bio-Fuel
Source: Neil Bowdler in *Science Reporter* BBC

Hydrogen produced could be used to fuel hydrogen fuel cells in cars, for example, while a combination of hydrogen and carbon monoxide can be used to create "syngas" for fuel. It is this harnessing the properties of ceria in the solar reactor which represents the major breakthrough, say the inventors of the device. They also say the metal cerium is readily available, being the most abundant of the "rare-earth" metals. Methane can be produced using the same machine, they say.

The prototype is grossly inefficient, the fuel is created harnessing only between 0.7% and 0.8% of the solar energy taken into the vessel. Most of the energy is lost through heat loss through the reactor's wall or through the re-radiation of sunlight back through the device's aperture. But the researchers are confident that efficiency rates of up to 19% can be achieved through better insulation and smaller apertures. Such efficiency rates, they say, could make for a viable commercial device. The chemistry of the material is really well suited to this process, says Professor Sossina Haile of the California Institute of Technology (Caltech). She says, "the reactor could be used to create transportation fuels or be adopted in large-scale energy plants, where solar-sourced power could be available throughout the day and night."

Daniel Davies, chief technology officer at the British photovoltaic company Solar Century, said, "the research was very exciting. I guess the question is where you locate it – would you put your solar collector on a roof or would it be better off as a big industrial concern in the Sahara and then shipping the liquid fuel?" he said. Solar technology is moving forward apace but the overriding challenges remain ones of efficiency, economy and storage. New-generation solar tower plants have been built in Spain and the United States which use an array of mirrors to concentrate sunlight onto tower-mounted receivers which drive steam turbines. A new Spanish project will use molten salts to store heat from the Sun for up to 15 hours, so that the plant could potentially operate through the night.

Wind Energy

Wind Energy is one of the largest sources of renewable energy globally today. In the later chapter on wind energy, further details will be furnished. In Fig. 1.3, we show a large wind turbine blade under installation in Hebei Province of China. China has announced to increase wind power from 20 gigawatts in 2009 to 90 gigawatts. This is great jump in capacity and foreign firms are up for orders of wind turbines.

In Fig. 1.4 we show a very small wind turbine under development by Wind-Smile company of Japan, quiet, safe and compact, Wind-Smile generators are easy to install and maintain. Their design has been optimised as to minimize impact on the surrounding landscapes. The Wind Smile Generator is an independent-type general purpose small wind-turbine.

Fig. 1.3. Workers paint wind turbine blades at a factory in China's Hebei province. Photograph: Alexander F. Yuan/AP

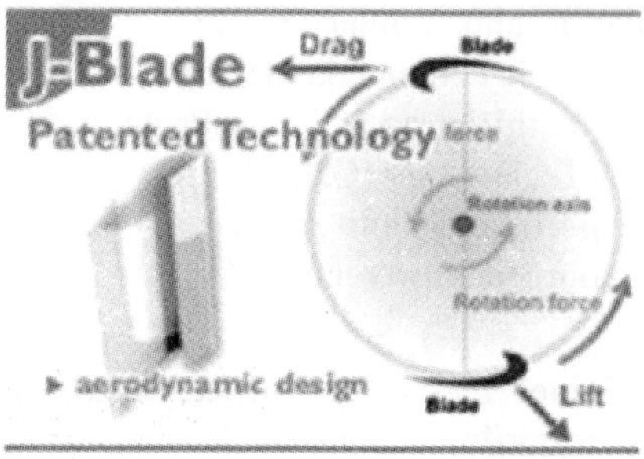

Fig. 1.4. Small Wind Energy Turbine (Wind-Smile Co Ltd. Japan)

IV

Green Corporates Tata, ITC and Fujitsu

The Green Corporates are those who offer green products which use green materials and green energy optimising the use of all resources. Their manufacturing processes are energy efficient, economising water and other resources and further emiting minimum or no GHG. We briefly introduce three well-known corporates– Tata and ITC of India and Fujitsu of Japan.

Tatas, India

In Fig. 1.5, Tata Motors shows its prototype Nano electric vehicle to European audience at the Geneva Auto show. It is now known that Nano, with petrol drive, was the world's smallest and cheapest car offered by Tatas about two years ago; the electric version now displayed will also accommodate four individuals comfortably in its spacious interiors. It will have a predicted range of 160 kms and an acceleration of 0-60 kmph in less than 10 seconds, using super polymer lithium ion batteries.

Tata Motor Electric vehicle strategy is that Miljo Greenland AS, Norway, a Tata subsidiary, will produce the electric version of Tata Models **Nano**, **Indica** and Sedan **Prima.** The vehicles are undergoing extensive trials in Europe.

Fig. 1.5 Prototype Tata Nano Electric Vehicle

ITC India

ITC is India's most celebrated Green Company; it is a large diversified, multi-business conglomerate of FMCG, Hotels, Paperboards, Agribusiness and Information Technology.

In India, it is the first company to have strictly introduced its famous *Triple Bottom line* strategy of management, where all businesses are evaluated under three heads Economic, Environment and Social consideration.

- Economic: The Company does aspire achieving leadership in each of its business segments.
- Environment: In its operations, it has achieved positive carbon balance, that is, its net carbon release to the atmosphere is nil or negative by virtue of using low carbon technology and heavy foresting; similarly, its water balance is positive using zero discharge recycling and water harvesting.
- Social: It maintains zero accident rate, promotes woman empowerment and imparts supplementary education and conserving natural resources.

Its Hotel business uses all Green Buildings duly certified 'LEED' by USGBC. The ITC Green Centre at Girgaum is the largest Platinum LEED (see Fig. 1.6) building in India.

In manufacturing of Paperboard, it has successfully deployed several green technologies including ECF (Elemental Chlorine Free) technology and ozone bleaching which are all global feats.

Fig. 1.6: ITC Centre (Gurgaon, India) Largest Platinum LEED Building in the world in 2004

Fujitsu Environment Care

Fujitsu environment care is symbolised by the green sapling in Fig. 1.7. To contribute to the creation of a sustainable environment for future generations, Fujitsu has made environmental protection a top management priority. With goals set for all business areas, the company conducts all business activities in a well-planned and sustainable manner. The Fujitsu Green Procurement Policy includes a commitment to implementing Green Procurement throughout the supply chain, which includes business partners.

Fig. 1.7. Symbolising Fujitsu Enviro-Care

V

Recipes for Greening India

A noted Journal *Echoworld* of UK has presented a note on how to green India in the coming decade by providing optimum level of energy and clean water to the growing population of India. For this purpose, we have collected updated primary data for population, GDP, primary energy and fresh water from the World Bank, FAO and World Resource Institute and introduced them in Table 1.1; somewhat on the lines of *Echoworld,* but using the new data, we have revised recommendations as to how to green India, as shown in Box 1.4.

Box 1.4: Greening India: Problem and Possibilities

Comparative Energy Scenario

The integrated Energy Policy of the Planning Commission of India estimates that to sustain the current Indian economic growth of 8 per cent, India will need primary energy of 1536 mtoe in 2031-32 from a level of 513 mtoe in 2005. This projected increase amounts to per capita energy of 53.5 million BTU on population base of 1.31 billion in 2032. The corresponding target by Echo World was 75 million BTU, calculated at 50% of current UK primary energy consumption per their data. It may thus take another decade or so to reach 50% level of European per capita power.

The total Indian electric power installed capacity of 152 GW provides per capita of about 700 KWH compared to 2000 KWH for China and the world average of 2600 KWH. It has been projected that power supply would have to grow to 3000 KWH per capita by 2031-32 (Ghosh Roy 2010) to provide the basic minimum economic wellbeing.

Here we would like to add a comment on the present energy efficiency of the Indian economy based on the comparative values of (Energy/GDP) ratio. From the Table 1.1, we observe that just at this present time the Indian efficiency at 0.389 mtoe per bUSD, which is somewhat higher than China (0.241) and Europe (0.244) and more higher than several other developed countries. This is mainly due to the very low base value of the Indian GDP, which form the denominator of the efficiency ratio. The efficiency ratio has already improved considerably in the last 5 years or so and we hope in the next ten years the value should compare favourably with both China and Europe.

Water Resource Scenario

From the Table1.1, we see that the gross Indian water availability at 1754 cu. m. per capita is satisfactory by global standards. But the net water usability at 635 cu. m. per person is on the low side, though the similar figure for China is also as low as 494 cu. m. What is not shown in the Table 1.1 is that the Indian agricultural consumption is 70 to 80 % of the total water withdrawal and hence the sectoral allocations for all other users are rather poor.

The other negative feature of the Indian scenario is the very uneven geographical distribution of water; the western, and the southern region are highly water-deficient, while the Eastern region is water surplus. Hence, for the last several decades the water planners have been dreaming of transferring a sizable quantity of water from the eastern Ganga-Brahmaputra basin to the southern Krishna/Cauvery basins. The *Echo World* recommends transferring 50 cu. Km. water over head of 250 m from Ganges Basin to Krishna basin, electricity requirement is only 5 GW about 3 % of India's 2005 power generation

Greening India

In spite of earlier deforesting, India is now increasing its forest cover during the last 20 years; if India receives compensatory financial allocation from the UNFCCC, India would be able to undertake comprehensive reforesting under a proactive Ministry of Environment and Forest. Indian reforesting programme has now been commended by UNEP and India has been named as the Host Country for 2011 Environment Day celebration with the theme "Forests: Nature at your Service".

Table 1.1: Multi-Nation Strategic Comparisons of Economic Parameters

Sources	*Resources*	*China*	*India*	*USA*	*Europe*	*Germany*	*UK*	*World*
WB2009	**Population**(millions)	1331	1115	307	459	81.4	52.6	
WB2009	**GNP** (USD billion)	4,985	1,310	14,119	6,978	3,330	2,174	
	GNP/Pop (USD/cap)	3,650	1,180	46,360		42,450	41,370	
WB2009	**GNP/Pop (PPP)** (USD/cap)	6 838	3237	46,436				
FAO2005	**Total Water Available** (Cubic Kilometre)	2,830	1,897	3,069	7,771	154	147	55,273
FAO2005	**Water/Pop** (Cu.m./capita)	2,206	1,754	10,333	10,655	1,866	2,474	8,549
FAO2005	**Total Water Used** (million cu.m.)	630,83 7	645,837	479,29 3	400,266	47,052	9,541	3,807,320
FAO2005	**Water used/Pop** (Cu.m./capita)	494	635	1682	581	572	183	683
	Water used /GNP (cu.m/USD)	0.474	0.579	1.561	0.872	0.578	0.181	
2002WRI -CAIT	**Primary Energy** quadrillion BTU (million toe)	54 (1200)	23 (510)	103 (2300)	76 (1700)	16 (350)	9 (200)	
	Energy/ GNP (BTU per 1000 USD) (mtoe/bUSD)	10,832 (0.241)	17,557 (0.389)	7295 (0.163)	10891 (0.244)	4804 (0.105)	4140 (0.092)	
	Energy/Pop (million BTU /capita)	40	21	333	164	196	171	

Echoworld recommends a strong reforesting programme –

- Reforestation of 90% of forest cover lost;
- Reforestation of desert along the sea coast with Mangroves;
- Reforestation of deserts including planting of bio-fuel energy crop for bio-energy.

VI

India vows on Green Technology and Climate Change

The Prime Minister of India, Dr. Manmohan Singh has often reiterated that the country must invest in its own environmentally friendly green technologies; our growth strategies can be different; rather they must be different, referring to the western world's energy intensive industrial policies and luxurious lifestyle that is blamed for climate change.

Global negotiations for a new U.N. agreement on climate change are stuck on the question of how much cash or technology rich nations will provide the poorer countries. Besides, they also want the developing and least developed nations to commit themselves to carbon emission cut limit like the developed countries. Therefore the policy statement of the Indian Prime Minister fits in the overall global aspiration to control emission of green house gas (GHG). The Kyoto Protocol of 1997 had required only the developed countries (Annex I Parties) to commit themselves to define target of GHG emission, but the developing countries were exempted from specific commitment of CO2 emission limit. The Kyoto protocol laid down the principle of 'common but differentiated responsibilities' and 'respective capabilities'. India being the fifth largest GHG emitting country of the world, the Prime Minister's statement showed India's sense of common but differentiated responsibility to put in money to develop expensive clean technologies to supplement what it might get from rich countries on the basis of their superior respective capabilities. The Indian Prime Minister also said India's energy use will rise sharply in the coming decades as it tries to lift the multitude of people out of poverty, but stressed a different development path must be walked. He further corroborated, "For this we need access to new technologies that are already available with developed countries. We must also make our own investments in new environment-friendly technologies" in a national conference on environment and forests in New Delhi.

VII

Forests: Nature at your Service

The United Nations designated 2011 as the Year of Forests. At the same time the United Nations Environment Programme (UNEP) named India as the Global Host country to celebrate Environment Day on 5 June 2011 with theme "Forests: Nature at Service". The forests serve mankind in manifold ways as shown in Box 1.5.

Box 1.5: Forest at Your service

Forests cover 31% of total land area.

The livelihoods of 1.6 billion people depend on forests.

Forests provide a home to more than 300 million people worldwide.

The total global trade in forest products was valued at around $379 billions in 2005.

Forests are home to 80% of terrestrial diversity.

Preserve and Grow Forests

Change lifestyle: Forests play multiple roles in our lives, including providing a source of livelihood, refuge for many species, and clean air for all. As a result of the growing global pollution levels, forests have often come to be referred to as the 'lungs of the earth'. This is particularly because deforestation and forest degradation account for nearly 20% of global greenhouse gas emissions, which forests would absorb if carefully managed. Thus, by changing our life style, we should endeavour to save forest.

Avoid Deforestation: Deforestation is the reduction of forest cover, notably viewed by the loss of trees. Commercial logging and fires are examples of causes of deforestation. It is however possible to use forests in a properly managed way that maintains their existence. There is therefore no deforestation if there is a guarantee of continuity in maintaining the forest cover.

Avoid Forest Degradation: Broadly, there are three main types of forest degradation: commercial logging, fires, and gathering wood for fuel. Insects and pests also cause considerable forest degradation.

Degradation refers to the loss of quality of the forests, rather than coverage. The quality of a forest can be observed through monitoring the survival rates of its ecosystem, for example vegetation layers, soil, flora and fauna. Some of the causes of forest degradation are the gathering of wood for fuel, and insects and pests.

Benefits of forests

Forests provide many important natural resources, such as timber, fuel, rubber, paper and medicinal plants. Forests also help sustain the quality and availability of freshwater supplies. More than three quarters of the world's accessible freshwater comes from forested catchments. Water quality declines with decreases in forest condition and cover, resulting in natural hazards such as floods, landslides, and soil erosion.

Climate change: It is well known that forests play a key role in our battle against climate change; storing carbon and sucking in carbon dioxide from the atmosphere and locking it into their biomass.

Beneficial Services: Some of the beneficial services provided by the forests are not that well-known, but are essential to every aspect of life. By regulating water for many of the world's rivers, they help secure water quality, and supply nearly half of the world's largest cities from Caracas to New York. They also help decrease the impacts of storms and floods, whilst helping control erosion.

Biodiversity: As the most biologically diverse ecosystems on land, forests are home to more than half of terrestrial species, from the great apes to the smallest of creatures. Use of the genetic diversity within forests enables the development of new medicines; progress in healthcare and science.

Economics and Livelihoods: They also provide homes, security and livelihoods for 60 million indigenous peoples, whilst contributing to the livelihoods of 1.6 billion people worldwide. The impact of forests reaches even further. In many developing countries more than 80% of total energy such as fuel wood and charcoal consumed by people and industry is derived from forest. Trade in timber and other forest products, is estimated at almost 330 billion US Dollars /year. Its value multiplies as it is processed into a myriad of products used globally every day.

UN International Year of Forests Initiative UN-REDD + Programme

In UN International Year 2011 of Forests, the theme of preserving forests throughout the world was to be instilled in our collective consciousness so as to change our lifestyles.

Beyond supporting the natural habitat, forests help to sustain economic growth. In 2004 trade in forest products was estimated at $327 billion. Continued and uncontrolled deforestation therefore not only has devastating consequences for the environment, the wildlife and communities, but for economies around the world. It is therefore strange that 36 million acres of natural forest are lost each year. World Environment Day (WED) chose 2011. So in theme, 'Forests: Nature at Your Service', to encourage forest conservation and sustainable consumption for green growth.

In September 2008, United Nations launched a collaborative initiative on Reducing Emissions from Deforestation and forest Degradation (REDD) in developing countries. The *UN-REDD Programme* assists developing countries to prepare and implement national REDD+ strategies, and builds on the convening power and expertise of the Food and Agriculture Organisation of the United Nations (FAO), the United Nations Development Programme (UNDP) and the United Nations Environment Programme (UNEP).

The Programme currently has 29 partner countries spanning Africa, Asia-Pacific and Latin America. REDD+ is seen as one of the most cost-effective ways of stabilising the atmospheric concentration of greenhouse gas (GHG) emissions to avoid a temperature rise of two degrees Celsius. But standing forests also conserve carbon while supporting the livelihoods of a large number of indigenous peoples and forest-dependent communities as well as providing essential ecosystem services such as habitat for biodiversity and provisioning clean water supplies.

A quick and easy way to get involved is to make others aware of the forests issues by sharing Green Living website, organising an event, or participating in this years WED celebrations. Governments should develop and implement policies that encourage sustainable use of forests. They should consider cordoning off areas inhabited by endangered species and promote forest restoration where they have been depleted.

Private companies have an opportunity to invest wisely into the new Green Economy whilst developing a socially responsible status with its consumers. They can develop procurement processes that buy only into sustainably managed forests, such as products certified by the Forest Stewardship Council (FSC). One should be the first in his organisation to shift your company policies towards green growth! Civil society can play a significant role by independently monitoring all parties involved, raising awareness on forests and supporting grassroots initiatives.

Like private companies, individuals can make wise premeditated choices over what products they will buy and only purchasing forest products that originate from sustainable sources. This means checking that furniture, wood, paper and other products you buy one buy are verified as coming from legal sources. A quick and easy way to do this is checking for an FSC certification logo.

Most importantly, taking action on forests requires an ongoing commitment to changing lifestyle and therefore is not a one-off action. Your new lifestyle demands that you are forest-conscious in all your choices, consumption and actions.

References

Echoworld on Green technology: ed. Redwood Ring, "Terra Green", March 2007; http://www.echoworld.com/water-supply/technoogy-sunlight-indias-green-future-html

FAO 2009

Fujitsu: http://ts.fujitsu.com/energy-eficiency-form-germany.info/EIE/Navigation/en/root.html.

Green Building: Mike Valenti; www.envirocitizen.org

Green Corporates Tata Motors; http://bit.ly/cYGevy (Nano)

Green Corporates ITC:

LEED 2009: Certification and Credentials (LEED), Feature 'State of LEED' by Kristie Dispenza 14 September 2010.

Small Wind Turbine J-bladed; http://www.wind-smile.com/products/technology.

Solar energy: Direct photosynthesis of Biofuels: Neils Bowler, *Science Reporter*, BBC;http://www.bbc.co.uk/news/science-environment-12051167#story_continue.

Solar Tower: Neils Bowder, *Science Reporter*, BBC; http://www.bbc.co.uk/news/science-environment-12051167#story_continue.

Wikipedia Chicago Center for Green Technology; http://www.egov.cityofchicago.org/city/webportal/portal.

Wind Turbine Chinese: reported in *Guardian* by Jonathan Watts; http://www.guardian.co.uk .

World Bank 2009

World Resource Institute/ CAIT 2005

Times of India; http://timesofindia.indiatimes.com/home/environment/developmental-issues/india-must-invest-in-green-technology-PM/articleshow/4907321

2

Technology Solution Towards Green Sustainability

Some Indian Resource Case Studies

I

UN Sustainable Development Programme

World Commission on Environment and Development

In realisation of the accelerating global environmental degradation and the international responsibility for affirmative action, United Nation set up the World Commission on Environment and Development (WCED) in 1983. The Commission, headed by Gro Harlem Brundtland of Norway, after arduous work, submitted a Report "Our Common Heritage" in 1987 (popularly known as "Brundtland Report 1987). For the first time, the Commission Report boldly presented the concept of Sustainable Development as an alternative to the policy of only economic growth.

"Sustainable Development is the development that meets the needs of the present without compromising the ability of the future generation to meet their needs."

Brundtland Report: Our Common Heritage

The Brundtland Report created a profound effect on the concept of economic development globally. It had for the first time highlighted two pioneering concepts.

Needs of the Poor: "The concept of needs, in particular the essential needs of the world's poor, to which overriding priority should be given"; and

Carrying Capacity of Environment: "The idea of limitations, imposed by the state of technology and social organisation that acts on the environment's ability to meet the present and the future needs". (WCED 1987:43)

Earth Summit 1992, Rio de Janeiro (UNCED)

Influenced by the highly revealing recommendations of the Brundtland Report, the UN General Assembly called for the UN Conference on Environment and Development (UNCED) at Rio de Janeiro in 1992 (popularly known as Earth Summit).

After prolonged deliberations by over a hundred participating nations, the Earth Summit arrived, inter alia, at the following major international agreements:

(i) Very well-known Agenda 21 – a very comprehensive set of action programmes towards sustainable development.

(ii) Rio Declaration of a series of twenty seven Principles defining the rights and responsibilities of individual state countries towards sustainability of environment and development;

Besides the above Agenda 21 and the Rio Declaration, following two very important Conventions were signed at the Earth Summit:

1. Framework of the Convention on Climate Change signed by 153 countries (UNFCCC);
2. Convention on Biological Diversity, signed by 153 countries.

II
UNDERSTANDING SUSTAINABILITY OF RESOURCES

In simple language, sustainability can be looked upon as a system where basic life-supporting resources have a surplus balance over consumption over a long period without deterioration of the earth's environment.

When we talk of surplus balance of resources, we have to identify also the prime resources other than food and nourishment on which the present technological structure of human civilisation rests. In Table 2.1 is given a list of very important natural resources, whose availability is limited. Now, Petroleum oil will be exhausted, say, in 40 years. Will then the wheels of the global industries and services come to a halt, which is very unlikely. Secondly, iron will be exhausted, say, in 150 years, what will happen to global transportation, bridges, building etc. Now, when fossil fuel gets depleted in 40 years, a new source of energy will have to be innovated; such process of innovation is discussed later.

Table 2.1: Some Non-Renewable Global Resources

Resource Name	*Known Global Resource*	*Projected Availability (years)*
Aluminum	1.17x 10 e 9 tons	100
Chromium	7.75 x 10 E 8 tons	420
Coal	5 X 10 E 12 tons	230
Cobalt	4.8 x 10 E 9 tons	60
Copper	308 x 10 E 6 tons	36
Iron	1 x 10 E 11 tons	240
Lead	91 x 10 E 6 tons	26
Manganese	8 x 10 E 8 tons	97
Molybdenum	10.8 x 10 E 9 lbs	79
Natural Gas	1.14 x 10 E 15 cu.ft	38
Nickel	147 x 10 E 9 lbs.	150
Petroleum	155 x 10 E 9 bbls	31
Silver	5.5 x 10 E 9 troy ozs.	16
Tin	4.3 x 10 E 6 tons	17
Tungsten	2.9 x 10 E 6 lbs.	40
Zinc	123 z 10 E 6 tons.	23

Source: Banerjee 2005

III

Exhaustible Resources: Theories of Resource Sustainability

Innovation of New Resources: Equations 1 and 2

In view of apparent limitation of current stock of Resources, we must innovate new resources by application of new Technologies formulated by the following two equations:

Unused Materials = Materials (Elements, Compounds, Alloys)
+ Material Composites ——— Eq. 1.0

Future Resources = (Unused Materials) × (Technological Innovation) ——— Eq. 2.0

The theory of equation 1 stipulates –

(i) That we must re-examine all known but still unused materials which could be put to newer uses as viable substitutes, for example, to reinvent refractory as metals substitutes.

(ii) We must explore new materials to replenish dwindling stock of existing materials. For example, explore bio-resources to replace petro-fuels.

(iii) The stock of viable unused materials shall be kept greater than the decrease in the stock of existing materials.

The theory of equation (2) assures –

(i) That by technologically innovating unused materials, we would be able to transform them into future useable resources. For example, it may be possible to convert a titanium alloy into a constructional material or develop or convert a polymer with carbon matrix into a structural material.

(ii) The growth in the stock of the new resources shall be more than the deficit of existing resources arising out of current consumption.

Conservation of Resources: Equations 3 and 4

In the short or medium terms, the sustainability will greatly depend on the efficient use, maintenance and conservation of the existing resources. The Conservation theory is well formulated by the following equations:

"Resource available" - "Resource surplus"
= "Resource utilised" + "Wastage"

———— Eq. 3.0

"Wastage"
= "Useless Byproduct" + "Product Degraded"
+ "Pollution (uncontrolled loss)"

———— Eq. 4.0

The theory of Eq. 3.0 mandates following Conservation regimes:

(i) We must conserve as much resources as possible i.e. maximize "Resource Surplus".

(ii) We must maximize the "Resource Utilised".

(iii) We must minimize the "Wastage"

The theory of Eq. 4.0 urges us to think on the following lines:

(i) To convert "Useless Byproduct" into useful product by secondary processes or by catalytic conversion or otherwise.

(ii) To prevent degradation, toxicity and contamination of other systems.

(iii) To prevent pollution by way of uncontrolled gaseous emission or liquid effluent discharge (and in some rare cases radioactive contamination)

Resource Recycling: Equation 5

The third option is recycling of "Resource utilised" in Eq. 5, after the materials have been used beyond their utility and are to be thrown out. There are substantial amounts of throwaways of old materials, aluminum utensils and furniture, Iron/Steel scraps, glasses, plastics, papers, textiles, etc.

These can be easily reconverted into newer applications of use.

New Resources = (Discarded/used Materials) × (Recycling Technology)

——— Eq. 5

The theory of Eq. 5 reiterates that new resources developed by recollection and recycling old scraps and throwaways are more economical and environmentally less pollutant than freshly reprocessing fresh ores or raw materials.

IV
Renewable Resources

There is a newer class of materials, mostly Bio-resources, which are limited in supply at any point of time but renewable on a time-scale of a year or season or a breeding cycle. These resources are agricultural food crops, energy crops, fisheries, animal foods, and other biomasses. As long as the quanta of consumption remain within the regenerating capacity of the earth ecosystem, this resource system is sustainable. Further, the production level at any particular level can be enhanced by adoption of appropriate technology already in existence or to be innovated. In course of time, perhaps newer engineering materials can be developed from bio-composites.

Technological Evolution

In equation 2.0, we have postulated how with the application of the magic wand of technology, new resources are created from unused materials. Technology has converted coal and petroleum into energy resource to feed and sustain the industrial infrastructure of the modern world.

Technology has converted mineral ores into iron, steel and other metals which are resources of construction to build bridges, ships, and railway which transports resources across the globe.

Though the green activists and sustainability practitioners have often condemned technology having created the toxic gases, hazardous wastes, radioactive materials, these are wastes under Equation 3.0 theory. They could possibly be reprocessed, by catalytic conversion or otherwise as already enumerated into useful or harmless byproducts or sequestrated and captured.

Further, like everything else in this world, technology is in a flux, it is constantly evolving and transforming over time. Table2.2 shows some main variants of old, current, and emerging technologies in today's world.

Possibly, at a foreseeable future, there could be a transition from macro-mechanical process of combustion to biotechnological fermentation aided by nanotechnologically controlled separation in lieu of chemical and metallurgical processes, the most notorious sector contaminating our environment.

Table 2.2: Evolution of Technologies

S. No	*Variants of Technologies*	*Some Conspicuous Products*
1	Macro-Mechanical Technologies	Prime-Movers (Steam/I.C. Engine, Electrical Generator/Motor, Turbine etc.) Ship, Aircraft, Smoke-Stack Industries, etc.
2	Micro-Electronics Technology	Transistor-Radio, TV, Telephone, Mobile Telephone
3	Digital Technology	Computer, MRI Scanner, Digital Photography etc.
4	Information Technology	Software-based knowledge products.
5	Biotechnology and Genetics	Newer medicines, new varieties of transgenic food, Recombinant DNA, including genetic modification.
6	Nanotechnology	Semi-permeable Membrane Technology, Ni/ C60 Batteries.

V
Emerging Technologies: Biotechnology

Biotechnology is the most powerful of modern technologies; it promises revolutionary empowerment in meeting the global challenges of sustainability, raising immensely food productivity by transgenic agriculture, ridding humans of fatal diseases by gene targeting, saving the environment of severe toxicity by substituting biotechnological fermentation in place of energy-guzzling combustion processes. It will help raise forestry by fast replicating tissue-culture. It will generate bio-energy. In short, Biotechnology will lead to quantum technological transformation of the world in a benign environment.

Biotechnology I: Fermentation Resources

Fermentation or Bioprocess, in its many forms, involves a multitude of complex enzyme-catalysed reactions engineered by specific micro-organism in carefully controlled environment at low temperature to generate new bioprocess products:

Three process-specific product groups that are identified:

(i) Primary Metabolic products: acetic acid, lactic acid, glycerol, acetone, butyl alcohol, amino acids, vitamins, polysaccharides;

(ii) Secondary metabolic products: penicillin, streptomycin, etc.;

(iii) Enzymes: (Exocellulars): amylases, pectinases, and proteases; (Intracellulars): invertase, asperaguinase, restricted endonucleases.

The large variety of fermentation products belong to a wide range of industries really:

Table 2.3: Fermentation Products

Industries	*Industrial Fermentation Products*
Organic Chemicals	Ethanol, acetone, butanol, Enzymes, Perfumeries, polysaccharides.
Pharmaceuticals	Antibiotics, steroids, Vaccines
Energy	Ethanol, methane, biomass
Food	Dairy products (cheese, yogurts etc.), Mushroom, glucose, Beverages (alcohols, tea, coffee)
Agriculture	Animal feed (SCP), Tissue-culture, transgenic fruits and vegetables

Biotechnology II: Genetic Engineering Resources

Transgenic products (see Fig. 2.1) i.e. products obtained by gene modification or recombinant DNA technology, for example, Bt Cotton, Bt Brinjal are now engineered bio-products, even known to farmers.

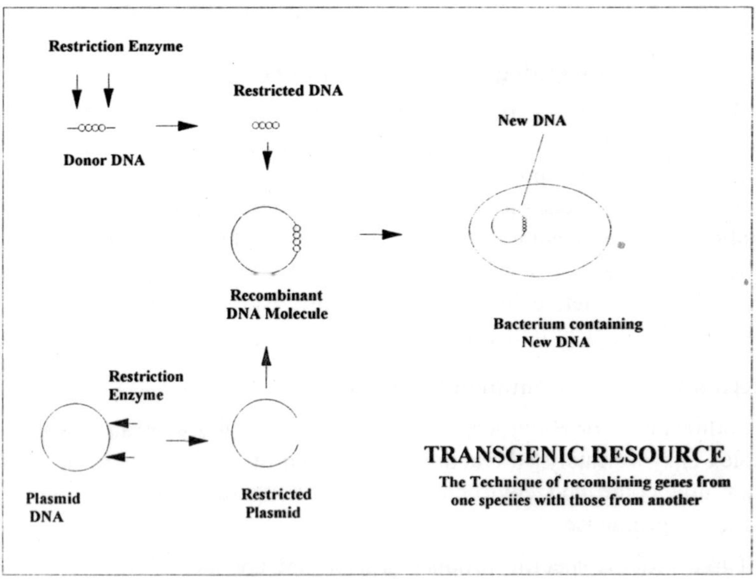

Fig. 2.1 Transgenic Resources

The birth of above processes has really started from the seminal studies by pioneers like James Watson and Francis Creek in the fifties of the last century. Today the industrial genetics have opened up vast scope of new or enhanced products. At another level. Considerable concern has been generated by the ill-effects of transgenic molecules on national health.

Biotechnology III: Ecological Resources

Wastes, the civilisational byproducts, representing the rejects of high consumption and elaborate production systems, are a source of environmental pollution. Biodegradation of solid wastes through microbial action with or without aerial oxidation is one of the primal contributions of bioprocess, generating usable resources.

Water is the most important resource for human survival, which gets polluted by all kinds of bio-wastes. Here again, microbiological processes in aerated lagoon and ponds, in activated sludge reaction, in aeration tanks help to greatly salvage the precious water and enable reuse after necessary disinfection. The process saves the ecology and preserves the natural resource.

But the organic manure is one of the direct ecological resources we obtain from land–fill biodegradation and composting. Organic manure is today highly valued being favourable to soil ecology in comparison to chemical manure which adversely affects the soil quality, if used over long period.

Bioremediation of the ocean water and sea beaches polluted with huge oil spill from ship wreck and otherwise is one of the current microbial process technology, which is serving the modern society.

Emerging Technologies: Nanotechnology

Nanotechnology: Concept

There are more than one ways to define nanotechnology

First Definition (*Koruga, Hameroff, Withers and Sundareshan, 1993*): "Nanotechnology is the newest scientific and engineering paradigm which enables the manipulation of atoms and molecules for building new materials and or molecular devices."

The basic concept of nanotechnology was proposed by Toniguchi of Japan in 1974. Nanoscale fabrication principles were proposed later by Drexler (1981), and Schneiker and Hameroff (1987).

Alternative Definition (*McGraw-Hill Encyclopedia of Science and Technology*, Vol.11, 1997): "Systems for transforming matter, energy and information, based on nanometer-scale components with precisely defined molecular features".

The term nanotechnolgy has also been used more broadly to refer to techniques that produce or measure features less than 100 nanometer in size; this meaning embraces micro-fabrication and metrology.

Nanotechnology: Health Hazards

Of late, we have heard of many apprehensions in India of using transgenic

Brinjals. Similar fear of risks from nanotechnological products have been expressed in US.

Because ordinary materials display unique properties at nano-scale, nano-size bits of seemingly benign materials might turn out to be noxious. As John Young and Jan Martel report in the Rise and Fall of Nano-bacteria, even naturally occurring nanoparticles can have deleterious effects on human body. Now, if naturally occurring nanoparticles can harm us, we would be wise to carefully consider the possible actions of engineered nanomaterial. In response to the risk, the US Environmental Protection Agency recently announced a grand research strategy to study the health and environmental effects of nanomaterials.

Superiority of Sustainable Development

If economic development is for immediate survival, sustainable development is for long-term enrichment of life, economically, socially, culturally. Sustainable Development, therefore calls for higher awareness or conceptual preparedness. It demands creating newer policy delineation or a new set of instruments beyond simple market forces. Apart from introducing enabling features in governance and economic instruments, the educational systems should be remodeled focusing on social and ecological consciousness, rights and duties of an enlightened citizenry, conservation of global resources, wild life, forestry, water, air, and biodiversity in general.

In conclusion, Sustainable Development is a Holistic Philosophy; it is a whole new way of life.

CASE STUDIES OF THREE VITAL INDIAN RESOURCES AFFECTING SUSTAINABILITY

Let us glance at following three important Indian resources facing challenges

- Water,
- Energy,
- Economic wellbeing.

Indian Fresh Water Availability

Water is a vital resource on which depends our precious life. It is the life line of our agriculture. Our industry cannot survive without fresh water. At independence, our population was about 400 million and it has grown to more than a billion (see Fig. 2.2), but the availability of fresh water remains more or less same. Thus, the per capita availability of fresh water has come down from 5000 cu. m. to 1000 cu. m (see Fig. 2.3), which is the stress level as per UN norm. How do we overcome this serious shortage of fresh water?

From the Table 2.4, we see that globally, the available fresh water is a very small percentage (less than 1%) of the total available water system.

Table 2.4: Global Fresh Water Availability

Sources	*Quantity (Million Cu Km)*	*Percentage %*
Ocean Water	1362.2	97.3
Fresh Water	37.8	2.7
Total Water	**1400.0**	**100.0**
Locations of above Fresh Water		
Icecaps and Glaciers	29.18	
Groundwater & Soil Moisture	8.47	
Lakes & Swamps	0.13	
Atmosphere	0.02	
Streams & Rivers	0.01	
Total Fresh Water	**37.81**	**2.7**
Available Fresh Water		
Groundwater and Swamps	8.47	
Lakes & Swamps	0.13	
Streams & Rivers	0.01	
Total Available Fresh Water	**8.61**	**0.62**

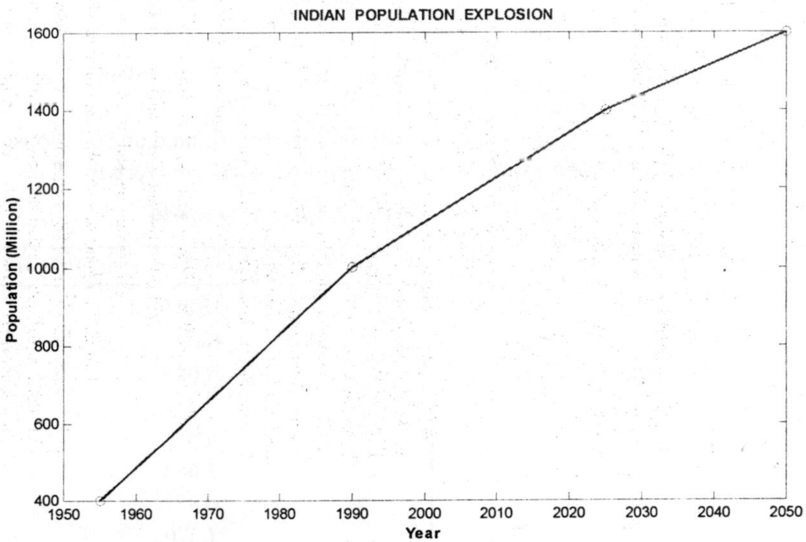

Fig. 2.2. Indian Population Explosion

There are many ways to alleviate the situation. Indian agriculture is a great guzzler of water. By modernising agriculture, a great saving of water can be made. There is great scope of rationalising the consumption of industrial water and saving water. Rain water harvesting and recharging of ground water are other very effective measures. Finally, the call is on revitalising the various river basin authorities, shorn of water politics. No less important is the better management of water resources by way of proper balancing of demand and supply, avoiding wastages.

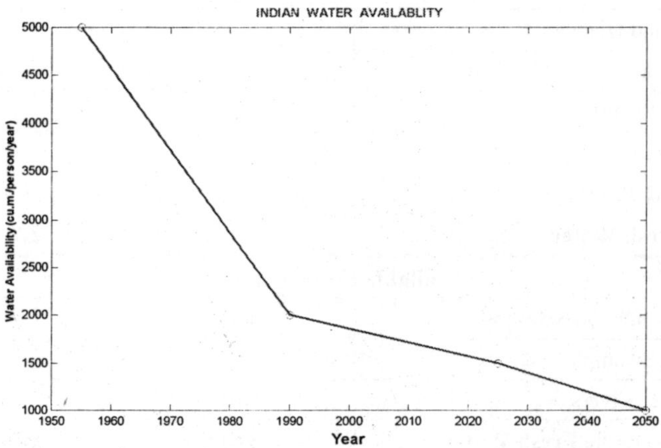

Fig. 2.3. Indian per Capita Water Availability

Indian Energy Availability

Most of us are conversant with load shedding. The problem of energy famine is of course much deeper and extensive. Unless we do improve our energy availability, our industrial growth will remain stunted and millions of our people will remain condemned to unending severe poverty.

Table 2.5: Global Electrical Energy Scenario

Country	*Annual Per Capita Consumption (kwh)*
Canada	18,309
USA	12,900
France	8,089
Japan	7,211
Germany	6,693
U.K.	5,660
Singapore	6,353
Brazil	1,570
China	2000
India	700

Look at the per capita power consumption in Table 2.5. We are behind the whole host of countries at 700 kwh per capita per annum. Fifty per cent of our household are without electricity, remedying which could trigger a burst of educational efforts and other activities in our Indian homes.

It has been recently estimated to arrive at the reasonable level of prosperity of our people, the per capita electric consumption needs to be upscaled to 3000 kwh. The aggregate national electrical generation capacity is 150,000 MW, which must be augmented to 800,000 MW in 20 years.

Indian Economic Scenario

In spite of recent galloping GDP growth, India remains an abysmally poor country. Its average per capita income is comparable with only sub-Saharan Africa and few other neighbouring countries of our South Asia (see Table 2.6). About half the population have annual income of less than two US dollars per day, of which about 50 per cent earn just one US dollar a day.

Of course, great lot of work is going on at the national level. But the task of eradicating poverty is complex; we not only need high and sustained GDP growth but also wider dissemination of education, skill upgradation, health and housing; all these together will raise the quality of human resource which can bring about real well-being and prosperity of the people in India.

Table 2.6: Income per Capita

Country	*2002 $ (PPP)*	*Global Ranking*
United States	34,557	4
Canada	28,728	7
Australia	27,256	11
Germany	26,241	13
France	26,090	14
Japan	25,788	15
Italy	25,453	17
United King.	25,139	18
E U - 25	22,917	21
Spain	20,777	25
South Korea	16,570	33
Saudi Arabia	11,994	43
Argentina	10,664	46
Poland	10,299	48
South Africa	9,750	51
Mexico	8,662	59
Russia	7,993	61
Brazil	7,480	64

(contd...)

Country	*2002 $ (PPP)*	*Global Ranking*
Iran	6,277	73
Turkey	6,145	76
Ukraine	4,719	97
China	4,379	100
Indonesia	3,057	118
India	2,572	121
Pakistan	1,941	138
Developed	22,254	
Developing	3,806	
World	6,980	

Source: World Resource Institute

Conclusion

We may close this discourse by concluding that given the human ingenuity and endeavour, the world can overcome its resource challenges and march on to greater heights of excellence.

References

Banerjee, B.P., *Handbook of Energy and Environment in India*, Oxford University, 2005.

Ghosh Roy, M.K., *Sustainable Development – An Engineering Critique*, Indian Science Congress Association, Sept. 2007.

Ghosh Roy, M.K., *Biomass Energy – An Outline of Futuristic Fuels*, MIEEM Energy Summit, Kolkata, Oct 2007.

Ghosh Roy, M.K., *Sustainable Development – Environment, Energy and Water Resources*, Ane Books, New Delhi, 2011.

McGraw-Hill, *Encyclopedia of Science and Technology*, 10th Edition.

Shankar, U. (Ed.), *Environmental Economics*, Oxford University, 2004.

Smith, E. John, *Biotechnology*, Cambridge University, 1996.

3

Food Security, World Hunger and Millennium Development Goals

I

Indian Food Security

Food Security may be defined as to enable people to obtain safe and nutritious food in adequate quantity, and preferably of their choice at all times. This definition rings similar to the definition provided by the UN for Energy Security.

The Government of India has now promised to introduce a Food Security Bill (FSB) in Parliament with the avowed objective of giving the citizens an enforceable right to food. The bill when it becomes a law should provide every citizen comprehensive food security as defined above.

Guaranteeing the food security in the above manner is a gigantic task for the Indian Government to undertake. For the International Food Policy Research Institute (IFPRI) awarded India a very poor Global Hunger Index, placing India 67th with a score of 24.1, far below China and below Sri Lanka and Pakistan and even Nepal. According to FAO, 25 per cent of world's undernourished people live in India. It will be the great task of the Food Security Act to create an appropriate instrument of intervention to remedy this dismal situation.

The Global Hunger Index is based on three important but very sensitive parameters:

1. Percentage of the population that are undernourished, recorded at 22% in India;
2. Percentage of underweight children under five is recorded at 43.5 % in India;
3. Percentage of children dying before five recorded at 6.9% in India.

Despite doubt about the legitimacy of data, the fact remains that the

Indian nation is highly undernourished on the whole and the abject poverty of a large part of the Indian people is matter of shame.

We will deal with the huge problem of hunger at the global and Indian contexts in the later part of this Chapter.

Nature of National Intervention

The food production in India in the last two decades shows an upward trend of sizeable proportion; there is still persistent hunger and malnutrition in the country for a large section of economically weaker population. That calls for national intervention by the Government to provide subsidised food to this section through public distribution system (PDS). From the experience gathered so far, it appears that the PDS has not been effective enough, for the subsidised food does not always reach the targeted poor. The system suffers from 'delivery deficit' in respect of diversion or theft and poor quality. Thus, it is very much essential that the proposed Food Security Act does take into consideration preventive measures to make the PDS effective. There is already much awareness that an appropriate technology intervention should be devised including well-operated monitoring system to support the technology. The Government of India is already implementing a comprehensive biometric identification known as UID (Universal Identification Device) popularly known as Adhar under the leadership of iconic technologist Nanadan Nilekeni of Infosys fame. It has been assured that the UID technology shall be able to eradicate the diversion of PDS Food to very large extent. In smaller towns, UID may be used to issue food coupons so that the holders may buy food from the market and the PDS will not need to store food.

In the procurement of rice and wheat in the rural areas, India may also follow the practice of UN World Food Programme named "Purchase for Progress" (abbreviated "P4P") which helps the farmers to participate in the market and thereby also improve his agricultural productivity.

The other innovation that remains to be attempted is to discontinue the present centralised procurement by distributed procurement and distribution at the district level and if not at the block level but with foolproof integrated monitoring.

II

UN World Food Programme (WFP)

Purchase for Progress (P4P)

UN World Food Programme (WFP) is working to ameliorate the global hunger. A brief resume of one of their main activities known as Purchase for Progress (P4P) is presented here.

As the world's largest humanitarian agency, the UN World Food Programme (WFP) is a major food buyer. WFP spends hundreds of millions of dollars in food every year and around 82 per cent of these purchases are made in developing countries. Local procurement for WFP's food assistance programmes requires cost-efficiency, quality commodities and timeliness of deliveries.

Through the Purchase for Progress **(P4P)** pilot initiative, WFP's uses its demand in innovative ways to support the development of agricultural markets and smallholder farmers' engagement in these markets. As the majority of the world's poorest people lives in rural areas and relies on agriculture for their livelihood, P4P's contribution to agricultural market development is an important step towards reducing hunger and poverty.

P4P is based on three critical components: 1) WFP's demand, 2) supply side support through partners and 3) learning and sharing as explained below:

The contour of the WFP in 21 countries is briefly described in Box 3.1.

1. WFP's Demand-Buying food in new ways

The normal WFP is to buy food through large competitive tenders. Through P4P, WFP is testing new procurement methods more suited to small-scale farmers:

- *Tendering for regular WFP procurement* processes are adjusted by:
  (a) reducing the quantities involved; (b) removing the need for performance bonds for guaranteed delivery; (c) providing pre-printed WFP bags.
- *Competitive purchases* through emerging commodity exchanges, which serve as platforms for farmers to trade their produce (in Zambia, WFP is buying 78% of its food this way)
- *Direct Contracting*, with farmers' organisations — almost half the food contracted under P4P so far has been purchased this way.
- Through the *Warehouse Receipt System* in Uganda and Tanzania, smallholder farmers deposit their crops in a certified warehouse, receiving a receipt for about 60% for the market value of the commodities which can be exchanged for cash at a local financial institution. The final balance is paid once the commodities are sold.
- **WFP purchases** which allow farmers to deposit the harvest, receive a payment for a portion of value and collect the remainder once the product is sold.
- In Mali and Burkina Faso, WFP is pioneering *forward contracting* that commits WFP to buy food commodity from farmers' organisations in the future at an agreed price.
- In Guatemala, Kenya, Mozambique and Zambi, P4P is helping smallholder farmers to access the private sector *food processing* market.

2. Partnership support in Supply Side

Having WFP as a committed buyer serves as an incentive for smallholder farmers to increase productivity, raise quality and diversify crops. However, critical to the success of P4P is the wide range of partnership offered to farmers in gaining access to new markets, for example:

- Improving their skills to increase the quantity and quality of their crops- e.g., training in farming techniques, and use of the best seeds and fertilizers for their land;
- Reducing post harvest loss- e.g., training in storage techniques and rehabilitation/construction of storage facilities;
- Enhancing farmers' knowledge of markets- e.g., quality, pricing, packaging;
- Strengthening the farmers' organisations-e.g., training in management and negotiation of contracts, strengthening account and fiscal systems;
- Improving business skills- e.g., preparing business plans and access to credit.

3. Learning and sharing

Through P4P, WFP is identifying mechanisms that help develop secure markets for small-scale farmers. The best practices identified will be mainstreamed into WFP's long-term policies and programme practices. The lessons drawn will be shared with national governments and other public and private-sector partners in the agricultural sector to serve as a model.

WFP has established many ways to capture and share valuable information from P4P through a comprehensive monitoring and evaluation (M&E) system that analyzes data on farmer organisations, smallholder farmers, and traders. Annual surveys and case studies will provide a detailed understanding of how P4P affects farmers, farmers' organisations and traders.

P4P Achievements

Since P4P's launch in September 2008,

- *630 farmers' organisations* representing more than 800,000 farmers have been identified to participate in P4P—so far, those farmers' organisations have contracted to sell food commodities to WFP.
- Over 50,000 people have been trained in skills including organisation management, farming technique, quality control and post harvesting handling.
- Working relationships specific to P4P have been established with approximately 160 different Partners including Government (ministries and other agencies), UN Agencies, local and international NGOs.
- Over 12,000 million tons (mt) of food through P4P in 20 countries have been contracted.

Box 3.1: Contour of WFP Project in 21 Pilot Countries

Africa: Burkina Faso, Democratic Republic of Congo, Ethiopia, Ghana, Kenya, Liberia, Malawi, Mali, Mozambique, Rwanda, Sierra Leone, Sudan, Tanzania, Uganda, Zambia

Asia: Afghanistan and Laos

Latin America: El Salvador, Guatemala, Honduras, Nicaragua

Beneficiaries: 500,000 smallholder farmers

Duration: 5 years (Sept 2008 — Sept 2013)

Total funding: US$137 million for technical capacity, including sub-grants, for 5 years (food not included)

Key donors – Bill and Melinda Gates Foundation, Howard G. Buffett Foundation, European Commission, Governments of Belgium, Canada, Ireland, Luxembourg and the United States of America and the Kingdom of Saudi Arabia

III

Review of World Hunger and Possible Solutions

It is sometimes difficult to perceive that the global hunger is so pervasive and there are 925 million people hungry out of the total population of about 7 billion i.e. 1 in 7 persons are hungry.

To understand the world hunger problem and eventually follow the FAO initiatives to try to ameliorate or eliminate them, we seek their definitions.

Hunger concepts and definitions

Hunger is a term which connotes the want of food in a country (*Oxford English Dictionary* 1971). World hunger refers to the hunger aggregated to the world level.

Malnutrition, related to hunger, is a general term that indicates a lack of some or all nutritional elements necessary for human health. There are two basic types of malnutrition:

(i) The first and most important is protein-energy malnutrition—the lack of enough protein (from meat and other sources) and food that provides energy (measured in calories) which all of the basic food groups provide. This is the type of malnutrition that is referred to when world hunger is discussed.

(ii) The second type of malnutrition, also very important, is micronutrient (vitamin and mineral) deficiency. This is not the type of malnutrition that is referred to when world hunger is discussed, though it is certainly very important.

Protein-energy malnutrition (PEM) is the most lethal form of malnutrition/hunger. It is basically a lack of calories and protein. Food is converted into energy by humans, and the energy contained in food is measured by calories. Protein is necessary for key body functions including provision of essential amino acids and development and maintenance of muscles.

The extent of global hunger including child hunger is presented in the Boxes 3.2A and 3.2B and the incidence of malnutrition is described in Box 3.3.

Box 3.2A: Extent of Global Hunger (FAO 2010)

- 925 million people do not have enough to eat — more than the populations of USA, Canada and the European Union; (see Fig. 3.1)
- 98 per cent of the world's hungry live in developing countries;
- Asia and the Pacific region is home to over half the world's population and nearly two thirds of the world's hungry people;
- Women make up a little over half of the world's population, but they account for over 60 per cent of the world's hungry. *(ECOSOC, 2007)*
- 65 per cent of the world's hungry live in only seven countries: India, China, the Democratic Republic of Congo, Bangladesh, Indonesia, Pakistan and Ethiopia.

Box 3.2B: Extent of Child Hunger (UNICEF 2006 and 2007)

- More than 70 per cent of the world's 146 million underweight children under age five years live in just 10 countries, with more than 50 per cent located in South Asia alone; *(UNICEF, 2006)*
- 10.9 million children under five die in developing countries each year. Malnutrition and hunger-related diseases cause 60 per cent of the deaths; *(UNICEF, 2007)*
- The cost of under-nutrition to national economic development is estimated at US$20-30 billion per annum; One out of four children — roughly 146 million — in developing countries are underweight;
- Every year WFP feeds more than 20 million children in school feeding programmes in some 70 countries. In 2008, WFP fed a record 23 million children.

Box 3.3: Incidence of Malnutrition

- It is estimated that 684,000 child deaths worldwide could be prevented by increasing access to vitamin A and zinc (WFP, 2007).
- Under-nutrition contributes to 53 per cent of the 9.7 million deaths of children under five each year in developing countries.(UNICEF, 2006).
- Lack of Vitamin A kills a million infants a year (UNICEF).
- Iron deficiency is the most prevalent form of malnutrition worldwide, affecting an estimated 2 billion people. Eradicating iron deficiency can improve national productivity levels by as much as 20 per cent. (WHO)

(contd...)

- Iron deficiency is impairing the mental development of 40-60 per cent children in developing countries (UNICEF)
- Vitamin A deficiency affects approximately 25 per cent of the developing world's pre-schoolers. It is associated with blindness, susceptibility to disease and higher mortality rates. It leads to the death of approximately 1-3 million children each year. *(UN Standing Committee on Nutrition. World Nutrition Situation 5th Report. 2005)*
- Iodine deficiency is the greatest single cause of mental retardation and brain damage. Worldwide, 1.9 billion people are at risk of iodine deficiency, which can easily be prevented by adding iodine to salt

(UN Standing Committee on Nutrition: 5th Report World Nutrition Situation)

The distribution of hungry people in the world is shown in Fig. 3.1 and the change in number of hungry people during the period 1969-2010 is depicted in Fig. 3.2.

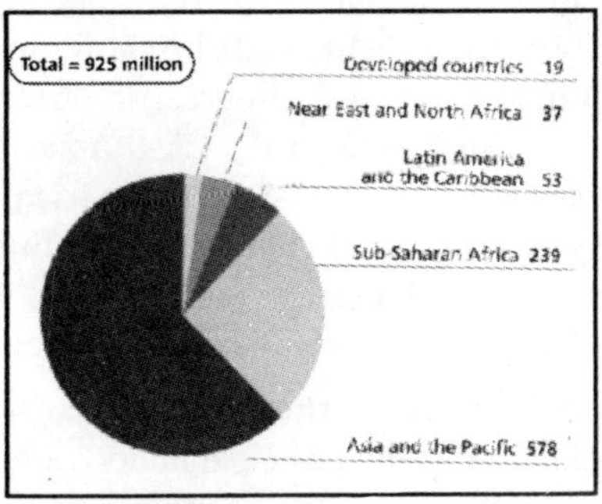

Fig. 3.1. 925 million hungry people in the world 2010

Source: FAO.

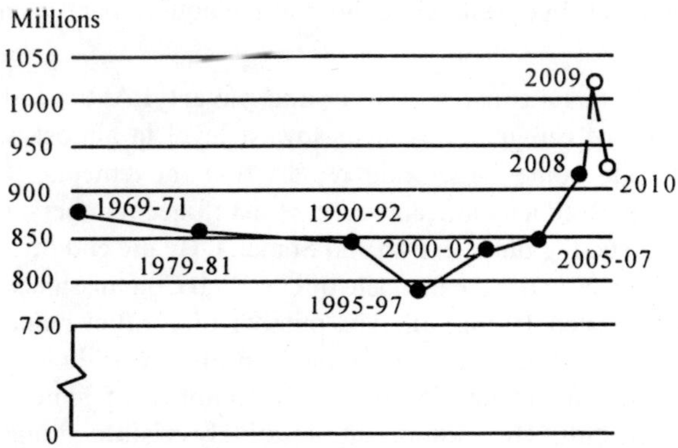

Fig. 3.2: Change in number of hungry people in the world 1969-2010

Source: FAO.

Towards Elimination of Hunger

The world produces enough food to feed everyone. World agriculture produces 17 per cent more calories per person today than it did 30 years ago, despite a 70 per cent population increase. This is enough to provide everyone in the world with at least 2,720 kilocalories (kcal) per person per day (FAO 2002, p. 9). The principal problem is that many people in the world do not have sufficient land to grow, or income to purchase, enough food. What are then the causes of hunger is a fundamental question, with varied answers.

(1) Poverty is the principal cause of hunger: The causes of poverty include poor people's lack of resources, an extremely unequal income distribution in the world and within specific countries, conflict, and hunger itself. As of 2008 (2005 statistics), the World Bank has estimated that there were an estimated 1,345 million poor people in developing countries who live on $1.25 a day or less. This compares to the later FAO estimate of 1.02 billion undernourished people. Extreme poverty remains an alarming problem in the world's developing regions, despite some progress that reduced "dollar—now $1.25—a day" poverty from (an estimated) 1900 million people in 1981, a reduction of 29 per cent over the period. Progress in poverty reduction has been concentrated in Asia, and especially, East Asia, with the major improvement occurring in China. In Sub-Saharan Africa, the number of people in extreme poverty has increased. The statement that 'poverty is the principal cause of hunger' is, though correct, unsatisfying. Why then are (so many) people poor? The next section summarizes Hunger Notes answer.

(2) Harmful economic systems are the principal cause of poverty and hunger: Hunger Notes believes that the principal underlying cause of poverty and hunger is the ordinary operation of the economic and political systems in the world. Essentially control over resources and income is based on military, political and economic power that typically ends up in the hands of a minority, who live well, while those at the bottom barely survive, if they do.

(3) Conflict as a cause of hunger and poverty: At the end of 2005, the global number of refugees was at its lowest level in almost a quarter of a century. Despite some large-scale repatriation movements, the last three years have witnessed a significant increase in refugee numbers, due primarily to the violence taking place in Iraq and Somalia. By the end of 2008, the total number of refugees under the mandate of UN High Commissioner for Refugees (UNHCR) exceeded 10 million. The number of conflict-induced internally displaced persons (IDPs) reached some 26 million worldwide at the end of the year. Providing exact figures on the number of stateless people is extremely difficult. But, important, relatively visible though it is, and anguishing for those involved conflict is less important as poverty as a cause of hunger. (Using the statistics above 1.02 billion people suffer from chronic hunger while 36 million people are displaced [UNHCR, 2008])

(4) Hunger is also a cause of poverty, and thus of hunger: By causing poor health, low levels of energy, and even mental impairment, hunger can lead to even greater poverty by reducing people's ability to work and learn, thus leading to even greater hunger.

(5) Climate change: Climate change is increasingly viewed as a current and future cause of hunger and poverty. Increasing drought, flooding, and changing climatic patterns may require a shift in crops and farming practices that may not be easily accomplished.

(6) Unsatisfactory Progress in reduction of hunger: The target set at the 1996 World Food Summit (WFS) was to halve the number of undernourished people by 2015 from their number in 1990-92. (FAO uses three year averages in its calculation of undernourished people.) The estimated number of undernourished people in developing countries was 824 million in 1990-92. In 2009, the number had climbed to 1.02 billion people. The WFS goal is a global goal adopted by the nations of the world; the present outcome indicates how marginal the efforts were in face of the real need. So, overall, the world is not making progress toward the world food summit goal, although there has been progress in Asia, and in Latin America and the Caribbean.

(7) Micronutrients are essential for hungry children: Quite a few trace elements or micronutrients—vitamins and minerals—are important for health. 1 out of 3 people in developing countries are affected by vitamin and mineral deficiencies, according to the World Health Organisation. Following three micronutrients are perhaps the most important in terms of health consequences for poor people in developing countries, as explained below:

Vitamin A: Vitamin A deficiency can cause night blindness and reduces the body's resistance to disease. In children Vitamin A deficiency can also cause growth retardation. Between 100 and 140 million children are vitamin A deficient. An estimated 250,000 to 500,000 vitamin A-deficient children become blind every year, half of them dying within 12 months of losing their sight. (World Health Organisation)

Iron: Iron deficiency is a principal cause of anemia. Two billion people—over 30 per cent of the world's population—are anemic, mainly due to iron deficiency, and, in developing countries, frequently exacerbated by malaria and worm infections. For children, health consequences include premature birth, low birth weight, infections, and elevated risk of death. Later, physical and cognitive developments are impaired, resulting in lowered school performance. For pregnant women, anemia contributes to 20 per cent of all maternal deaths (World Health Organisation).

Iodine: Iodine deficiency disorders (IDD) jeopardize children's mental health– often their very lives. Serious iodine deficiency during pregnancy may result in stillbirths, abortions and congenital abnormalities such as cretinism, a grave, irreversible form of mental retardation that affects people living in iodine-deficient areas of Africa and Asia. IDD also causes mental impairment that lowers intellectual prowess at home, at school, and at work. IDD affects over 740 million people, 13 per cent of the world's population. Fifty million people have some degree of mental impairment caused by IDD (World Health Organisation). (Updated March 6, 2011)

IV
Millennium Development Goals (Target 2015)

The concept of sustainability and the world without poverty but with equity were the core theme of the Brundtland Report 1987 under the aegis of the WCED (World Commission on Environment and Development) of the United Nations. At the advent of the 21st Century, the United Nations declared the Millennium Development Goals (MDGs) setting specific targets of poverty alleviations, health and sanitation, literacy and women's empowerment of the Third World, as presented in Box 3.4.

WFP and the Millennium Development Goals

Box 3.4: Millennium Development Goals

Goal 1: Eradicate Extreme Poverty and Hunger.

Target 1. Halve, between 1990 and 2015, the proportion of people whose income is less than a $1 a day.

Target 2: Halve, between 1990 and 2015, the proportion of people who suffer from hunger.

Goal 2: Achieve Universal Primary Education.

Target 3: Ensure that, by 2015, children everywhere, boys and girls alike, will be able to complete a full course of primary schooling.

Goal 3: Promote Gender Equality and Empower Women.

Target 4: Eliminate gender disparity in primary and secondary education, preferably by 2005, and in all levels of education no later than 2015.

Goal 4: Reduce Child Mortality.

Target 5: Reduce by two-thirds, between 1990 and 2015, the under-five mortality rate.

Goal 5: Improve Maternal Health.

Target 6: Reduce by three-quarters, between 1990 and 2015, the maternal mortality ratio.

Goal 6: Combat HIV/AIDS, malaria and other diseases.

Target 7: Have halted by 2015 and began to reverse the spread of HIV/AIDS.

Target 8: Have halted by 2015 and began to reverse the incidence of malaria and other major diseases.

Goal 7: Ensure Environmental Sustainability.

Target 9: Integrate the principles of sustainable development into country policies and programs and reverse the loss of environmental resources.

Target 10: Halve, by 2015, the proportion of people without sustainable access to safe drinking water and basic education.

Target 11: Have achieved, by 2020, a significant improvement in the lives of at least 100 million slum dwellers.

Goal 8: Develop a Global partnership for Development.

Target 12: Develop further an open, rule-based, predictable non-discriminatory trading and financial system (includes a commitment to good governance, development, and poverty reduction, both nationally and internationally).

Target 13: Address the special needs of the least Developed. Countries (includes tariff- and quota-free access for the Least Developed Countries' exports, enhanced program of debt relief for the heavily indebted poor

countries (HIPCs) and cancellation of official bilateral debt, and more generous official development assistance for countries committed to poverty reduction).

Target 14: Address the special needs of land-locked developing countries and small Developing States (through programs of action for the sustainable development of Small Island Developing States and 22nd General Assembly Provisional).

Target 15: Deal comprehensibly with the debt problems of developing countries (through national and international measures in order to make debt sustainable in the long term).

Target 16: In consultation with the developing countries, develop and implement strategies for decent and productive work for youth.

Target 17: In consultation with the pharmaceutical companies, provide access to affordable essential drugs in developing countries.

Target 18: In consultation with the private sector, make available the benefits of new technology, especially information and communication technologies.

WFP's role lies in achieving the first MDG is clear. Whether through its emergency operations or its relief or development projects, WFP's focus is on making sure food reaches hungry people. WFP's mission statement emphasises that WFP's activities aim to save lives, improve nutrition and quality of life among vulnerable people and help promote self-reliance.

But WFP's work also helps toward the second MDG - universal primary education. WFP school meals projects encourage parents to keep their children in school.

WFP's food assistance does also play an important role in meeting another five of the eight MDGs: empowering women, reducing child and maternal mortality, improving maternal health, combating AIDS and ensuring environmental sustainability. Thus WFP is really complimentary to MDGs.

Review of Progress of Millennium Development Goals

Most of the MDG targets have a deadline of 2015 and 1990 is the baseline against which the progress is gauged. Progress towards the eight Millennium Development Goals is measured through 18 targets and many official indicators. Table 3.1 mentions the gist of the progress of implementation of these goals as per the 2010 UN Report.

Though the progress has been made, it is uneven. Without major

Table 3.1: Progress of MDGs – UN Report 2010

Goal 1:	**T1:** The Global economic crisis has slowed down progress, but the world is still on truck to meet the poverty reduction target.
	T2: Hunger may have spiked in 2009, one of the many dire consequences of the global financial melt-down
Goal 2	**T3:** Hope has dimmed for universal primary education by 2015, even though many poor countries have made tremendous progress.
Goal 3	**T4:** For girls in some regions, education remains still elusive.
Goal 4	**T5:** Child deaths are falling, but not quickly enough to achieve the target.
Goal 5	**T6:** Giving births are especially risky in Southern Asia and Sub-Sahara Africa, where most women deliver without skilled care.
Goal 6	**T7:** The spread of HIV appeared to have stabilized in most regions and more people are surviving longer.
Goal 7	**T9:** The rate of deforestation shows signs of decreasing, but is still alarmingly high.
Goal 8	**General:** Aid continues to rise despite the financial crisis, but Africa is short-changed.
	T12: Developing countries gain greater access to the markets of developed countries.
	T13: Only five donor countries have reached the UN targets for official aid.
	T15: Debt burdens ease for developing countries and remain well below the historical levels.
	T18: Demand grows for information and communication technology.

collective international push, many of the MDG targets are likely to be missed in most regions.

The number of people who are under-nourished has continued to grow, while slow progress in reducing the prevalence of hunger stalled or even reversed itself.

One in four children under the age of five is underweight mainly due to lack of food and quality food, inadequate water and health services.

An estimated 1.4 billion are still living in extreme poverty in 2005. The global financial crisis will likely delay their amelioration of the poverty.

Gender equity and the empowerment of women are at the heart of MDGs and in fact are preconditions of overcoming poverty, hunger and disease. But the progress on the gender front – from education to political participation – has been less than satisfactory.

However, the overall progress so far shows that the goals are achievable when nationally owned development strategies, policies and programmes are supported by international development partners (Ban Ki-Moon, UN).

The world possesses the resources and knowledge to ensure that even the poorest countries and others held back by diseases, geographic isolation or civil strife, can be empowered to achieve the MDGs. (Ban Ki-Moon).

Bibliography

Black, R.E., Morris S.S., Bryce, J., "Where and why are 10 million children dying every year?"*Lancet*. 2003 Jun 28;361(9376):2226-34.

Shaohua Chen and Martin Ravallion. June 2004. "How have the world's poorest fared since the early 1980s?" World Bank Policy Research Working Paper 3341 Washington: World Bank.

de Onis, Mercedes, Edward A. Frongillo and Monika Blossner. 2000. "Is malnutrition declining? An analysis of changes in levels of child malnutrition since 1980." *Bulletin of the World Health Organisation* 2000 : 1222–1233.

Food and Agriculture Organisation, International Fund for Agricultural Development, World Food Program. 2002 "Reducing Poverty and Hunger, the Critical Role of Financing for Food, Agriculture, and Rural Development."

Food and Agriculture Organisation. 2006. State of World Food Insecurity 2006

Food and Agriculture Organisation. 2010. The state of food Insecurity in the World 2010

International Food Policy Research Institute. 2010. 2010 Global Hunger Index.

United Nations High Commissioner on Refugees. 2007. *Statistical Yearbook 2006* "Main Findings"

UNHCR 2008 *Global Report 2008* "The Year in Review" http://www.unhcr.org/4a2d0b1d2.pdf

World Bank. Understanding Poverty website.

References

FAO 2010: Purchase for Progress:

http//typo3.org/fileadmin/user_upload/fsn/docs/P4P_OVERVIEW_FACT_SHEET-FINAL-NOV09.pdf

http://www.fao.org/docrep/013/i1683e.pdf

http://www.worldhunger.org/articles/learn/world_hunger_facts%202002.htm

Sam Pitroda: file:///E:/sam%20Pitroda's%20new%20mission%20in%29 mitigating %20Hunger%20in%20in%20India.htm

File:///E:/SamPitroda's new mission in mitigating hunger in%20India.htm

T. Nandkumar, former Union Secretary, Food and Agriculture, Government of India, *Economic Times* of 14.02.2011, Kolkata; Lead Articles: Indian Food Security.

World Food Programme: http://en.wikipedia.org/wiki/World_Food_ Programme.

World Food Programme: http: www.facebook.com/pages/World-Food-Programme/103162203056957

World Hunger and Poverty Facts and Statistics 2011: fiile:///E:/World%20Hunger&%Food%20Security/2011%20World%20Hunger%20and%20Poverty%20Facts%20 and%20Statistics.htm

World Hunger: http://www.worldhunger.org/articles/Learn/world%20 hunger%20facts%2002.htm

4

Green Agriculture and Organic Foods

I
Concept of Green Agriculture Practices

If one were evolving a new agricultural system for the world in the new millennium that would not impact the social, economic and environmental costs that we may find in the current practice, what would one want to change? Improving water use efficiency; reducing pesticide use; promoting agro forestry, both to maintain nearby forest resources and to improve carbon sequestration; adopting conservation tillage; helping aquaculture, to cultivate fish, shrimp and the like as part of a larger integrated farm system. All these innovations add up to what we know as the Green Agriculture.

To expand the concept, the Green Agriculture may inter alia constitute of some or all of the following practices (Jules Pretty):

- Integrated pest management, while utilising ecosystem resilience and diversity for pest, disease and weed control, seeks only to use pesticides when other options are ineffective.
- Integrated nutrient management, which seeks both to balance the need to fix nitrogen within farm systems with the need to import

inorganic and organic sources of nutrients, and to reduce nutrient losses through erosion control.

- Conservation tillage, which reduces the amount of tillage, sometimes to zero, so that soil can be conserved and available moisture used more efficiently.
- Agro-forestry, which incorporates multifunctional trees into agricultural systems, and collective management of nearby forest resources.
- Aquaculture, which incorporates fish, shrimps, and other aquatic resources into farm systems, such as into irrigated rice fields and fish ponds, and so leads to increases in protein production.
- Water harvesting in dry land areas, which might mean formerly abandoned and degraded lands can be cultivated, and additional crops can be grown on small patches of irrigated land owing to better rainwater retention.
- Livestock integration into farming systems, such as dairy cattle and poultry, including using zero-grasing.

Over the four years of the study by Jules Pretty and his partners in Thailand, China, Sri Lanka, and Mexico, the number of farmers increased by 56% and the number of hectares under sustainable agriculture increased significantly by 45%, in randomly sampled areas. Of the various crops the group studied, a quarter — largely maize, potatoes and beans — saw yields double, and half saw increases of at least 20%. Only rice and cotton saw minor increases or slight declines.

The above research team sees three of the methods under study as particularly useful; water efficiency practices, including collective irrigation management, rainwater harvesting to cultivate formerly degraded lands, and watershed-level conservation; soil management to improve organic matter accumulation and carbon sequestration, including conservation tillage and no-till agriculture; and pest, weed and disease control emphasising "in-field biodiversity and reduced pesticide" use.

The team also points out benefits of carbon mitigation in these improvements as well; if 25% of developing world farmland adopted these techniques, the potential increase in carbon mitigation would amount to about100 million tonnes of carbon every year. As Pretty notes, "such gains in carbon may offer new opportunities to households for income generation under emerging carbon trading schemes." That is, not only does the adoption of sustainable agriculture practices mean more yields, it could also mean income from Clean Development Mechanism Credit available under the UN Framework Convention on Climate Change.

II

Developments in Indian Agriculture

Thrust Towards Modernisation, Growth and Sustainability Agriculture is the largest private enterprise in India; it is really the lifeline of the Indian economy. Agriculture provides food and ensures livelihood security for over a billion of the Indians and works as a catalyst to the economic growth and social transformation of the sub-continent. In 2008-09 the agricultural sector contributed approximately 15.7 per cent of India's GDP (at 2004-05 prices) and about 10.23 per cent of total exports, besides providing employment to around 58.2 per cent of the work force.

Table 4.1 Growth in Agriculture Sector vis-à-vis GDP

Year	*Total-GDP (Actual Growth) %*	*Agriculture Sector (Actual Growth) %*
2005-06	9.5	5.2
2006-07	9.7	3.7
2007-08	9.2	4.7
2008-09	6.7	1.6
2009-10	7.2	(-) 0.2

Source: Central Statistical Organisation (base year 2004-05).

In view of the above scenario, there is absolute necessity to increase investment in the agriculture sector. The overall target of GDP growth in India for the 11th Five Year Plan period 2007-2012 has been projected at 9% with an average growth rate of 4 per cent for agriculture, 10-11 per cent in industry and 9-11 per cent for the service sector. Because of the worldwide financial melt-down, the Indian plan has been adversely affected as seen in the Table 4.1. The fall in agricultural production in 2009-10 is particularly due to the deficit of 23% in rainfall during the Southwest Monsoon, which affected kharif (summer) crop production.

Reversing the Decline in the Investment in Agricultural Sector

Unfortunately the investment in the agricultural sector declined in the recent past leading to its apparent neglect. Public investment in Agriculture in real terms had witnessed a steady decline from the Sixth Plan (1980-85) as shown in Table 4.2. However, this lacunae was finally corrected by increasing Public investment in the 10th Plan period (2002-07).

Table 4.2: Plan Investment in Agriculture

Sixth Plan (1980-85)	Rs. 64,012 crores
Seventh Plan (1985-90)	Rs. 52,107 crores
Eighth Plan (1990-95)	Rs. 45,565 crores
Ninth Plan (1997-2002)	Rs. 42,226 crores
Tenth Plan (2002-2007)	Rs. 67,200 crores

However, because of the private sector contributions, the total capital formation in agriculture continued to grow faster in recent times as shown in Table 4.3.

Table 4.3: Investment in Indian Agriculture

Years	*Public Invest-ments (Rs. Crores)*	*% Share GDP*	*Private Invest-ments (Rs. Crores)*	*% Share GDP*	*Total Invest-ments (Rs. Crores)*	*% Share GDP*
2004-05	16183	2.9	62665	11.2	78848	14.1
2005-06	19909	3.4	73221	12.4	93121	15.8
2006-07	22978	3.8	71422	11.7	94400	15.4
2007-08	23039	3.6	86967	13.6	110006	17.2
2008-09	24452	3.8	114145	17.5	138597	21.3

National Food Security Mission (NFSM)

The implementation of NFSM since 2007-08 has helped to widen the geographical food basket of the country with significant contributions by the NFSM districts. Productivity gains of 25 % or more have been recorded in many of the NFSM districts. New farm practices have been promoted through 287,000 demonstrations of improved varieties of rice and wheat. Systems of Rice Intensification (SRI) for rice as well as hybrid rice have been demonstrated. Nearly, 8.4 million quintals (100 kg) of rice, wheat, pulses and hybrid seeds of rice have been distributed. An area of 5.6 million hectares has been supplied with soil ameliorants (gypsum/lime/micro nutrients) to restore soil fertility. An area of 2.2 million ha of farm lands has been treated under integrated pest management (IPM). Improved Farm machinery/water saving machinery has been distributed. Capacity building of farmers has been enhanced through 27834 farmer's field schools at the farm level. All these efforts through the last three years have resulted in increased production of rice and wheat, but the production of pulses has not unfortunately seen any significant increase in productivity.

National Agricultural Development Plan RKVY

RKVY (Rashtriya Krishi Vikas Yojana) is presently the biggest development scheme in the agricultural sector. It aims at incentivising the Indian States to increase outlays in agriculture and allied sectors so as to significantly contribute to the target of 4 per cent growth of agriculture outputs in the 11th Five Year Plan (FYP). An outlay of Rs. 25,000 crores has been earmarked in the 11th FYP for the RKVY programmes. It is an entirely grant-based programme.

The experience of past three years shows that the States have been undertaking a wide variety of projects under RKVY integrating with available resources from other schemes. Some States have already undertaken to prepare holistic States Agricultural Plan (SAP) and other States are also contemplating the same. The RKVY funds have enabled the States and district plans to scale up substantially.

National Horticultural Mission (NHM)

The horticulture sector covers a wide range of crops including fruits, vegetables, spices, flowers, medicinal and aromatic plants, mushrooms and plantation crops like cashew, coco and coconut. The National Horticulture Mission was launched in 2005-06 for the holistic development of this sector in the country. Under NHM in 2009-10, 201 new nurseries were set up and vegetable seeds were produced in 2062 ha areas. An area of 296,000 ha was brought under new gardens of various horticulture crops and an area of 37,000 ha of old and senile plantations were rejuvenated. Under the post-harvest management scheme component, 153 packaging houses, 157 storages, 1 refrigerated van, 101 primary processing units and 83 rural markets were set up. The infrastructure for collection, grading, etc was created, which helped in the proper handling and marketing of horticulture produces.

Implementation of NHM in North East India

In 2009-10, an additional area of 32,500 ha has been brought under cultivation of various horticulture crops in the North East India Himalayan States. The shares of various crops are as given in Table 4.4.

Table 4.4

Crops	*Areas Hectares*
Fruits	10,902
Vegetables	10,510
Spices	3,078
Plantation Crops	750
Aromatic Plants	1,368
Flowers	906

The infrastructure for post-harvest management, marketing and processing is also being strengthened to prevent or minimize losses of the produce. Twenty primary markets are being developed and two rural markets have already been commissioned. Six processing units are under development and three have been completed. Ropeways for improvement of accessability in difficult terrain are also being envisaged.

Seed Development Programme

For this purpose following action plans are under implementations:

- Seed Village Programmes to improve the quality of Farm saved seeds;
- Availability of quality seeds has increased from 1.4 million quintals (100 kg) in 2005-06 to 2.8 million quintals in 2009-10.
- Fifty two Seed Processing plants to be established by the Private sectors;
- Nineteen training programmes on seed quality control have been organised;
- Grants-in- aid will be released for strengthening of 25 notified seed testing laboratories;
- In response to changes in seed sector, the Seeds Act 1966 is proposed to be amended.

National Action Plan on Climate Change (NAPCC) and National Mission on Sustainable Agriculture (NMSA)

To meet the impact of the climate change on the country, National Action Plan on Climate Change (NAPCC) was proposed by the Indian Prime Minister's Council on Climate Change. The NAPCC constituted eight National Missions. Out of these eight Missions, the National Mission on Sustainable Agriculture (NMSA) devolves on the Ministry of Agriculture. The strategy of minimising the adverse impacts of climate change consists of research, adaptation, and mitigation including the following specific action programmes:

- Strategic research and application of latest agriculture technologies for increasing productivity and reducing vulnerability to climate change;
- Sustained growth in food grain production to overcome the reduced productivity arising out of increased temperature and occurrences of climatic extremes floods, draughts, and storms;
- Improvement in water use efficiency;
- Improvement in risk management systems to minimize the adverse impacts of climatic variability risks, pests and diseases;
- Other supportive measures for farmers and other stakeholders.

Growing Indian Food Production

The total geographical area of the country is 328.7 million ha, of which 193.7 million ha is the gross cropped area. The gross and net irrigation areas are 85.8 and 60.9 million ha areas respectively. Production of food grain during 2009-10 was 216.85 m tonnes.

In 2010-11 (April-March), the country's food production reached all time record of 235.88 m t (see Table 4.5). This is good news, for the country is plagued by consistent food inflation since the beginning of the last year. India is the world's largest importer and consumer of pulses. Though the production of pulses have risen to all time high at 17.3 m t (Table 4.5), the consumption has also risen and some import might still be required.

Table 4.5: Indian Food Grain Production

Food Grain Types	*Production (million tonnes)*
Food Grains	
Rice	94.1
Wheat	84.3
Coarse Cereals Maize	40.2
Maize	20.2
Pulses	17.3
Tur	3.2
Urad	1.8
Moong	1.4
Total Food Grains	**235.9**
Oil Seeds	
Soybean	12.6
Groundnut	7.1
Total Oilseeds	**30.3**
Sugarcane	**340.6**
Cotton	**33.9**

In Table 4.6 and Table 4.7 are shown the increasing rates of food production for different types of food grains.

Table 4.6: All India Area, Production and Yield of Food Grain and Irrigation Coverage

Year	*Area (million Ha)*	*Production (million tonnes)*	*Yield (Kg/Ha)*	*Area Under Irrigation %*
1950-51	97.32	50.82	522	18.1
1955-56	110.56	66.85	605	18.5
1960-61	115.58	82.02	710	19.1
1965-66	115.10	72.35	629	20.9
1970-71	124.32	108.42	872	24.1
1975-76	128.18	121.03	944	26.5
1980-81	126.67	129.59	1023	29.7
1985-86	128.02	150.44	1175	31.4
1990-91	127.84	176.39	1380	35.1
1995-96	121.01	180.42	1491	41.1
2000-01	121.05	196.81	1622	43.4
2005-06	121.65	217.26	1756	46.3
2007-08	124.07	230.76	1860	46.8
2010-11		235.9		

III

An Overview of Green Agriculture in India

Present Position

There has been a spurt in domestic and international demand for greening agriculture across the countries as a result of initiatives of green activists against chemical and synthetic fertilizers and pesticides.

Table 4.7: Different Types of Food Grain Productions

(million tones)

Year	*Rice*	*Wheat*	*Coarse cereals*	*Jowar*	*Bajra*	*Maize*	*Pulse*	*Gram*	*Tur*	*Total Food Grain*
1950-51	20.58	6.46	15.38	5.50	2.60	1.73	8.41	8.65	1.72	**50.82**
1955-56	27.56	8.76	19.49	6.73	3.3	2.60	11.04	5.42	1.86	**66.85**
1960-61	34.58	11.00	23.74	9.81	3.28	4.08	12.70	6.25	2.07	**82.02**
1965-70	30.59	10.40	21.42	7.58	3.75	4.82	9.94	4.22	1.73	**72.35**
1970-71	42.22	23.83	30.55	8.11	8.03	7.49	11.82	5.20	1.88	**108.42**
1975-76	48.74	28.64	30.41	9.50	5.74	7.26	13.04	5.88	2.10	**121.03**
1980-81	53.63	36.31	29.02	10.43	5.34	6.96	10.63	4.33	1.96	**129.59**
1985-86	63.83	47.05	26.20	10.21	3.66	6.46	13.36	5.79	2.44	**150.44**
1990-91	74.29	55.14	32.70	11.68	6.89	8.96	14.26	5.36	3.41	**176.39**
1995-96	76.98	62.10	29.03	9.93	5.38	9.53	12.31	4.98	2.31	**180.42**
2000-01	84.98	62.10	31.08	7.56	6.76	12.04	11.08	3.96	2.25	**196.81**
2005-06	91.79	69.35	34.07	7.24	7.68	14.71	13.39	5.60	2.74	**208.60**
2008-09	99.18	80.68	40.04	7.25	8.89	19.73	14.57	6.91	3.79	**234.47**
2009-10	89.13	80.71	33.77	6.98	6.50	16.68	14.10			**218.20**
2010-11	94.1	84.3	40.2			20.2	17.3		3.2	**235.88**

But the uses of chemical fertilizers and pesticides have played a positive role in increasing agricultural productivity and in making India self-sufficient in food grain production. Yield of food-grain in India increased from 644 kg per hectare in 1966-67 to 1636 kg per hectare in 2000-2001 (http://www.indiastat.com/), thus registering an impressive increase by around two and half times. This was mainly brought about by a more than 12 fold increase in the consumption of chemical fertilizers (from 1.1 million ton to 13.56 million tons) during the same period (Pawan Wadhwa, 2001). This apart, inorganic chemical use in agriculture has also contributed towards increasing productivity of cash crops.

Chemical Fertilizers

After independence the use of fertilizers in India in the last 50 years has grown nearly 170 times. In 1950 use of fertilizer per hectare in India was 0.55 kg but by 2001-02 this figure has increased to around 90.12 kg per hectare. Green revolution during 1960s and subsequent years led to intensification of agriculture being the major cause behind this growth. Fertilizers and pesticides have become major cost of production in India along with the cost of other input like seeds, and labor cost (www.etagriculture.com).

Pesticides

Other major inputs for Indian agriculture are use of various pesticides, like insecticides, weedicides, fungicides, rodenticides, etc. As the cropping pattern is becoming more intensive, use of these pesticides is also increasing. Consumption of insecticide in agriculture has increased more than 100% from 1971 to 1994-95. For instance, insecticide consumption in India, which was to the tune of 22,013 tonnes, has increased to 51,755 tonnes by 1994-95 (www.indiastat.com). Consumption of all of these pesticides in same duration has increased more than two times, that is from 24305 tonnes to 61357 tonnes.

But in recent past, changes have been observed in trends of pesticides consumption. As a consequence of adoption of bio-intensive Integrated Pest Management Programme in various crops, the consumption of chemical pesticide (Tech. Grade) has come down from 66.36 thousand tonnes during 1994-95 to 43.59 thousand tonnes during 2001-02 with a reduction of 27.69% (Thirty Seventh Report of Standing Committee on Petroleum and Chemicals, 2002).

One of the consequences of indiscriminate use of pesticide is the adverse health impact on society in general and vulnerable population like children in particular. Some of the well-known health effects of pesticide exposure include acute poisoning, cancer, neurological effects and reproductive and developmental harm (CSE, 1997). The major causes of concern in this respect are bio-accumulation of pesticides and the prolonged time period that it takes to express the negative health consequences.

Green Input Market - Spatial and Temporal Trends

It is very difficult to estimate the total size of green inputs market in India because of its diversity in terms of products and also due to the nature of it being unorganised market. Green inputs into agriculture include bio-fertilizers, bio-pesticides, compost, Farm Yard Manure (FYM), green manure, etc. As most of the inputs are either not traded and even if they are traded, it is only at informal levels, available information regarding production capacity, demand and sales is at best sketchy estimation and hence inadequate. The limited number well-established firms that have their presence in this market today; as a result, this market is predominately controlled by the small and local producers of bio-fertilizer, vermi compost and other input producers who are in large numbers.

Relatively speaking there is more rigour in estimation of bio-fertilizer market in India, because of the presence of some large producers in production of bio-fertilizers. Comparatively this is more organised than other green inputs market. Based on the gross cropped area in India (190 million hectare) and recommended doses of bio-fertilizers, potential demand is estimated to be about 6,27,000 tonnes. This demand can further be segregated into different categories of bio-fertilizer, such as Rhizobium, Azotobactors, Azospirillium, BGA, and Phosphate Solubilizer etc. with widely differing potential demands as shown in Table 4.8.

Table 4.8: Estimated Total Potential Demand for Bio-fertilizers in India

Category of Bio- Fertilizer	*Amount in Million Tonnes*
Rhizobium	35,730Mt
Azotobacter	162,610Mt
Azospirillum	77,160Mt.
BGA	267,510Mt
Phosphate solubilizer	275,510Mt
TOTAL	818,730Mt

Source: Abhay Phadke, 2001

Green Outputs Market Trends and Potentials in India

Demand for green agricultural products is a stimulant for growth for input market. In other words, if there is demand in market for organically produced farm products, that will encourage farmers to implement the organic farming practices and also to use organic input like bio-fertilizers, bio-pesticides, vermin-compost, green manure and FYM.

Estimating area under organic agriculture in India is a difficult task as there is no central agency that collects and compiles this information. Different agencies have estimated the area under organic agriculture differently, for instance, the study undertaken by FIBL and ORG-MARG

(Garibay S V and Jyoti K, 2003) estimates the area under organic agriculture to be 2,775 hectares (0.0015% of gross cultivated area in India). But there is another estimate by SOEL-Survey, which shows that the land area under organic cropping is 41,000 hectares. The total numbers of organic farms in the country as per SOEL-Survey are 5661 but FIBL and ORG-MARG survey puts it as 1426. Some of the major organically produced agricultural crops in India include crops like plantation, spices, pulses, fruits, vegetables and oil seeds etc (Table: 4.9).

Table 4.9: Major products produced in India by organic farming

Major products produced in India by organic farming	
Type of Product	*Products*
Commodity	Tea, Coffee, Rice, Wheat
Spices	Cardamom, Black pepper, white pepper, Ginger, Turmeric, Vanilla, Tamarind, Clove, Cinnamon, Nutmeg, Mace, Chili
Pulses	Red gram, Black gram
Fruits	Mango, Banana, Pineapple, Passion fruit, Sugarcane, Orange, Cashew nut, Walnut
Vegetables	Okra, Brinjals, Garlic, Onion, Tomato, Potato
Oil seeds	Mustard, Sesame, Castor, Sunflower
Others	Cotton, Herbal extracts

Source: Garibay S V and Jyoti K, 2003

Export Market

Organic agricultural export market is one of the major drivers of greening of agriculture in India. The current production of organic crops is around 14,000 tonnes (Garibay S V and Jyoti K, 2003). Out of this production, tea and rice contributes around 24% each, fruits and vegetables combined makes 17% of this total production. From India around 11,925 tons of organic products are exported, that makes around 85% of total organic crop production. Major export market for Indian producers are Australia, Belgium, Canada, France, Germany, Italy, Japan, Netherlands, Sweden, Singapore, South Africa, Saudi Arabia, UAE, UK, and USA. Estimated quantity of various products that are exported from India in 2002 is shown in Table 4.10. This shows that around 3000 tons of tea was exported and in quantity term it was the highest, next major exports are rice (2500 tons), fruits and vegetables (1800 tons), cotton (1200 tons) and wheat (1150 tons) (Garibay S V and Jyoti K, 2003).

Table 4.10: Major Organic Products exported from India

Product	*Sales (Tonnes)*
Tea	3000
Coffee	550
Spices	700
Rice	2500
Wheat	1150
Pulses	300
Oil Seeds	100
Fruits/Vegetables	1800
Cashew Nut	375
Cotton	1200
Herbal Products	250
Total	**11,925**

Source: Org-Marg, 2002

Domestic Market

Given the unorganised nature of the domestic organic agriculture market in India, it is difficult to estimate the magnitude and trends in this growing market. In general, the sale of organic produces is limited to metros like Mumbai, Delhi, Kolkata, Chennai, Bangalore and Hyderabad. To a large extent this sale is also based on individual initiative of the farmers, Non-Government Organisations and some entrepreneurial traders etc. The current domestic green products market demand is mainly for fruits, vegetables, rice and wheat. Other products include tea, coffee and pulses (ORG-MARG Survey, 2002). The market prospects for other commodities like organic spices, fruits, herbal plants and cotton are relatively high. For next five years it is projected that organic spices would grow by 14%, fruits 8% and herbal plants and cotton estimated to be around 7% (ORG-MARG Survey, 2002). Markets for different range of organic agricultural products are as shown in Table 4.11. It is estimated to reach up to 1568 tonnes in 2006-07.

Table 4.11: Growth forecast for specific Products

Product	*% Projected Growth in the next Five Years*
Spices (all)	14
Pepper	5
Turmeric	4.5
Tea	13
Rice	10
Fruits (all)	8
Banana	15
Mango	5
Orange	5

(contd...)

Product	*% Projected Growth in the next Five Years*
Pineapple	5
Herbal extracts	7
Cotton	7
Coffee	5
Oil seeds	5
Honey	5
Groundnut	5
Baby food	5
Coconut	5

Source: Org-Marg, 2002.

Government Role in Promotion of Green Agriculture in India

There has been limited allocation under 9th National Five Year Plan for the popularisation of bio-fertilizer and bio-pesticide in the agriculture sector; the provision for this programme was included in expanded scope of activities for the promotion of organic agriculture as part of the activities targeted in the 10th National Plan. As part of 10th Five year Plan, Government of India has earmarked about Rs. 100 crores for the promotion of organic agriculture in the country. The main components of this initiative include framing of standards, negotiating with different countries and putting in place a system of certification for organic products.

Central Government is also promoting the production and use of bio-fertilizers to make it popular. Government has initiated a project *"National Project on Development and Use of Bio fertilizers"* for this purpose. Main objectives of this project are as following:

(i) Production and distribution of Bio fertilizers (BFs)
(ii) Developing Standards for different BFs and Quality control
(iii) Releasing of grants for setting up BF units
(iv) Training and Publicity

To promote the organic agriculture in India government has also taken some initiative in recent past. APEDA (Agriculture and Processed Food Products Export Development Agency) has been made the nodal agency to promote the Indian organic agriculture and its exports opportunities. Under this program, National Organic Standards have been evolved. This apart, it has also developed criteria for accreditation of certification agencies, Accreditation Procedure and Inspection and Certification Procedures. In developing these standards and procedures due attention is paid to the guidelines as enumerated by international organisations such as International Federation for Organic Agricultural Movement (IFOAM), EU Regulations and FAO Codex Standards. As part of this program, a National Logo for organic products on behalf of Govt. of India has also been developed (http://www.apeda.com/organic.htm).

Some of the other efforts towards promotion of organic exports include attempts to collaborate with all the major organic importing countries. Towards this APEDA is deliberating with European Union for inclusion of India in the list of third world countries under Article 11 of the EU regulations No 2092/91 so that India's National Programme for Organic Production gets the required recognition under the EU regulations.

Facilitating Factors for Growth of Organic Agriculture

Organic agriculture provides economic opportunities for different stakeholders. Some of the drivers that facilitate growth of organic agriculture in India are (Source: Mongillo 2004, Ghosh Roy 2011)

- Growing export market for organically produced crops (Kortbech-Olesen, 2003).
- Price premium for organically produced agriculture products from10% to 100% (V. Garibay, K. Jyoti, 2003).
- Diverse agro-climate regions across the country that provides environment for wide range of crops that can cater to different market demands.
- Increasing awareness and health consciousness especially among certain sectors of domestic consumers.
- Availability of comparatively cheap labor for labor-intensive organic agriculture.
- Huge numbers of small farmers those who do the traditional farming with very limited capacity not able to pay for most of the chemical inputs into agriculture ((Planning Commission, 2001).
- Presence of Non-Government Organisations (NGOs) as active promoters of Organic farming in different agro-climatic regions (Donthi N. Reddy, 2001).
- Increasing involvement of private companies in field of agricultural extension, trade, consultation and other services.
- Enhanced Government attention and support for organic agriculture through various policy initiations and action programs.

Roles of APEDA in Export of Processed/Organic Food

The Agricultural and Processed Food Products Export Development Authority (APEDA) was established by the Government of India with the objectives including registration of exporters, standardisation, inspection and quality control, improving and assisting marketing, training in food processing and export promotion with the additional responsibility of promoting Green agriculture.

APEDA has the responsibility of export promotion and development of the following products:

- Fruits, Vegetables and their Products.
- Meat and Meat Products.
- Poultry and Poultry Products.
- Dairy Products.
- Confectionery, Biscuits and Bakery Products.
- Honey, Jaggery and Sugar Products.
- Cocoa and its products, chocolates of all kinds.
- Alcoholic and Non-Alcoholic Beverages.
- Cereal and Cereal Products.
- Groundnuts, Peanuts and Walnuts.
- Pickles, Papads and Chutneys.
- Guar Gum.
- Floriculture and Floriculture Products
- Herbal and Medicinal Plants

In addition, APEDA has been entrusted with the responsibility to monitor import of sugar.

Membership of APEDA

The following reputed Institutions are nominated members of APEDA to enrich its activities:

- Indian Council of Agricultural Research
- National Horticultural Board
- National Agricultural Cooperative Marketing Federation
- Central Food Technological Research Institute
- Indian Institute of Packaging
- Spices Export Promotion Council and
- Cashew Export Promotion Council.

The following Industry bodies are also nominated members of APEDA:

- Fruit and Vegetable Products Industries
- Meat, Poultry and Dairy Products Industries
- Other Scheduled Products Industries
- Packaging Industry

IV
Outlook on World Agriculture Till 2030

Population

World population will grow from around 6 billion people today to 8.3 billion people in 2030. Population will be growing at an average of 1.1 per cent a year up to 2030, compared to 1.7 per cent annually over the past 30 years. At the same time, an ever increasing share of the world's population will be well-fed. As a result, the growth in world demand for agricultural products is expected to slow down, from an average 2.2 per cent annually over the past 30 years to 1.5 per cent per year until 2030. In developing countries, the slowdown will be more dramatic, from 3.7 per cent for the past 30 years to an average of 2 per cent until 2030. However, the developing countries with low to medium levels of consumption, accounting for about half of the population, would see demand growth slowing only from 2.9 to 2.5 per cent per year, but per capita consumption will rise.

Food Consumption

The world population will be increasingly well-fed by 2030, with 3050 kilocalories (kcal) available per person, compared to 2360 kcal per person per day in the mid-1960s and 2800 kcal today. This change reflects above all the rising consumption in many developing countries whose average will be close to 3000 kcal in 2030. The number of hungry people in developing countries is expected to decline from 777 million today to about 440 million in 2030. This means, that the target of the World Food Summit in 1996, to reduce the number of hungry by half from its level in 1990-92 (815 million) by 2015, will not even be met by 2030.

Sub-Saharan Africa is cause for serious concern, because the number of chronically undernourished people will only decrease from 194 to 183 million. Patterns of food consumption are becoming more similar throughout the world, shifting towards higher-quality and more expensive protein foods such as meat and dairy products. Meat consumption in developing countries, for example, has risen from only 10 kg per person annually in 1964-66 to 26 kg in 1997-99. It is projected to rise to 37 kg per person per year in 2030. Milk and dairy products have also seen rapid growth, from 28 kg per person per year in 1964-66, to 45 kg now, and could rise to 66 kg in 2030. FAO expects increases in meat and dairy consumption to be less dramatic than in the past.

Food Production

Cereals are still by far the world's most important sources of food, both for direct human consumption and meat production. An extra billion tonnes of cereals will be needed by 2030. The developing countries will become increasingly dependent on cereal, meat and milk imports, their production will not keep pace with demand. By 2030 they could be producing only 86 per cent of their own cereal needs, with net imports rising from currently 103 million tonnes to 265 million tonnes by 2030. Traditional grain exporters such as the US, the EU, Canada, Australia and Argentina, and the transition countries as emerging exporters, are expected to produce the surpluses needed to fill this gap. "If real food prices do not rise and exports of industry products and services grow as previously, then most countries will be able to afford to import cereals to meet their needs. However, the poorest countries tend to be the least able to pay for imports."

The use of cereals as animal feed does not contribute to hunger and undernutrition. Globally, some 660 million tonnes of cereals are used as livestock feed each year. This represents just over a third of total world cereal use. If these cereals were not used as feed, they would probably not be produced at all, so would not be available as food in many cases, according to the report. More likely, the lack of demand for cereals for livestock production would lead to lower crop production.

Much of future food production growth will come from higher productivity. In developing countries, almost 70 per cent of the increase in crop production will come from higher yields, around 20 per cent from an expansion of arable land and around 10 per cent from multiple cropping and shorter fallow periods.

The expansion of farmland for food production will be slower than in the past. In the next 30 years, developing countries will need an additional 120 million ha for crops; this means, less new land will be opened up than in the past. The expansion will mainly take place in sub-Saharan Africa and Latin America. A considerable part of this extra land will probably come from forest clearance. In other developing regions, almost all suitable land is already in use. Some countries and communities will face problems related to land scarcity.

During the 1990s, the world lost a forest area of 9.4 million ha per year, about three times the size of Belgium. However, the rate of deforestation was slower than in the 1980s, and, globally, deforestation will probably continue to slow down in future, although much of the cropland expansion will have to come from forests and world consumption of industrial roundwood is expected to rise by 60 per cent over current levels.

Irrigation is crucial to the world's food supplies. The developing countries are likely to expand their irrigated area from 202 million ha today to 242 million ha by 2030.

Water

At global level there is enough water available, but some regions will face serious water shortages. A 14 per cent increase in water withdrawals for irrigation is expected for developing countries by 2030. One in five developing countries will be suffering water scarcity. Two countries, Libyan Arab Jamahiriya and Saudi-Arabia, are already using more water for irrigation than their annual renewable resources, by drawing on fossil groundwater. In large areas of India and China, ground-water levels are falling by 1 to 3 metres per year. These regions will need to use water more efficiently. Agriculture is responsible for about 70 per cent of all fresh water withdrawn for human use. Saving water in agriculture means that more water is available for other sectors.

Application of Technologies

Modern biotechnology offers promise as a means to improving food security. If the environmental threats from biotechnology are addressed, and if the technology is affordable and geared towards the needs of the poor and undernourished, genetically modified crop varieties could help to sustain farming in marginal areas and to restore degraded lands to production. To address the concerns of consumers FAO called for improved testing and safety protocols for genetically modified organisms.

Other promising technologies have emerged that combine increased production with improved environmental protection. These include no-till/ conservation agriculture and integrated pest or nutrient management. Locally, organic agriculture could become a realistic alternative to traditional agriculture over the next 30 years.

Livestock

Future demand for livestock and dairy products can be met, but the consequences of increased production must be addressed. Production will shift away from extensive grasing systems towards more intensive and industrial methods. "This could pose a threat to the estimated 675 million rural poor whose livelihoods depend on livestock. Without special measures, the poor will find it harder to compete and may become marginalised, descending into still deeper poverty. If the policy environment is right, the future growth in demand for livestock products could provide an opportunity for poor families to generate additional income and employment." Environmental and health problems of industrial meat production (waste disposal, pollution, the spread of animal diseases, overuse of antibiotics) also need to be addressed.

Climate Change

Climate change could increase the dependency of some developing countries on food imports. The overall effect of climate change on global food production by 2030 is likely to be small. Production will probably be boosted in developed countries. Hardest-hit will be small-scale farmers in areas affected by drought, flooding, salt water intrusion or sea surges. Some countries, mainly in Africa, are likely to become more vulnerable to food insecurity.

Fish Production

With many marine stocks now fully exploited or overexploited, future fish supplies are likely to be constrained by resource limits. The share of capture fisheries in world production will continue to decline, and the contribution of aquaculture to world fish production will continue to grow. The capacity of the global fishing fleet should be brought to a level at which fish stocks can be harvested sustainably, FAO said. "Past policies have promoted the build-up of excess capacity and incited fishermen to increase the catch beyond sustainable levels. Policy makers must act to reverse this situation."

References

APEDA; www.apeda.com ; http://www.apeda.com/organic.htm

FAO: World Agriculture 2030

File:///E:/World%20Agriculture/FAO%20%20%20world% 20agriculture%202030% 20%220 br%20%29Main%20findings.htm

Green Agriculture; www.indiastat.com

Greening Agriculture in India - An Overview of Oppportunities and Constraints, Thilotham R kolanu (trkolanu@ascu.org.in) and Sunil Kumar (sunilkumarg@yahoo.com)

Jules Pretty, "Resource-Conserving Agriculture Increases Yields in Developing Countries," University of Essex; http://pubs.acs.org/cgi-bin/abstract.cgi/esthag/asap/abs/es051670d.html.

Mongillo, John 2004, *Environmental Sciences*, Vol 5, Creating a Sustainable Society, Greenwood Press, Westport, Connecticut, USA.

M.K. Ghosh Roy, *Sustainable Development - Environment, Energy and Water Resources*, 2011, Ane Books, New Delhi, India.

Ministry of Agriculture, Dept of Agriculture, Annual Report 2010.

www.agricoop.nice.in Ministry of Agriculture, GOI; http://agricoop.nice.in/compendium7410.pdf.

TOI;www.gogreenindia.co.in Abode; [greenindia.co.inPlanet%20 Earth%202011%20 Circularr.pdf]

5

Sustainability of Global Fisheries Resources

I

Indian Fisheries Development

Indian Fish Production

The Department of Animal Husbandry, Dairying and Fisheries has been undertaking various production, input supply and infrastructure development programmes and welfare oriented schemes besides formulating/ initiating appropriate policies to increase production and productivity in the Fisheries Sector.

Fish production since 1980-81 is shown in the table 5.1:

Table 5.1: Indian Fish Production

Fish production (Lakh tonnes) since 1980-1981			
Year	*Marine*	*Inland*	*Total*
1980-81	15.55	08.87	24.42
1990-91	23.00	15.36	38.36
1995-96	27.07	22.42	49.49
2000-2001	28.11	28.45	56.56
2005-2006	28.16	37.55	65.71
2006-2007	30.24	38.45	68.69
2007-2008	29.19	42.07	71.26

1 lakh = 10^5

The fisheries sector has been one of the major contributors of foreign exchange earnings through export. Export of fish and fishery products has grown manifold over the years. From about 15,700 tonnes valued at Rs. 3.92 crore in 1961-62, exports have grown to 5.41 lakh tonnes valued at Rs. 7621 crore in 2007-08 (1 crore = 10 million).

Inland and Coastal Fishing

Inland Fisheries and Aquaculture

The ongoing scheme of Development of Freshwater Aquaculture and Integrated Coastal Aquaculture have been combined with four new programmes on Development of Coldwater Fish Culture, Development of water-logged Area and Derelict water bodies into aquaculture estates, Use of Inland Saline/Alkaline Soil for Aquaculture and programme for augmenting the Productivity of Reservoirs. This scheme broadly has two components: Aquaculture and Inland Capture Fisheries.

Development of Freshwater Aquaculture: FFDAs

The Government has been implementing an important programme in inland sector, viz., Development of Freshwater Aquaculture through the Fish Farmers Development Agencies (FFDAs). A network of 429 FFDAs covering all potential districts in the country are in operation. During 2007-08, about 24,752 ha of water area was brought under fish culture and 35,000 fish farmers were trained in improved aquaculture practices through FFDAs.

Development of Brackish-water Aquaculture: BFFDAs

With the objective of utilising the country's vast brackish-water area for shrimp culture, an area of about 30,889 hectares was developed for shrimp culture till 2007-08 through 39 Brackish-water Fish Farmers Development Agencies (BFFDAs) set up in the coastal areas of the country. The agencies have also trained 31,624 fishermen in improved practices of shrimp culture till 2007-08. Presently about 50 per cent of the shrimp exported from the country is from aquaculture.

Marine Fishing

Development of Marine Fisheries

The Government is providing subsidy to poor fishermen for motorising their traditional craft, which increases the fishing areas and frequency of operation with consequent increase in catch and earnings of fishermen. About 46,223 traditional crafts have been motorised so far. The Government has also been operating a scheme on fishermen development rebate on HSD oil used by fishing vessels below 20 meter length to offset the operational cost incurred by small mechanised fishing boat operators.

Fishing Harbour

The Government has been implementing a scheme with the objective of providing infrastructure facilities for safe landing and berthing to the fishing vessels. Since inception of the scheme, six major fishing harbours viz., Cochin, Chennai, Visakhapatnam, Roychowk, Paradip and Sassoon Dock (Mumbai), 62 minor fishing harbours and 190 fish landing centres have been taken up for construction in various coastal States/UTs.

Welfare Programmes for Traditional Fishermen

Important programmes for the welfare of traditional fishermen are:

- Group Insurance Scheme for active fishermen

- Development of Model Fishermen Villages
- Saving-cum- relief Scheme, and
- Saving cum component financial assistance is provided to the fishermen during the lean fishing season.

About 3.5 lakh fishermen were assisted under the saving-cum-relief programme in 2008-2009.

Developmental Institutes for Fisheries

The Central Institute of Fisheries, Nautical and Engineering Training, Kochi with units at Chennai and Visakhapatnam, aims at making available sufficient number of operators of deep-sea fishing vessels and technicians for shore establishments. Integrated Fisheries Project, Kochi, envisages processing, popularising and test marketing of unconventional varieties of fish. The Central Institute of Coastal Engineering for Fisheries, Bangaluru, is engaged in techno-economic feasibility study for location of fishing harbour sites. Fishery Survey of India (FSI) is the nodal organisation responsible for survey and assessment of fishery resources under the Indian Exclusive Economic Zone (EEZ).

National Fisheries Development Board

National Fisheries Development Board was established to work towards blue revolution with a focus on increasing the fish production of the country to a level of 10.3 million tonnes, to achieve double the exports from Rupees 7,000 crores to Rupees14,000 crores and direct employment to an extent of 3.5 million by extending assistance to the various agencies for implementation of activities under Inland, Brackish water and Marine sectors. It will become a platform for public-private partnership for fisheries, a mechanism for an end approach for ensuring self-availability to efficient marketing etc.

It is an autonomous organisation under the administrative control of the Department of Animal Husbandry, Dairying and Fisheries, Ministry of Agriculture, Govt. of India. It has been registered with the Office of the Registrar of Societies, Hyderabad on 10th July 2006. The Board was inaugurated on 9th September 2006. The office was established at Hyderabad.

The period of implementation of the various activities of the Board is 6 years (2006-12):

- To bring major activities relating to fisheries and aquaculture for focused attention and professional management.
- To coordinate activities pertaining to fisheries undertaken by different Ministries/Departments in the Central Government and also coordinate with the State/Union territory Government.
- To improve production, processing, storage, transport and marketing of the products and culture fisheries.
- To achieve sustainable management and conservation of natural aquatic resources including the fish stocks.

- To apply modern tools of research and development including biotechnology for optimising production and productivity from fisheries.
- To provide modern infrastructure mechanisms for fisheries and ensure their effective management and optimum utilisation.
- To generate substantial employment.
- To train and empower women in the fisheries sector, and
- To enhance contribution of fish towards food and nutritional security.

Major activities to be taken up by the National Fisheries Development Board:

- Intensive aquaculture in ponds and tanks.
- Enhancing productivity from Reservoir Fisheries
- Brackish water Coastal Aquaculture
- Mariculture i.e. Marine aquaculture
- Sea Ranching
- Seaweed Cultivation
- Infrastructure for Post Harvest Programmes
- Fish Dressing Centres and Solar Drying Fish
- Domestic Marketing
- Other Activities

The total budget provision for implementation of various activities of National Fisheries Development Board for the period 2006-12 is Rs.2100.00 crores and subsequently reduced to Rs.1500.00 crores.

II

Perspectives of Global Fisheries

Brief Overview: Fish Resources

Total trends in fishing trade comprising production, utilisation, capture fishing and aquaculture resulted in the availability of about 110 million tonnes of food fish in 2006, providing an effective per capita fish of 16.7 kg, which is the highest on record.

Since 1992, the aquaculture accounted for 47 per cent or more of total catch and the growth in aquaculture made up the offsets in production of capture fishing.

Overall fish provided more than 2.9 billion people with at least 15 per cent of their protein intake. The share of the fish protein in total world annual protein supply grew from 14.9 per cent in 1992 to peak of 16 per cent in 1996, declining again to about 15.8 per cent in 2005.

China remains by far the largest producer, with reported fisheries production of 51.5 million tonnes (17.1 mt by captive fisheries and 34.4 mt from aquaculture respectively), providing an estimated domestic food supply of 29.4 kg per capita as well as production for export and non-food purposes. But FAO believes that there are genuine concerns that the Chinese statistics are overly high and likely to be incorrect.

Table 5.2 World Fishing: Capture and Aquaculture Production and Utilization

Year	2002	2003	2004	2005	2006
PRODUCTION (million tonnes)					
INLAND					
Capture	8.7	9.0	8.9	9.7	10.1
Aquaculture	24.0	25.5	27.8	29.6	31.6
Total Inland	32.7	34.5	36.7	39.3	41.7
MARINE					
Capture	84.5	81.5	85.7	84.5	81.9
Aquaculture	16.0	17.2	18.1	18.9	21.1
Total marine	100.9	98.7	103.8	103.4	102.0
Total Capture	93.2	90.5	94.6	94.2	92.0
Total Aquaculture	40.4	42.7	45.9	48.5	51.7
Total World Fishing	**133.6**	**133.2**	**140.5**	**142.7**	**143.6**
UTILILZATION (Million tones)					
Human Consumption	100.7	103.4	104.5	107.1	110.4
Non-food Uses	32.9	29.8	36.0	35.6	33.3
Population billion	6.3	6.4	6.4	6.5	6.6
Per capita food fish supply (kg/capita)	16.0	16.3	16.2	16.4	16.7

Marine Capture Fisheries

According to FAO data, global marine capture production in 2006 was about 92 million tonnes, a decrease of 2.2 million tonnes in comparison to 2005 (Table 5.2). This fall was attributed to adverse environment fluctuations. However, while marine capture production fell in 2006, the inland capture increased in 2006.

China has remained by far the global leader with more than 17 million tonnes in 2006 and a very stable capture production. Compared with 2004, the ranking of top ten producer countries (Table 5.3) remained unchanged with only two exceptions. Chile ranked two places lower and Philippines replaced Norway in the tenth position.

Table 5.3: Capture (Marine and Inland): Top 10 Countries (2006)

Rank	*Countries*	*Production million tonnes*
1	China	17.1
2	Peru	7.1
3	United States	4.9
4	Indonesia	4.8
5	Japan	4.2
6	Chile	4.2
7	India	3.9
8	Russian Federation	3.3
9	Thailand	2.8
10	Philippines	2.3

Global marine capture production was 81.9 million tonnes in 2006, third lowest since 1994. Only in 1998 and 2003, the production was lower.

Although the ranking of the first eight principal marine fishing areas in 2006 (Table 5.4) was same as in 2004, overall catches in the Western Central Pacific and the Western Indian Ocean continued to increase.

Table 5.4: Capture Fishing: Principal Sea Areas (2006)

Rank	*Principal Sea Areas*	*Production million tonnes*
1	North-West Pacific	21.6
2	South-East Pacific	12.0
3	Central-West Pacific	11.2
4	North-East Atlantic	9.1
5	Eastern Indian Ocean	5.8
6	Western Indian Ocean	4.5
7	Eastern-Central Atlantic	3.3
8	North-East Pacific	3.1
9	South-West Atlantic	2.4
10	North-West Atlantic	2.2

Inland Capture Fisheries

In 2006, reported global inland water catches exceeded 10 million tonnes for the first time. Almost all the increase in the last two years has come from Asia. This continent now accounts for two-thirds of total global inland capture production (Table 5.5). Africa is the clear second.

Table 5.5: Inland Capture Fishing by Continent (2006)

Oceania	*Europe*	*Americas*	*Africa*	*Asia*	*Total*
0.2%	3.5%	5.9%	23.5%	66.9%	100% (10.1 mt)

China and other developing countries together now represent 95 per cent of global inland capture production (Table 5.6). Generally, in the developing countries, inland fisheries constitute primary sources of animal protein.

Table 5.6: Inland Capture Fisheries: Top ten Producer Countries (2006)

Countries	*Inland Capture Production thousand tonnes*	*Inland Capture Production%*
China	2544	25.3
Bangladesh	957	9.3
India	868	8.5
Myanmar	631	6.3
Cambodia	422	4.2
Uganda	367	3.6
Indonesia	301	3
Tanzania	293	2.9
Egypt	256	2.5
Brazil	251	2.5
Other Countries	3500	31.6

Aquaculture

The contribution of aquaculture to global supplies of fishes and other aquatic animals has continued to grow, increasing from 3.9 per cent of total production by weight in 1970 to 36% in 2006. In the same period, production from aquaculture easily outpaced population growth, with per capita supply increasing from 0.7 kg in 1970 to 7.8 kg in 2006, an average annual growth rate of 7.0 per cent.

Aquaculture accounted for 47 per cent of world's fish food supply in 2006. World aquaculture has grown dramatically in the last 50 years. From a production of less than 1 million tonnes in early 1950, production in 2006 had risen to 51.7 million tonnes (Table 5.2) with a value of USD 78.8 billion.

Table 5.7: Aquaculture Production by Region (%)

China	*Rest of Asia Pacific*	*Rest of World*				
		10.5				
		Europe	*Latin America & Caribbean*	*Africa*	*North America*	*North East*
66.7	22.8	4.2	3.0	1.5	1.2	0.6

In 2006, countries in the Asia and Pacific regions accounted for 89.5 per cent of production by quantity (Table 5.7). China accounted for 66.7 %. The Latin America and Caribbean regions showed highest average growth rate at 22 % followed Near East (20 %) and Africa (12.7 %).

Table 5.8 shows the top ten producing countries in terms of annual growth in aquaculture production for the period of two years 2004-2006.

In 2006, more than half of global aquaculture production was fresh water finfish. Output amounted to 27.8 mt, worth USD 29.5 billion.

Fishers and Fish Farmers

Fisheries and fish farming play, either directly or indirectly, an essential role in the livelihood of millions of people around the world. In 2006, 43.5 million people were directly engaged, part time or full time in primary production of fish either in fishing or fish farming (Table 5.9). They accounted for 3.2 per cent of the 1.37 billion people actively engaged in agriculture worldwide.

In recent decades, major increases in the total number of people engaged in fisheries have come from the development of aquaculture activities. Aquaculture can provide an important source of livelihood for the rural poor.

Table 5.10 compares fresh productions by continent with number of people employed in the primary sector. The highest concentration of people employed is in Asia, but the average production per person there is only 2.5 tonnes per year, whereas it is 21 tonnes per year in Europe and nearly 20 tonnes in America. The high figure for Oceania might be due to incomplete reporting. The higher figures on production per person indicate the degree of industrialisation of fishing activities and also the key role played by small-scale fisheries in Asia and Africa.

While the number of people employed in fisheries and aquaculture has been growing steadily in most low-income and middle-income countries, employment in Japan and Norway have about halved since 1970, down 61 and 42 per cent respectively. In many industrial countries, the decline has occurred mainly in capture fisheries, while the number of fish farmers has increased.

Table 5.8: Aquaculture Production - Top Ten Countries

	China	*India*	*Vietnam*	*Thailand*	*Indonesia*	*Bangl-adesh*	*Chile*	*Japan*	*Norway*	*Philippine*
2004 (mt)	30.61	2.79	1.19	1.26	1.04	0.91	0.66	0.77	0.64	0.51
2006 (mt)	34.43	3.12	1.66	1.38	1.29	0.89	0.80	0.73	0.71	0.62
% Growth	6.05	5.71	17.6	4.87	11.23	-1.25	9.81	-2.78	5.5	10.32

mt: million tonnes

Table 5.9: People employed in Fish & Fish Farming by Continent (million)

Continents	1990	2000	2005	2006
Africa	1.77	3.63	3.59	3.64
North & Central America	0.77	0.89	1.03	1.04
South America	0.73	0.0.71	0.70	0.71
Asia	23.77	34.78	36.65	27.34
Oceania	0.06	0.05	0.05	0.06
World	27.74	40.87	42.76	43.5

Table 5.10: Fishery Production per Fisher/ Fish Farmer (2006)

Continents	*Production (Capture + Aquaculture) million tonnes*	*Percentage of Total Production %*	*No. of Fishers & Fish Farmers (No. millions)*	*Percentage of Persons (%)*	*Production per Person (Tonnes/year)*
Africa	7.684	5.3	3.637	8.4	2.1
Asia	94.300	65.6	33.337	85.8	2.5
Europe	15.552	10.8	0.725	1.7	21.4
North America	6.778	4.7	0.334	0.8	19.7
Latin America	17.832	12.4	1.401	3.2	12.7
Oceania	1.393	1.0	0.055	0.1	25.1
Total	143.647	100	43.501	100	3.3

In industrialised countries, younger workers seem reluctant to go to sea on fishing vessels. For many young people, neither the salaries nor the quality of life aboard the fishing vessels compares favourably with those of land-bound industries. In Europe, fishers from economies in transition or from developing countries are starting to replace local people. In Japan, foreign workers have been allowed to work on Japanese distant-water fishing vessels under the "Maru-ship system".

In 2006, in addition to the estimated 43.5 million part-time and full-time fishers, about 4 million occasional fishers and fish-farmers were reported by FAO (2.5 million from India).

Climate Change Effect on Fisheries

Climate change is a compounding threat to the sustainability of capture fisheries and aquaculture developments. Impacts have been experienced as a result of gradual warming at the global scale and associated physical changes, as well as consequences of the increased frequency of extreme weather events.

In terms of physical and biological impacts, climate change is modifying the distribution of marine and freshwater species. In general, warm-water species are being displaced towards the poles and experiencing changes in habitat size and productivity. In warmed world, ecosystem productivity is likely to decline in lower latitudes. Increased temperatures will also affect fish physiological processes, resulting in both positive and negative effects on fisheries and aquaculture systems. Climate change is already affecting the seasonability of particular biological processes, altering marine and freshwater food webs with unpredictable consequences for fish production.

Differential warming between land and oceans and between polar and tropical regions will affect the intensity, frequency and seasonability of

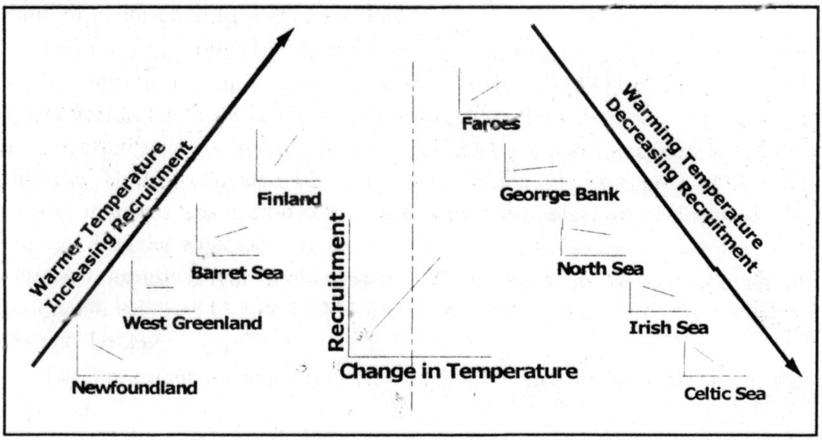

Fig. 5.1: Varying Impacts of Warm Water

climate patterns (e.g. El Nino) and extreme weather events (e.g. floods, droughts, and storms) and hence the stability of marine and freshwater resources adapted to or affected by them (Fig 5.1)

With respect to Fig. 5.1, it may be noted that owing to Bell-shaped relationship between changes in stock recruitment and bottom temperature for various Cod stocks in North Atlantic, populations at the near-pole extents of their ranges tend to increase in abundance with warmer temperature, whereas those in the equatorial parts of their range tend to decline as temperature warms up.

Global Outlook of Fishing Industry

Will land and water suffice for agriculture to feed a growing human population? The question about human ability to feed itself is old and recurring. However, fairly recently fish has also been included in this concern. As late as the first half of the twentieth century, the sea was considered virtually inexhaustible reservoir of fish to exploit. It was only in the mid-century that the marine biologists could convince that the stocks of wild fish were finite and could be fished too heavily. These concerns could become serious early in the second half of the century, when the capacity to overfish wild stocks became apparent. However, aquaculture started to grow at about the time and for many this was reassuring. It sustained the hope that there would be enough fish to eat in the future.

In the last three decades, aquaculture has grown rapidly. In the 1970s, it accounted for about 6% of fish available for human consumption; in 2006, the figure of growth was 47% of total consumption. However, overall the rate of growth in aquaculture has started to slow. For the world as a whole, while the average yearly growth rate had been 11.8 per cent in the period 1985-94, it was 7.2 per cent in the following period.

In the late twentieth century, when the capture fishery production leveled off and aquaculture production increased rapidly, most observers tended to conclude that any supply shortfall would be filled by aquaculture production. This position is still held.

References

Indian Fisheries 2008, Ministry of Agriculture, India; http://india.gov.in/sector.php

Mongillo John 2004, *Environmental Science*, Vol. 5, Creating a Sustainable Society, Greenwood Press, Westport, Connecticut, USA

Pauly, D and Chritensen 1995, *Primary Production required to sustain global fisheries,* International Centre for Living Aquatic Resources Management, Makati, Philippines; Sustainable GlobalFisheriesPrimaryProductionRequired.pdf

Wild Fish Catch hits limits – Ocean decline offset by Increased Fish Farming, June 22 2005, Earth Policy Institute.

The State of World Fisheries and Aquaculture 2008, FAO.

6

Conserving Forests Resources for Ecological Sustainability and Mitigation of Global Warming

Hinglajgarh Forest

I

Introduction

Indian Forestry and Some Aspects of Indian Social Forestry

Some 500,000 square kilometres, about 17 per cent of land area of India, were regarded as Forest Area in the early 1990s. In year 1987, however, actual forest cover was 640,000 square kilometres. But, because more than 50 per cent of this land was barren or bush land, the area under productive forest was really less than 350,000 square kilometres, or approximately 10 per cent of the country's land area. The high demand of the growing population for forest resources continued to lead to the destruction and degradation of forests through the 1980s, taking a heavy toll on the soil. An estimated 6 billion tons of topsoil were lost annually. However, India's 0.6 per cent average annual rate of deforestation for agricultural and non-lumbering land uses in the decade beginning in 1981 was one of the lowest in the world and on a par with Brazil. Many forests in the mid-1990s are found in high-rainfall, high-altitude regions, areas to which access is difficult. About 20 per cent of total forestland is in Madhya Pradesh; other states with significant forests are Orissa, Maharashtra, and Andhra Pradesh (each with about 9 per cent of the national total); Arunachal Pradesh (7 per cent); and Uttar Pradesh (6 per cent). The variety of forest vegetation is large: there are 600 species of hardwoods, sal (*Shorea robusta*) and teak being the principal economic species.

Measures to Grow Forest Cover

India's long-term strategy for forestry development reflects three major objectives: to reduce soil erosion and flooding; to supply the growing needs of the domestic wood products industries; and to supply the needs of the rural population for fuel wood, fodder, small timber, and miscellaneous forest produce. To achieve these objectives, the National Commission on Agriculture in 1976 recommended the reorganisation of state forestry departments and advocated the concept of social forestry. The commission itself worked on the first two objectives, emphasising traditional forestry and wildlife activities; in pursuit of the third objective, the commission recommended the establishment of a new kind of unit to develop community forests. Following the leads of Gujarat and Uttar Pradesh, a number of other states also established community-based forestry agencies that emphasised programs on farm forestry, timber management, extension forestry, reforestation of degraded forests, and use of forests for recreational purposes.

Such socially responsible forestry was encouraged by state community forestry agencies. They emphasised such projects as planting wood lots on denuded communal cattle-grasing grounds to make villages self-sufficient in fuel wood; to supply timber needed for the construction of village houses, and to provide the wood needed for the repair of farm implements. Both individual farmers and tribal communities were also encouraged to grow trees for profit. For example, in Gujarat, one of the more aggressive states in developing programs of socio-economic importance, the forestry department distributed 200 million tree seedlings in 1983. The fast-growing eucalyptus is the main species being planted nationwide, followed by pine and poplar.

IA
Aspects of Indian Social Forestry

Social forestry means the management and protection of forests and afforestation on barren lands with the purpose of helping in the environmental, social and rural development. The term 'Social forestry' was first used in 1976 by The National Commission on Agriculture, Government of India. It was then that India embarked upon a social forestry project with the aim of taking the pressure off currently existing forests by planting trees on all *unused and fallow lands.*

Social Forestry Programme

Government forest areas that are close to human settlement and have been degraded over the years due to human activities needed to be afforested. Trees were to be planted in and around agricultural fields. Plantation of trees along railway lines and roadsides, and river and canal banks were carried out. They were planted in village common land, Government wasteland and Panchayat land.

Involvement of Common People

Social forestry also aims at raising plantations by the common man so as to meet the growing demand for timber, fuel wood, fodder, etc, thereby reducing the pressure on the traditional forest area. This concept of village forests to meet the needs of the rural people is not new. It has existed through the centuries all over the country but it was now given a new character.

With the introduction of this scheme the government formally recognised the local communities' rights to forest resources, and is now encouraging rural participation in the management of natural resources. Through the social forestry scheme, the government has involved community participation, as part of a drive towards afforestation, and rehabilitating the degraded forest and common lands.

Need of Social Forestry

This need for social forestry scheme was felt as India has a dominant rural population that still depends largely on fuel wood and other biomass for their cooking and heating. This demand for fuel wood will not come down but the area under forest will reduce further due to the growing population and increasing human activities. Therefore, the government managed the projects for five years and then gave them over to the village panchayats (village council) to manage them for themselves and generate products or revenue as they saw fit.

Social forestry scheme can be categorised into groups - farm forestry, community forestry, extension forestry and agro-forestry.

Farm forestry

At present in almost all the countries where social forestry programmes have been taken up, both commercial and non-commercial farm forestry is being promoted in one form or the other. Individual farmers are being encouraged to plant trees on their own farmland to meet the domestic needs of the family. In many areas this tradition of growing trees on the farmland already exists. Non-commercial farm forestry is the main thrust of most of the social forestry projects in the country today. It is not always necessary that the farmer grows trees for fuel wood, but very often they are interested in growing trees without any economic motive. They may want it to provide shade for the agricultural crops, as wind shelters, or as soil conservation or to use wasteland. Farm Forestry is another name for Agro-forestry forming a part of Social Forestry.

Community forestry (Rural Forestry)

Another scheme taken up under the social forestry programme is the raising of trees on community land and not on private land as in farm forestry. All these programmes aim to provide for the entire community and not for any individual. The government has the responsibility of providing seedlings, fertilizer but the community has to take responsibility of protecting the trees. Some communities manage the plantations sensibly and in a sustainable manner so that the village continues to benefit. Some others took advantage and sold the timber for a short-term individual profit. Common land being everyone's land is very easy to exploit. Over the last 20 years, large-scale planting of Eucalyptus, as a fast growing exotic, has occurred in India, making it a part of the drive to reforest the subcontinent, and create an adequate supply of timber for rural communities under the augur of 'social forestry'.

Extension forestry (Urban Forestry)

Planting of trees on the sides of roads, canals and railways, along with planting on wastelands is known as 'extension' forestry, increasing the boundaries of forests. Under this project there has been creation of wood lots in the village common lands, government wastelands and Panchayat lands. Schemes for afforesting the degraded government forests that are close to villages are being carried out all over the country.

Agro-forestry (Rural Forestry)

In agro-forestry, silvicultural practices are combined with agricultural crops like leguminous crop, along with orchard farming and live stock ranching on the same piece of land. In lay man language agro-forestry could be understood as growing of forest trees along with agriculture crop on the same piece of land.

In a more scientific way agro-forestry may be defined as a sustainable land use system that maintains or increases the total yield by combining food crop together with forest tree and live stock ranching on the same unit of land, using management practices that takes care of the social and culture characteristic of the local people and the economic and ecological condition of the local area.

Objectives of Social Forestry and Environment

Social forestry schemes that have been started all over the country have made a considerable difference in overall forest cover in a short time. Afforestation outside the conventional forest area for the benefit of rural and urban communities serves the following main objectives:

1. Improve the environment as well as protecting agriculture from adverse climatic factors,

2. Increase the supply of wood fuel for domestic use, small timber for rural housing, fodder for livestock, and minor forest produce for local industries,
3. Increase the natural beauty of the landscape; create recreational forests for the benefit of rural and urban population,
4. Provide jobs for unskilled workers and
5. Help land rehabilitation
6. Finally, its help raise the standard of living and quality of life of the rural and urban people.

Other Important Missions

- To carry out a need based and time bound programme of afforestation with special emphasis on fuel wood and fodder development on all degraded and denuded lands/forests.
- Afforestation of abandoned jhum lands and mined areas.
- Linear strip plantation of fast growing species on sides of public roads, rivers, streams and irrigation canals.
- Afforestation on unutilised lands under State/Corporate, institutional or private ownership.
- Green belts in urban/industrial areas.
- Shelter belt (generally more useful than the wind breaks) for the purpose of shelter from wind and sun covering areas larger than a single farm on a planned pattern.
- Farm forestry in the form of raising rows of trees on bund or boundaries of fields and individual trees in private agricultural land as well as creation of wind breaks round a farm or orchard by raising one or two lines of trees.
- Raise flowering trees and shrubs mainly to serve as recreation forests for the urban and rural population.
- Elicit people's participation involving women and young people in conservation of forests, wildlife and environment.
- Environmental awareness generation and celebration vanamahotsava, environment day, wildlife week etc.

IB

Indian Forests Institutions

The **Ministry of Environment and Forests (MoEF)** is an Indian government ministry. The Minister of Environment and Forests holds cabinet rank as a member of the Council of Ministers. Mr. Jairam Ramesh, is the Union Minister of Environment and Forests.

The ministry is responsible for planning, promoting, coordinating, and overseeing the implementation of environmental and forestry programmes in the country. The main activities undertaken by the ministry include conservation and survey of the flora and the fauna of India, forests and other wilderness areas; prevention and control of pollution; afforestation, and land degradation mitigation. It is responsible for the administration of the national parks of India.

Subordinate Organisations

The following organisations work directly under the Ministry of Environment and Forests:

- Botanical Survey of India (BSI), Kolkata
- Directorate of Forest Education (DFE), Dehradun
- Forest Survey of India (FSI), Dehradun
- Indira Gandhi National Forest Academy (IGNFA), Dehradun
- National Institute of Animal Welfare
- National Museum of Natural History (NMNH), New Delhi
- National Zoological Park (NZP), New Delhi
- Zoological Survey of India (ZSI), Kolkata

Autonomous institutions

The following organisations work autonomously, though administratively they are under the Ministry of Environment and Forests:

- Central Pollution Control Board (CPCB), Delhi
- Central Zoo Authority (CZA), New Delhi
- G.B. Pant Institute of Himalayan Environment and Development, Almora
- Indian Council of Forestry Research and Education (ICFRE), Dehradun
- Arid Forestry Research Institute (AFRI), Jodhpur
- Forest Research Institute (FRI), Dehradun
- Centre for Forestry Research and Human Resource Development (CFRHRD), Chhindwara
- Institute of Forest Genetics and Tree Breeding (IFGTB), Coimbatore
- Institute (TFRI), Jabalpur
- Centre for Social Forestry and Eco-habitation (CSFE), Allahabad
- Wildlife Institute of India WII

II
Conserving Forest for Ecological Sustainability

The role of forests in the national economy and in ecology was further emphasised in the 1988 National Forest Policy (See Box 6.1), which focused on ensuring environmental stability, restoring the ecological balance, and preserving the remaining forests. Other objectives of the policy were meeting the need for fuel wood, fodder, and small timber for rural and tribal people while recognising the need to actively involve local people in the management of forest resources. Also in 1988, the Forest Conservation Act of 1980 (see

Box 6.1: National Forest Policy 1988

Forests constitutes one of the most productive renewable natural resources. From 74.18 million ha recorded in 1951, the area under forest cover has dwindled progressively and in 1993, the area under forest cover did not exceed more than 64 million ha, of which hardly 19 million was effective forest with crown density of 40 per cent or more. The problem is further compounded by the fact that the forest at the rate of 47,500 ha, is disappearing unabated per annum.

The National Forest Policy of 1988, which replaced the National Forest Policy of 1951 vintage, recommended a number of innovative measures to revamp the situation. The Policy inter alia recommended planting trees on all vacant land including waste land through social forestry and agro-forestry in rural areas, recreational forestry in urban areas, planting trees in the watershed areas, river and canal banks, sides of National and State highways, railway lines, etc.

Another landmark is the enactment of National Forest Act 1980 which prohibits diversion of Forest lands for non-forestry purposes by the State Government without the written permission of the Government of India.

Biotic factors (grasing, pests and diseases and forest fires) have a significant effect in shaping the character and composition of forest vegetation in India.

Consequent to the decline of forests both qualitatively and quantitatively, there is now an ever increasing gap between the demand and supply of the wood materials. The projected demand for industrial wood (required for pulp and paper, safety match, plywood and panel products industry) is 45 million cu.m. as against the current supply for firewood is bare 12 million cu.m. In the energy sector, the projected demand for firewood is 153 million metric tonnes against meager supply of 19 million tonnes from all sources (1979 estimate). Regarding leaf-fodder, the picture is equally gloomy. There is about 550 million livestock in the country. Current availability of leaf-fodder is about 350 million tonnes per annum, which is incredibly lower than the actual requirement.

Box 6.3) was amended to facilitate stricter conservation measures. A new target was to increase the forest cover to 33 per cent of India's land area from the then-official estimate of 23 per cent. In June 1990, the Indian Government adopted resolutions that combined forest science with social forestry that is, taking into account the sociocultural traditions of the local people. The cumulative area afforested during the 1951-91 period was nearly 179,000 square kilometres. However, despite large-scale tree planting programs, forestry is one area in which India has actually regressed since independence. Widespread pilfering by villagers for firewood and fodder also represents a major decrement. In addition, the forested area has been shrinking as a result of land cleared for farming, inundations for irrigation and hydroelectric power projects, and construction of new urban areas, industrial plants, roads, power lines, and schools.

Chipko Movement: Since the early 1970s, as people realised that deforestation threatened not only the ecology but their livelihood in a variety of ways, people have become more interested and involved in conservation. The best known popular activist movement is the Chipko Movement, in which local women under the leadership of Chandi Prasad Bhatt and Sunderlal Bahuguna, decided to fight the government and the vested interests to save trees. The women of Chamoli District, Uttar Pradesh, declared that they would embrace—literally "to stick to" (chipkna in Hindi)—trees if a sporting goods manufacturer attempted to cut down ash trees in their district. Since initial activism in 1973, the movement has spread and become an ecological movement leading to similar actions in other forest areas. The movement has slowed down the process of deforestation, exposed vested interests, increased ecological awareness, and demonstrated the viability of people power.

The Government has also enacted the Forests Right Act 2006 recognising the rights of forest dwelling communities (see Box 6.2). Forestry is a major government enterprise in India which faces the challenges of dwindling forest cover area due to overpopulation, farming and environmental factors.

Box 6.2: Forests Right Act 2006

In 2006 forestry in India underwent a major change with the passage of the Forest Rights Act, a new legislation that seeks to reverse the "historical injustice" to forest dwelling communities that resulted from the failure to record their rights over forest land and resources. It also sought to bring in new forms of community conservation.

Box 6.3: Forest Conservation Act 1980

The Forest (Conservation) Act, 1980 came into force with effect from October 25, 1980. Under the provisions of this Act, prior approval of the Central Government is essential for diversion of forest lands for the non-forestry purposes. In the national interest and in the interest of future generations, this Act, therefore, regulates the diversion of forest lands to non forestry purposes. The basic objective of the Act is, to regulate the indiscriminate diversion of forest lands for non forestry uses and to maintain a logical balance between the developmental needs of the country and the conservation of natural heritage. The, guidelines have been issued under the Act from time to time, to simplify the procedures, to cut down delays and to make the Act more user friendly. To ensure this, recently, new rules under this Act have been framed and notified on 10 January 2003 by the Ministry of Environment and Forests.

The Act has succeeded in controlling the indiscriminate release of forest land for non forestry purposes. Prior to 1980, the rate of diversion of forest lands for non forestry purposes was about 1.43 lakh ha per annum. But, with the advent of the Forest (Conservation) Act, 1980, the rate of diversion of forest

(contd...)

(contd...)

lands has come down to around 15000 ha per annum and mostly diversion of forest land is allowed to meet the developmental needs for Drinking water projects, Irrigation projects, Transmission lines, Railway lines, Roads, Power projects, Defense related projects, Mining etc. For such diversions of forest lands for non forestry purposes, compensatory afforestation is stipulated and catchment area treatment plan, wildlife habitat improvement plan, rehabilitation plan, etc. are being implemented, to mitigate the ill effects of diversion of such vast area of green forests. To monitor the effective implementation of the compensatory afforestation in the country, an authority named as "Compensatory Afforestation Management and Planning Authority (CAMPA)" is being constituted at the national level. A monitoring cell is also being set up in the Ministry of Environment and Forests to monitor the movement of proposals at various stages and the compliance of the conditions stipulated in the forestry clearances by the user agencies.

The State Advisory Groups (SAGs) constituted by the Ministry of Environment and Forests, Government of India, under the Forest (Conservation) Act, 1980 shall meet according to the timetable. In case any date indicated in the time-table happens to be a Government holiday, the meeting shall take place on the next working day.

The senior-most Conservator of Forests/Deputy Conservator of Forests of the Regional Office shall act as the Chairperson of the concerned SAG in the absence of the Chief Conservator of Forests/Conservator of Forests. Nodal Officer of the State shall act as the Member-Secretary of the concerned SAG.

Plundering Forests (Timber Mafia): Protected forest areas in several parts of India, such as Jammu and Kashmir, Himachal Pradesh, Karnataka and Jharkhand, are vulnerable to illegal logging by timber mafias that have co-opted or intimidated forestry officials, local politicians, businesses and citizenry. Clear-cutting is sometimes covered-up by conniving officials who report fictitious forest fires.

III
Global Forest Policy for Climate Change Mitigation: UNFCCC and REDD plus Action Programmes

Acronyms

BAP	–	Bali Action Plan
COP	–	Conference of Parties
GHG	–	Green House Gases
UNFCCC	–	United Nations Framework Convention on Climate Change
CDM	–	Clean Development Mechanism
LULUCF	–	Land use, Land-use change and forestry
PES	–	Payment for Environmental Services

REDD	–	Reducing Emission from Deforestation and Degradation of Forests
SMF	–	Sustainable Management of Forests
A&R	–	Afforestation and Reforestation
REDD plus	–	REDD plus SMF and A&R
SBSTA		Subsidiary Body for Scientific and Technological Advice

GPG – Good Practice Guidance

Global Forestry and Climate Change

Global warming is the result of the thermal shield formed by the continuous emission of Green House Gases (GHG) in the atmosphere, particularly Carbon Dioxide in the industrial age. The Global warming then induces climate change, which has emerged as one of the most serious concerns of the 21st century. Due to these climate changes, there is likely to occur melting of glaciers and polar ice, sea level rise, draught, flood and storms, shifting of climatic seasons, adverse effects on agricultural systems, and increasing variability in weather patterns; thus, significant impacts in the developing world, where the world's poor are expected to face severe losses and uncertainties in a changing climate are expected.

Since the inception of the Kyoto Protocol 1997, the *'land use, land-use change and forestry'* (LULUCF) activities to combat climate change have been a topic of significant debate. Whether one is a protagonist or an antagonist on the issue, deliberations of the UN Framework Convention on Climate Change (UNFCCC) have significantly affected the policy making for the forest sector. Forests are both an emission source of carbon dioxide (CO_2) when they are destroyed or degraded and a sink when conserved, managed, grown or planted sustainably. Forest vegetation and soils currently hold almost 40% of all carbon stored in terrestrial ecosystems.

However, in the tropics, forest clearance and degradation are together acting as a "net source" of carbon emissions. Though growth in plantations is expected to absorb more carbon, the likely continuation of current deforestation rates means that the world's forests will remain a net source of carbon dioxide emissions and a contributor to global climate change. However, there are also significant opportunities to develop innovative carbon sequestration forestry projects that generate positive synergies between forest restoration, mitigation of climate change and livelihood improvements for the forest dwelling poor. These processes need to be explored through existing or extended mechanisms such as the Clean Development Mechanism (CDM) of the Kyoto Protocol and the Bio-Carbon Fund of the World Bank. However, the success of such carbon trading initiatives will ultimately depend on the resolution of many outstanding issues, including clearly defining the legal ownership of the carbon, and also on the outcome of the negotiations for the next commitment period under the Kyoto Protocol through Bali, Copenhagen, Cancun or other future conferences. Nonetheless, in many countries, the prospect of trading carbon credits generated by the forestry sector has started influencing operational decisions in the forest industries of those countries and this can also have effects on future timber markets.

India's forest and tree cover, contribution as carbon sink[3].

India is known as one of the twelve megadiverse countries in the world in the sense that Indian forests are the repository of most of India's rich biodiversity. The forests not only harbour unique wildlife, flora and fauna, nearly 200 million people in the country also depend on them for their livelihood, partly or fully. India has a great tradition of institutionalised scientific forest conservation dating back 147 years (Forest department was established in 1864). It is the result of these efforts that despite our rising population and the pressures of economic development, India is one of the few developing countries where the forest cover has increased over the last 20 years.

Forestry is at the centre-stage of global climate change negotiations. This is because forests have the potential to be a carbon sink as well as a source of carbon emission. India has been an active participant in the deliberations on forestry for the Bali Action Plan (BAP) and earlier for the United Nations Framework Convention on Climate Change (UNFCCC). It is India's view that we need an agreement on a comprehensive framework for compensation and positive incentives for forestry as part of the ongoing climate change negotiations.

It is important to note that any such agreement can provide for incentives not only for Reducing Emissions from Deforestation and Degradation (REDD), but also for Sustainable Management of Forests (SMF) and Afforestation and Reforestation (A&R). A "REDD Plus" approach that combines REDD with SMF and A&R is required in order to fulfill the principles of equity and efficiency.

The policy approach for REDD

At the deliberation on REDD started in the UNFCCC and various other international forums, India reiterated its position by acknowledging the seriousness of the threat of deforestation as a major contributor of GHG emissions, and participated actively in the international efforts to reduce deforestation at global level. The proposed range of actions that have emerged so far to address the issue of REDD are summarised below:

- Compensation for reducing deforestation
- Compensation for stabilising forest cover
- Compensation for conserving and increasing forest cover

The diverse national circumstances have to be taken into account while developing a policy framework for providing financial incentives under

REDD plus. Protection and sustainable management of forests need to be treated as positive practices to avoid deforestation.

The theme of "Reducing emission from deforestation in developing countries: Approaches to stimulate action" was first discussed in COP 11. COP 11 invited parties and accredited observers to submit their views on the issue and also asked the Secretariat to organize a workshop prior to 25th Session of SBSTA (Subsidiary Body for Scientific and Technological Advice). Accordingly, SBSTA 24 decided to organize a technical workshop in Rome in August – September 2006. The country submissions and recommendations of Rome workshop were deliberated upon in SBSTA 25 held in parallel with COP 12 at Nairobi in November 2006. India in furtherance of conservation and sustainable forest management based approach of Congo Basin Countries for providing positive incentives, proposed the concept of 'Compensated Conservation' to compensate the countries for maintaining and increasing the carbon stocks as a result of conservation and increase and improvement in forest cover. The Indian submission on the topic at the 27th Session of the SBSTA is as follows:

> *Protection and sustainable management of forests need to be considered as positive practices to avoid deforestation. Nations with history of continued deforestation and now committing to reduce deforestation rates, and those having already taken up strong conservation measures and thereby stabilising and increasing the forest cover against a pre-determined baseline, present a fit case to be rewarded under REDD, as these efforts reduce the emission of carbon in atmosphere or capture atmospheric carbon in the process. Besides the above approach, SBSTA/COP must be open to consider other successful range of actions that reduce or stabilize or conserve carbon in forest ecosystems.*

It will be advisable to examine different policy approaches suitable to various national circumstances through pilot projects. The countries with heavy deforestation rates must first identify the drivers of deforestation and then attempt to make preventive actions to reduce deforestation by addressing the drivers of deforestation. Such action will prevent the activity shift, and checking deforestation at one place will not cause deforestation at other place.

Developing REDDs Mechanism

A national level approach possibly working outside CDM, and financially supported by a specially designed and designated fund is being mooted to operationalize REDD concept. The comprehensive approach at national level can be developed by integrating existing IPCC 2006 guidelines, GPG, and use of remote sensing technologies for estimation, acceptance and disbursement of positive incentives.

Many Party nations including European Union (EU) and Australia are in favour of a national baseline, and reference scenario (FCCC/SBSTA/2007/Misc.2). For REDD accounting, India also favours a National Level Accounting Mechanism (as one single country project). COP/SBSTA may develop independent "Modalities and Procedure" for implementation of REDD mechanism under UNFCCC / Kyoto Protocol (on the analogy of decision 19CP.9 for CDM A/R Projects under Marrakech Accord).

To ensure uniform reporting at international level for the purpose of admissibility of incentives under REDD, there is need for recognising a general definition of forest in terms of crown density, which should include natural as well as industrial and short rotation plantations, or in the alternative, if technologically possible a minimum default biomass or carbon stocks per unit area.

Developing Financial Mechanism for REDDs Projects:

It is proposed to implement the REDD outside the CDM of Kyoto Protocol. Therefore, the existing market mechanism for CDM will not be applicable for disbursement of financial incentives for projects qualifying under REDDs. Developed country parties must deploy additional financial resources for the purpose. New financial arrangements within the Convention (UNFCCC) can be worked out to enable developing country parties to take admissible actions under REDD. UNFCCC may create different financial instruments or apportion funds under the same instrument for different range of actions to be implemented by the countries according to their national circumstances such as:

1. Reducing Deforestation Fund
2. Stabilisation Fund
3. Forest Carbon Conservation Fund

While developing any instrument of financial incentives to support REDD, strong need for capacity building expressed by most developing country parties at all stages of discussions on REDD should be kept in view. Many nations such as Bolivia, Central African Republic, Costa Rica, Democratic Republic of Congo, Chile, etc. have advocated for constitution of an 'Enabling Fund' to support capacity building and conducting pilot activities related to REDD activities for developing / least developed countries.

IV

Use of Forest Land for Industrial or Non-forest Purposes: Indian Experience

The task of finding suitable land for setting up industry, being adequate in extent, affordable in cost, causing little or no displacement of human beings and having easy access to raw materials, good communications to markets, availability of energy, skilled manpower, etc, has become more and more difficult over the years. In the initial few decades after Indian

Independence, when a fair amount of government waste land was available, it was possible for an industry to be set up exclusively on government allotted land; but, five or six decades later, with most government spare land already distributed or used up for various purposes, the only land that is available for setting up an industry, is land that is privately owned, or government land, that has a specific character or is earmarked for a specific purpose, e.g. forest land, wetlands etc. Privately owned land, can be, and usually is, purchased by the project promoter from the owners; however, if the extent of land required is large, and the land owners are many, the promoter prefers to get the land acquired, with the help of the Land Acquisition Act, 1894. There are significant problems associated with this, as anyone who has been following the happenings in the country for the past few years, would be aware. The use of the Land Acquisition Act to acquire land for industry or other projects has caused considerable resentment among private land owners, primarily because of the element of compulsion. The events in Singur[1] and Nandigram in West Bengal, and in Raigad[3], in Maharashtra, are cases in point, where public resentment and opposition forced the government to cancel the land acquisition. Aware of the public sentiment, government has proposed to amend the Land Acquisition Act to make acquisition of private land for industry permissible only as a last mile action, after the industry has already obtained seventy per cent of its land through purchase or other legitimate market means.

The use of the Land Acquisition Act for acquisition of private land for industry is not the theme in this chapter. Instead, we would like to examine, the action that needs to be taken when an industry (or other development project) finds that the only land that is suitable for it, in all respects, is government land – but land that is classified as forest.

Prior to 1980, if forest land had to be used for non-forest purposes, e.g. for an industry, or a road, or a school, an application needed to be made to the Divisional Forest Officer (DFO) concerned and he/ she would process the case and allot the land. Large areas of forest land did get allotted, through this process, for industries and other purposes in the years from 1947 until 1980. There was, as a consequence, considerable decrease in the overall extent of forest land. One estimate puts the diversion of forest land for non-forest use, in the 30 years from 1950 to 1980, at 45 lakh hectares. Alarmed by this rapid depletion of forest land, the government of India, in 1980, enacted the Forest (Conservation) Act 1980. As a consequence of this Act, the powers to allot forest land for non-forest purposes were withdrawn from the DFOs and even from the State governments, and were assumed by the government of India (specifically, by the Ministry of Environment and Forests). Subsequent to 1980, therefore, in all cases where development projects - industrial, or any other - needed to be established on forest land, clearance had to be obtained from the Ministry of Environment and Forests of the Government of India.

As a result, the procedure for obtaining such clearance has become a long and cumbersome one. The promoter of the project first has to identify the land in the field. If all, or some of it is forest land, he/she needs to apply to the State government, which will verify the land, the area, the coordinates, whether there are any trees on it, whether the land can be spared or not, and various other aspects. If convinced that the land can be permitted to be used for non-forest purposes, the State Government will send a proposal to the Ministry of Environment and Forests (MoEF), Government of India, requesting that the land be diverted for non-forest use by the project proponent. The proposal is then examined by a statutory committee, viz. the Forest Advisory Committee (FAC), set up by the MoEF, under the Forest (Conservation) Act. If the committee finds the proposal acceptable, it will recommend to the MoEF that the land be permitted to be given for non-forest use, subject to certain conditions. This is known as the Stage I clearance.

Once the conditions specified have been complied with by the project promoter, the Stage II clearance is given, finally permitting the use of the forest land for non-forest purposes. What are these conditions that are set out in a Stage I clearance? Over the years the conditions have evolved, and as on today, the principal conditions are (i) the provision of an equal amount of non-forest land for converting into forest land; (ii) the provision of compensatory afforestation costs to plant this land with trees; and (iii) the payment of net present value (NPV). Once the money and the land is provided, and are certified to have been received by the State government, the Stage II forest clearance is granted.

It is interesting to note that the Forest (Conservation) Act, 1980, does not specifically prohibit the diversion of forest land for non-forest purposes; it merely subjects the proposal to a stringent scrutiny by experts, at a high level, to ascertain whether or not such diversion of forest land is really required and justified. Consequently it is left to the Forest Advisory Committee, entirely, to advise the MoEF, whether to divert the forest land or not. There are no guidelines for the FAC to go by, but, no doubt, the ecological value of the forest land concerned is a major consideration, with clearance being more easily granted where the forests are not dense, natural forests, and the impact on the environment and ecology is minimal. In the past, however, very few cases for forest clearance have been refused permission, unless a wild life national park or sanctuary was likely to be affected. It is only recently that the MoEF has been talking of classifying forests into 'go' and 'no go' areas, depending on their ecological value, albeit, so far, only for proposals relating to coal mining[6]. A demand has been raised in some quarters, to apply this principle to projects other than mines as well. Perhaps, in due course, it will be applied to many other categories of development projects.

Until 2008, the only issue that project proponents, who wished to locate their projects on forest land, had to be concerned about was the grant of clearance under the Forest (Conservation) Act, 1980. However, this changed with effect from the end of 2007. In December, 2006, Parliament passed a historic Act (which came into force from 31 December 2007)[7]. The Act recognised that many people, particularly those belonging to the Scheduled Tribes, were forest dwellers; had, in fact, been so, for generations; and, even today, depended on forest land for their livelihood, and even survival. It also emphasised that this fact had so far not been legally recognised by the state. The new Act recognised the rights of forest dwellers, particularly of Scheduled Tribe forest dwellers, to live in, and earn their livelihood from, the forests. The Act, known formally as the Scheduled Tribe and Other Traditional Forest Dwellers (Recognition of Forest Rights) Act, (FRA), makes it mandatory for a process to be set in motion to identify the persons who had rights over the forests, to ascertain the nature of such rights (viz. cultivation rights, rights to collect minor forest produce, grasing rights, fishing rights, etc.) and to grant this right or rights to them through a legal document. Clearance to use forest land for projects, even under the Forest Conservation Act, could, therefore, from January 2008 onwards, only be given, after the process for grant of forest rights under the FRA had been completed. This has made it more difficult for projects to be put up on forest land. A recent case in which this issue has attracted a great deal of attention is the case in Jagatsinghpur district in Orissa, where the State had proposed to grant a large chunk of forest land to the Korean steel company, POSCO, to establish a steel plant. People living in the surrounding villages claimed that their traditional rights over the forest land had not been recognised by the State Government before handing over the forest land to the company. The MoEF had to step in and resolve the issue, but the matter still remains conflict ridden.

A third Act which should also be taken into consideration by all those who need to situate development projects on forest land is the Panchayat (Extension to the Scheduled Areas) Act, 1996, commonly known by its acronym PESA. This Act does not apply to forest areas specifically, but to Scheduled Areas listed in the Fifth Schedule of the Indian Constitution (in other words, areas occupied by Scheduled Tribes, mostly in and around central India). However, since much of the land in Scheduled Areas is forested, the Act becomes a relevant and important one for development projects proposed to be located in forest areas. According to the PESA Act, no land can be acquired in Scheduled Areas, or people relocated, without consulting the Gram Sabha(s) concerned[8] of the concerned area or district.

References

http://en.wikipedia.org/wiki/Forestry_in_India

http://en.wikipedia.org/wiki/Social_forest_in_India

http://www.indianetzone.com/24/forestry_india.com

1. *India's forest and tree cover, Contribution as carbon sink,*. 27 April 2010, Ministry of Environment and Forests - P. Singhal.
2. Forest Environmental Services Published: August 22, 2008, 3:09 am. Edited: August 22, 2008, 3:09 am, Lead Author: IUCN, the International Union for Conservation of Nature.
3. India's Forest and Tree Cover Contribution as a Carbon Sink, August 29 2009, Ministry of Environment and Forests Government of India.
4. Views from ICFRE, Dehra Dun, India (an Observer organisation) to UNFCCC on REDD – Sanjiv Tripathi.
5. Subrata Palit, Forest, its Multiple Services and its Primal Roles in Climate Change Mitigation - *Journal Green Earth*, Vol. 2, No. 1, published by Millennium Institute of Energy and Environment Management, Kolkata, India.
6. Meena Gupta, Forest Land and Development Projects Legal issues and Implications, *Journal Green Earth*, Vol. 2, No. 1, published by Millennium Institute of Energy and Environment Management, Kolkata, India.
7. The site in West Bengal where Tata Motors had initially planned to manufacture their Nano car.
8. The site where the West Bengal Government had proposed to have a chemical hub set up by the Salim Group of Indonesia, but which had to be abandoned because of strong opposition from the local population.
9. The site in Maharashtra where the Reliance (Mukesh Ambani) Group had proposed to set up a Special Economic Zone but which had to be given up because the land could not be acquired.
10. The Land Acquisition (Amendment) Bill 2009, clause 5 (f) (v) – definition of public purpose.
11. Statement of Minister, Environment and Forests in Parliament (Lok Sabha) on May 10, 2005.
12. Press release of the Ministry of Environment and Forests dated 18 July 2010.
13. Ministry of Tribal Affairs notification published in the Gazette of India, Extraordinary (31.12.2007).
14. The Panchayats (Extension to Scheduled Areas) Act 1996, Sec 4 (i).

7

Green Corporates : Issues of Sustainability and Governance

Part I

Corporate Transformation to Green Philosophy

Around a decade or so ago, the corporate initiatives to Greenism meant planting a few trees, funding a community water harvesting project, landscaping a greenery, beautify a public place or sponsor a few water fountains. As time rolled by the Kyoto Protocol, touching Bali and Copenhagen and now post- Cancun, there is new transformation in corporate green thinking that they must live and operate within the sustainable bounds of the environment to survive and prosper. The realisation is so overwhelming that you take any product from any industry in any category, be it a house, an office, an airconditioner, a refrigerator, wall paints, there is great corporate effort to make it conform to a green product; could the product be made more energy-efficient, organic and eco-friendly, non-toxic or biodegradable if applicable be.

The fact that earth's resources are exhaustible and hence scarce and are not going to last forever is a scary thought for all; more so for the industries that are hugely dependent on the same resources for their survival. A business that follows sustainable practices will stand to gain in the long run, beating competition. The main driver for carrying out the environmental agenda by the Indian corporates starting 80s and 90s was essentially to manage the business threats and risks in the aftermath of the Brundtland Report of the World Commission on Environment and Development, 1987. The Brundtland Report, famously known as "Our Common Future" lays down for the first time the principle of sustainable development in the following lines:

"Sustainable development is the development that meets the needs of the present without compromising the ability of the future generation to meet their needs."

The scope of the above sustainable development includes two following key concepts within it:

- The concept of the needs in particular, the essential needs of the world's poor, to which overriding priority should be given;
- The idea of limitations, imposed by the state of technology and social organisation on the environment's ability to meet the present and future needs. (WECED1987:43)

After considering the 1987 Brundtland Report, the UN General Assembly called for the UN Conference on Environment and Development at Rio de Janeiro in 1992. The Rio meet, also known as 'Earth Summit', was unprecedented in its size of participation attended by 108 Heads of State or representatives and many thousands of ordinary delegates, journalists and other participants.

Besides enunciating the internationally acclaimed Agenda 21, the Summit declared a series of 27 Principles of sustainable development, which greatly influenced the world nations including the corporate world. The guiding philosophy of the declarations may be summarised as follows:

- That human beings are at the centre of concern for the sustainable development. They are entitled to a healthy and productive life in harmony with the nature;
- That the right to development must be fulfilled so as to meet the development and environment needs of the present and the future generations;
- That the states have a sovereign right to exploit their own resources pursuant to their own development and environmental needs of the present and the future generations;
- That eradicating and reducing economic disparities of living standards among people of the world are indispensible requirements for sustainable development;
- That women having as vital a role in the environment management, their full participation is essential for achieving sustainable development.

Thus, propelled by the heightened consciousness of the global consumers, the corporates are opting to integrate their 'green agenda' into their business objectives. Corporates have found that there are significant economic incentives for reducing wastes and moving towards sustainable business models of operation. We refer to some live examples of reputed Indian corporates changing to green thinking.

- **Nokia India**: During 2006-2008, the company reduced the size of their packaging and used more recycled materials. This enabled them to save 100,000 tonnes of packaging materials and reduced also transportation by 12,000 truck loads.

- **Infosys Technology**: Infosys campuses are implementing solar photo-voltaic in few of their campuses equivalent to 400 KW. They have also installed one of the largest solar heating which can meet about 600,000 litres per day. In the last three years they have reduced their per capita electricity consumption by about 20%, saving about USD 5.00 per year.
- **Wipro Technologies**: Their switch to Green IT infrastructure in combination with new initiatives in virtualisation helped to reduce electricity consumption in 2009-10 by nearly 7.5 million units (Kwh) and thus save 6000 tonnes of carbon annually. Greenpeace has adjudged Wipro as number 1 Indian Green Brand.
- **Accelor Mittal India:** This is the Indian subsidiary of the world's largest Steel maker. Their R and D is working on finding the sustainable low carbon technical and product solutions concerning steel making. In Europe, AccelorMittal is leading the EU's Ultra-low CO2 steel-making initiative. Already they are the largest recycler of scrap steel in the world. They have made huge investments in cleaning and desulphurising the coke oven gas.
- **ITC-WelcomGroup:** This Indian MNC is perhaps largest Indian conglomerate to have extensively implemented the Green Corporate strategy in all activities. Their now famous Three-Bottom-Line Corporate Management will be briefly elaborated in the next section. But here we will enumerate some of their grand green achievements in their WelcomGroup Hotel division:
  - All hotels in ITC-WelcomGroup are ISO 1400 certified since 2005;
  - ITC Sonar in Kolkata is the only hotel in the world which has earned CRUs (Carbon Reduction Units) under CDM carbon trading regime under the UNFCCC.
  - ITC Royal Gardenia is the world's largest platinum rated green hotel;
  - ITC Maurya is the first hotel in the world in the existing building category to have acquired platinum very recently;
  - ITC is perhaps only company in the world which is both carbon and water positive.

Part IIA
UNILIVER- Global FMCG Leviathan
Recent Thrust to Sustainable Business

Unilever has so far done business the conventional way. But now, Paul Polman, their CEO has a 10-year plan to turn the leviathan multinational into a sustainability champion. By integrating the sustainability into the very core of Unilever's business and strategy, Polman also plans to double the turnover, so that the Sustainability programme itself is also commercially viable. In the Table 7.1 is shown a brief glimpse of Unilever's SLP programme.

Table 7.1: Unilever 2020 Sustainability Living Plan (SLP)

Health Improve the health and wellbeing of more than a billion people	**Environment** Decouple growth from environment impact; halve the environment footprint of making and using products	**Livelihoods** Improve the livelihoods of farmers and distributors
Global Target Bring clean drinking water to 500 million people. Improve levels of four nutrients in products; salt, sugar, saturated fat and trans fat.	**Global Target** Help 400 million consumers reduce green house gas (GHG) emissions while washing and showering.	**Global Target** Link 500 000 small farmers and distributors into supply chain and convert them into Shakti entrepreneurs.
India Initiatives Pureit water –purifier homes show 50% lower incidence of diarrhea; tea with vitamins launched to address the nutrition needs of low-income consumers	**India Initiatives** Reduce co^2 from energy in manufacturing operations by 28%, water usage by 31%,. Ice-cream business replacing HFC with hydrocarbon refrigerant.	**India Initiatives** Funding eight drip-irrigation trials gherkin farmers; yields increased by 84%, profits by 245%, water use reduced by 70%.

On taking over as CEO in 2009, Polman laid the grounds for his grand sustainability plan. In direct confrontation with the financial community, he suspended earning guidance and quarterly .results. He could not have them breathing down his neck every quarter, baying for growth and profits, when his goal is long-term. Unilever shares fell by 6%. But Polman soldiered on. The world incentive drivers are too short-term, he lamented.

Since then, though, the Unilever SLP has been hailed as ground-breaking. Unilever's SLP is league ahead of the corporate responsibility framework that continues to obsess many industry captains. Sustainability is not about what a company does with its profit, but how it makes its profits; it is the business process itself.

Nevertheless, some of the targets are really stiff. For example, by 2020 Unilever wants to buy all its agri-raw materials from sustainable sources; it is 10% today. If it achieves the target, it can bring about a significant change in global agriculture and livelihoods. Unilever already buys 12% of world tea, 6% of its tomatoes and 3% of its palm oil.

As shown in the Table 7.1, the company also wants to link over 500 000 marginal farmers and small-scale distributors into its supply chain; help a billion people to change their hygiene habits; bring safe drinking water to

500 million people; halve the green house gas impact its products, across the lifecycle.

Part IIB

Many-Splendoured Socio-Economic Paradigms of Sustainability ITC – An Indian Green Corporate Giant

ITC is a very large conglomerate of Indian Cigarette, Hospitality and many other products diversified as indicated in Box 7.1. Over more than a decade, the ITC group company has been engaged in converting its diverse manufacturing and service businesses into high quality green processes and have thus won many awards and international accreditations.

Box 7.1: ITC
An Indian Green Corporate Giant

• Market cap	Over Rs.1,00,000 Crores.
• Directly employs	Over 26,000
• Supports Turnover	Over Rs. 26,000 Cr.
• Livelihood of	Over 5 million

A diversified Conglomerate with multiple businesses and large Product Lines

Cigarettes	Education and Stationery products	Hotels
Safety Matches, Incense Sticks	Information Technology	Foods
Life Styling	Infotech Parks	Agri-Business
Home and Personal Care	Paperboards and Packaging	Exports

The first and overall business strategy is to take all decisions by trying to integrate the interests from three dimensions - economic, environment and social, which is famously known as the *'Triple Bottom Line'* principle.

1. Philosophy of Triple Bottom Line

In 2002-03, ITC formally embraced Triple Bottom Lines, evaluating the overall operations in three accounting heads - Economic, Environment and Social. The objectives in each of these Performance Heads were laid down as under:

1. Economic

- Sustaining ITC position as one of India's most valuable corporations.
- Achieving leadership in each of the business segments of ITC.

2. Environment

- Moving beyond compliance
- Adopt low carbon path — make ITC Carbon positive

- Maximise water usage efficiencies, rainwater harvesting — make ITC Water positive
- Minimise waste generation, maximise waste recycling, use external waste as raw material — make ITC Waste recycling positive

3. Social

- Zero Accidents
- Natural resource conservation
- Sustainable Livelihood
- Woman Empowerment
- Supplementary Education

2. Mitigating Climate Change : Multitude Green Initiatives, Agri-Development, Social Forestry, Water Harvesting and Community Development

ITC's engagement in the social sector since 2000 has addressed these major challenges through interventions on two simultaneous fronts:

- *Diversifying Farming Systems*, as a strategy for climate change mitigation by broad-basing the farm-based livelihoods portfolio of the poor. Initiatives have been spearheaded in five major areas: e-Choupal, wasteland development, soil and moisture conservation, value-added crop rotation and animal husbandry programme.
- *Expanding off-farm Income Opportunities*, to reduce the disproportionate dependence of rural households on land for supporting livelihoods with the help of two initiatives: economic empowerment of women and community development.

The number of above projects implemented during 2009-10 stood at 66, spread across 51 districts of 11 States : Andhra Pradesh, Tamil Nadu, Karnataka, Kerala, Orissa, West Bengal, Bihar, Madhya Pradesh, Maharashtra, Rajasthan and Uttar Pradesh. Together, these programmes reach out to more than 2.20 lakh households and over 1.20 million individuals who are directly benefited by the various interventions spread over 3787 villages.

Wasteland Development: Social Forestry

A total of 2082 hectares were brought under social forestry plantations during the year. An area of 933 hectares of timber was felled by 1,174 beneficiaries during the year, yielding 19,996 tonnes of debarked wood. The poor tribal farmers, who reaped the benefits of the social forestry programme, earned Rs. 3.60 crores at an average of Rs. 31,000 per farmer. The total village development fund increased to Rs. 94 lakh. The social forestry programme has created sustainable livelihood opportunities for 19,376 poor households to date by transforming 16,442 hectares of land into productive plantations. In addition, it has augmented the raw material base for ITC's Paperboards business, besides contributing to large scale carbon sequestration. A part of the total acreage covered under this programme was registered as a CDM project, the first of its kind in India.

Soil and Moisture Conservation: Water Harvesting

This programme, designed to help farmers in identified moisture stressed districts through scientific soil and moisture conservation projects, continues to bring precious water resources to several rural communities. A total of 278 large and small water-harvesting structures were constructed during the year, taking the cumulative total to 2,813 structures to date. These projects created 4.83 lakh person-days of work during the current year for the landless and marginal farmers, taking the total to 13.8 lakh person-days of employment. As a result, the total area covered under the watershed programme touched 51,294 hectares, covering some of the country's severely moisture-stressed districts. These structures provide critical irrigation security to 28,301 hectares and have contributed to the Company's 'Water Positive' status.

These meaningful interventions in watershed development were further strengthened by several public-private partnerships that were forged successfully during the year. The total area to be covered under collaboration with NABARD over five years in the states of Andhra Pradesh, Maharashtra, Rajasthan and Bihar stands at 34,000 hectares. Further areas of 49,000 hectares will be covered under the MoUs (Minutes of Meetings) signed with the Governments of Maharashtra and Rajasthan. As a result of the MoUs signed during the year, the total area targeted for watershed development under various partnerships increased to 83,000 hectares.

Integrated Agricultural Development

These interventions aim to improve farm productivity and contribute to higher incomes by promoting a package of farm practices and technologies. During 2009-10, 79 group irrigation projects were implemented, 230 composting units were constructed and 474 demonstration plots were promoted, covering 1,002 farmers. In addition, 205 hectares under Integrated Pest Management (IPM) spices and Medical Aromatic Plants (MAP) were promoted during the year.

Livestock Development

After land, livestock is the second most important asset for a vast majority of the rural population. Owned by more than 70%, it is one of the primary sources of livelihood. The Company's Integrated Animal Husbandry Programme targets this source of livelihood through an artificial insemination programme aimed at increasing yields of milch animals. 40 additional Cattle Development Centres (CDCs) were established during the year, taking the total to 161 CDCs. 1.31 lakh artificial inseminations (AIs) were performed and 35,881 live births took place during the year, taking the total number of cross-bred calves born to date to 1.04 lakh. There was significant improvement in the average number of artificial inseminations performed per Cattle Development Centre, a trend that has continued since 2006-07. This was largely possible because of the robust performance of 51 financially viable CDCs, up from 18 CDCs last year.

Community Development

The primary education programme has so far covered over 2.29 lakh primary school goers. In addition, 577 government primary schools have been provided infrastructure support consisting of benches/chairs, compound walls/gates/roofs, additional classrooms, teaching aids and sports kits. A total of 2,937 low cost sanitary units have been constructed so far, with one-third of the cost being met by the beneficiaries.

3. ITC is Energy-Efficient and Promoter of Renewable Bio-energy Improving Specific Energy consumption and Installing Renewable Energy Power Plant

ITC's focused energy conservation initiatives continued to yield significant benefits to a large number of Business Units during the year. Most Units achieved significant reduction in specific energy consumption (energy required per unit of production) through a rigorous process of third party audits and benchmarking.

ITC's paperboards factory at Bhadrachalam, already highlighted as the most energy efficient factory of its type in the country (reference "All about Paper" by Centre for Science and Environment, 2004) further reduced its specific energy consumption by 2.5% over last year. The specialty papers factory at Tribeni also achieved a reduction of 9% in its specific energy consumption. These two factories account for 82.7% of ITC's total energy consumption.

Specific energy consumption at the Haridwar Units of the Personal Care and Foods businesses reduced substantially as illustrated in the chart below:

Business	*Product Portfolio*	*% Reduction in Specific Energy*
Personal Care	Soap	20.4
	Shampoo/Cream	31.0
Foods	Biscuit	21.3
	Snack Foods	16.8

Similarly, specific energy consumption was reduced by 9.7% and 16.3% at the leaf threshing Units in Chirala and Anaparti, 8.7% at ITC's largest cigarette factory at Bengaluru, 22.6% at Hotel ITC Windsor and by 17.3% at Hotel Sheraton, New Delhi.

Flowing from its commitment to the triple bottom line philosophy, ITC has chosen renewable energy as a focus area for enhancing its positive environmental footprint. During the year, it commissioned an additional capacity of 6MW in wind energy and installed a 90 TPH (tonnes per hour) bio-mass residue fired boiler. These investments have helped offset the enhanced energy consumption of 11.2% as a result of increased production. Around 30.9% of ITC's energy consumption is now from renewable sources and carbon neutral fuels.

ITC businesses have 8 CDM Projects under the Kyoto Protocol, registered with the UNFCCC (United Nations Framework Convention on Climate Change). The Certified Emission Reductions, generated by these projects have helped offset some of the investments made towards reducing specific energy consumption and enhancing the renewable energy portfolio. ITC's large scale social forestry project, the first of its kind in India, has also been registered with the CDM Executive Board of the UNFCCC. This project has ensured that poor tribals in rural India will also be able to benefit from carbon credits traded across the world. Several other projects are in the process of being registered.

4. ITC is 'Carbon-positive', 'Water-positive', and Waste-positive'

The 'Triple Bottom Line' principle of ITC has over the years evolved into a strong Eco-Sustainability model of business which is simultaneously 'carbon-positive', 'water-positive' and 'waste-positive'. In other words, the overall processes emit less carbon, consume less water, and produce less waste than respectively what carbon ITC sequesters, than what water ITC harvests and than the amount of waste ITC recycles or reuses.

'Carbon Positive' - 5 Years in a Row and CO_2 Emissions and Sequestration

Total CO_2 emissions from ITC's operations (manufacturing and freight) amounted to 1709 Kilotonnes in 2009-10 (1,572 Kilotonnes in 2008-09). Higher CO_2 emissions were on account of significant growth in most businesses and increased pulp production which substitutes imported pulp. The new 6 MW wind energy plant and the 90 TPH bio-mass residue fired boiler have contributed to a reduction in CO_2 emissions for only part of the year and the impact of these initiatives are expected to be much larger during the course of 2010-11.

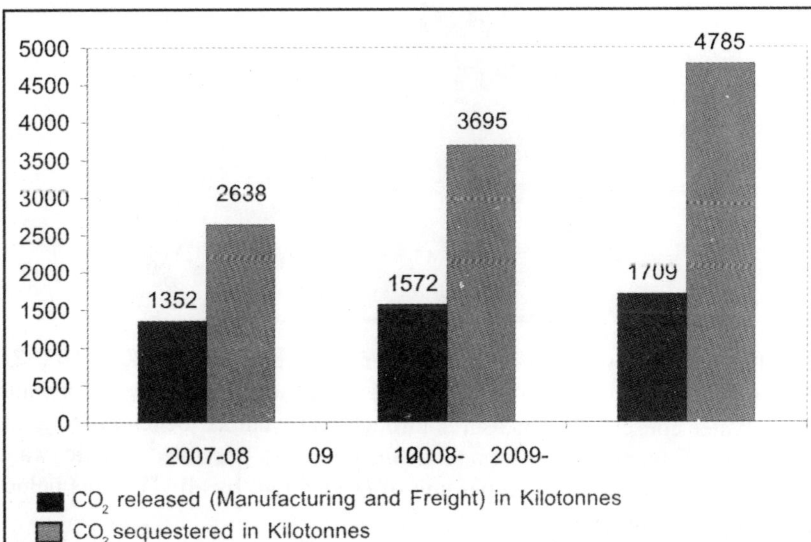

ITC's farm and social forestry initiatives added 13,333 hectares of plantations during 2009-10. Total plantations, as on March 31, 2010 covered over 1,00 000 hectares. The social and farm forestry initiatives have not only contributed to a sustainable source of raw material for the Paperboards business, but also helped sequester 4785 Kilotonnes of CO_2, thus consolidating ITC's status as a 'Carbon Positive Company' for the fifth year in a row. The forestry programme has also created over 46 million person days of employment for marginal farmers and tribals.

ITC 'Water Positive'- 8th Year in a Row

With nearly 17% of the global population and less than 4% of global fresh water resources, water availability is one of the most serious sustainability concerns for India.

ITC Units drew 29.96 million kilolitres of fresh water in 2009-10 (32.7 million kilolitres in 2008-09), a substantial reduction of 8.4% over the previous year. Water conservation measures through water audits, benchmarking and adoption of leading practices to achieve lowest possible specific water consumption (water required per unit of production) by all ITC units helped achieve this reduction.

Though a large number of business Units have achieved significant reduction in specific water consumption, what needs special mention is that the Paperboards and Specialty Papers business, which accounts for over 90% of ITC's fresh water usage, has achieved significant reduction in specific fresh water intake over the previous year.

Table 7.3: Reduction of water intake by the Paperboards and Specialty Papers business across all Units

	Specific Fresh Water Intake (KL/tonne)		*Percentage Reduction*
	2008-09	*2009-10*	
Bhadrachalam	72.2	54.7	24.3
Kovai	15.3	13.1	14.5
Bollarum	1.2	0.9	25.6
Tribeni	120.1	116.6	2.9

Some of the other Units that achieved significant savings in specific water intake during the year were: Anaparti Leaf Processing Plant (20.2%); Foods Business Unit at Haridwar (40.6%); Packaging and Printing Unit at Tiruvottiyur (10.3%); ITC Hotel Grand Central (12.7%), Maratha (10.4%), Sheraton Hotels Chola (16.6%) and Rajputana (11.9%).

Zero Effluent Discharge: The Cigarette Units at Bengaluru and Saharanpur, Leaf Processing Units at Anaparti and Chirala, Packaging and Printing Unit at Tiruvottiyur, Research Centres at Bengaluru and Rajahmundry, Paperboards Unit at Bollarum and ITC Green Centre at Gurgaon recycled all their treated effluents within the Unit.

Rainwater Harvesting: ITC continued to invest in rainwater harvesting potential (RWH) both within the Company premises and through external watershed development projects in socially relevant areas. Total rainwater harvesting potential so far created by the Company is over three times the total water consumed by its operations.

'Waste Recycling Positive'- 3rd Year in a Row

Waste management in India is an area of serious concern, with large amounts of waste being disposed through land filling/dumping. Most landfills/ dumping sites create serious health hazards to people living in the neighbourhood and are liable to contaminate soil and water. ITC Units recycle practically all the waste generated by their operations. The paperboards Unit at Bhadrachalam which generated 75% of the total waste of ITC, recycled more than 99% of the waste generated in 2009-10, yet another noteworthy achievement. Other Units which recycle/reuse more than 99% of waste generated out of their operations include: Cigarette Units at Bengaluru, Saharanpur, Munger and Kolkata, Leaf Processing Units at Anaparti, Paperboards and Specialty Paper Unit at Tribeni, the ITC Green Centre at Gurgaon, Surya Nepal's Cigarette Unit at Simra and ITC Hotels - Maurya, Maratha, Grand Central, Windsor, Mughal, Kakatiya, Sheraton New Delhi and Rajputana.

Table 7.4: Positive Waste Recycling

Waste Recovery Processes	*2007-08*	*2008-09*	*2009-10*
Total Waste (tonnes)	352,970	490,180	578,865
% Recycled	98.9%	98.8%	99.8%
Un-recycled Waste (tonnes)	3706	5893	1099
External Wastes Used as Raw Materials (tonnes)**	163,245	125,337	125,931
Waste Recycling Footprint (%)*	145	124	122

* Waste recycling footprint (%): (waste recycled + external waste used as raw material) x 100/total waste generated in ITC.

** The Paperboards and Specialty Papers business collects / procures significant amounts of post consumer waste.

Part III
Green Governance in Corporate Environment

General Principles

In traditional business, efforts are generally concerned with the reduction of environmental hazards at the processing plants or from some identified products. A total approach for controlling all potential environmental damages likely to be caused during production and later from the use of the products is normally overlooked. Often the corporate governance for sustainable development is limited to efforts to control GHG emission, which is no doubt also a significant contribution. For, metallurgical industries like aluminum, cement, and steel are very serious emitter of GHGs, severely polluting the environment. However, the Wuppertal Institute (Ruud 2004/von Weizsacker et al., 1998) has framed up a comprehensive framework of Corporate Governance norms, which go beyond specific or discrete actions:

- Reduce material intensity;
- Reduce energy intensity;
- Reduce harmful or toxic emission;
- Enhance recyclability;
- Maximize use of renewable resources;
- Extend product durability;
- Increase in service intensity;
- Enhance efficiency, and reduce wastes in all operations.

The last panacea, added by us, is an overarching aim of the corporate governance philosophy for "green business".

Developing a Green Business -Three Possible Strategies

This section lays down some guidelines to Corporates which are trying how to lead their organisation to Green business practices. Generally in all businesses, there are three possible ways to develop a green business (Gregory 2010):

1. To *refocus* or highlight the genuine green attributes in an existing product portfolio;
2. To *acquire* or purchase a known green product from another known green business or developer;
3. To develop or *create* inhouse a green product from a zero stage.

The decision as to which strategy to adopt will depend on the company's greenability that is its internal capability to go green, its business culture, its range of current product portfolio. Such decision shall also be influenced by the expectation of its potential customers, competitors' capabilities and the attitude and reactions of the social activists expressed by their acceptance or resentment. The logic can be presented graphically as shown in Fig. 7.1.

Green Attribute		
	Refocus	Any of three
	Acquire	Create

Fig.7.1: General Development Capabilities

Brita Filters started losing their hitherto popular water filter business as the bottled drinking water stormed the market into a billion dollar business. The social activists of course raised hell against the plastic bottles being dumped all over the land fill and choking the water drainages. Helped by this wave of severe criticism of the activist groups, Brita refocused the inherent superiority of their Water filters which do not create any environmental hazards of plastic bottles and regained fair share of the drinking water market.

Ben Jerry's Ice cream was acquired by Unilever. Using their very powerful marketing ability and wide customer base, Unilever increased the customer base of Ben and Jerry's by 70 % in one year. However, there is possibility of culture clash and strategic misfit between the existing product management and acquired product management.

Developing and creating a new green product inhouse from scratch is a gainful strategy only for companies with high innovation and R&D capability. Besides, each such innovative exercise helps to grow new competence and new capacity building. World Auto giant Toyota developed hybrid car Prius, though there was already another successful hybrid car

Insight of Honda, a finest auto engineering behemoth. Nevertheless Prius dominated the market over the Insight as a more fuel-efficient car.

Choose one of the above Green Strategies

We have briefly studied three possible strategies of going green in a business – accentuating an existing product, acquiring a new product or creating organically inhouse. To choose one or the other, a company must apply general management criteria studying Feasibility, Desirability and implementability, keeping all judgment focused on the underlying principles of Fig. 7.1.

References

www.gogreenindia.co.in 28.03.2011.

Go Green for the love of Business, Planet Earth Supplement, *Economic Times* 02.02.2011.

Sunil Kumar, UNILEVER, Bengal Chamber Seminar, ITC Sonar, January, 2011.

Gregory Unruh and Richard Ettenson, Growing Green – Three Smart Paths to Developing Sustainable Products, *Harvard Business Review*, June 2007.

www.itcportal.com/shareholders/annualreports/itc-report-2010/page-01/aspx.

8

Water Security for Sustainable Healthy Living

A. GLOBAL WATER RESOURCES

I
Challenges of World Water Resources

Water is a precious Resource

In the golden days of the Greek civilisation, philosopher Thales Miletus (640-546 BC) is believed to have proclaimed that "the best of everything is water". Down the centuries, Leonardo da Vinci (1452-1592) extolled water as the "driver of nature". It is difficult to conceive of any programme of human development that does not ensure a ready availability of water.

In 1930, Gartner (Furan 1930) stated that a jellyfish weighing 500 g reduces to 0.45 g after desiccation. For the humans, 70 per cent of an able-bodied man weighing 70 kg is made of water weighing 50 kg. Indeed water is at the source of all lives.

Because of its easy availability, water has traditionally been regarded as an inexhaustible free gift of nature; this has also led to extravagant and wasteful use, thus a life-saving resource may become scarce and threaten human life, livelihood, and development and progress.

The demands of water will significantly and continually increase because of growing world population, accelerating industrialisation and expanding agriculture and increasing economic well being of the people. It may consequently be inferred that there is likely to be critical shortage of water of required quality to sustain future growth and higher standard of living unless the water resource is adequately upgraded, scientifically managed, treated and vended as a valuable economic good.

Dimensions of Water Problems observed by UN

Globally there may potentially be enough water to meet the forthcoming human needs but unfortunately much of water resource is not available at the right place, right time and right quality to serve human needs. Thus, in one way or the other, all societies, rich or poor, are generally affected. A UN

Document (Mordel Plata 1977) summarises the worldwide water scenario as follows:

(a) Reasonably safe supplies of drinking water are unavailable for at least one fifth of the world's city dwellers and three-quarters of its rural people; in many countries, less than one half of the urban population and less than one tenth of the rural population are served with adequate and safe water supply.

(b) Increasing and unplanned concentrations of population and industry in large urban areas have strained water supply; this leads to problems of waste disposals, which in turn, degrade the environmental health and the quality of life;

(c) Proliferation of industrial processes, greater use of energy and increased use of agricultural activities are causing progressive and chronic degradation of the quality of available water by the increase of toxic compounds and other pollutants; the mutagenic and carcinogenic effect of these substances posses a potential threat to human life;

(d) Backwardness and relative isolation of rural areas where the great majority of the world population now lives aggravate the difficulty of providing adequate and safe supplies of drinking water, improved sanitation and waste disposal;

(e) Expansion of food production in water-short areas and in marginal lands has necessitated rapid development of irrigation and reclamation of lands to a degree that water and land resources have been exploited to their limit in many areas;

(f) Ever-growing land degradation from such causes as water logging, salinisation and erosion is leading to losses in production potential, investment and employment;

(g) Ground water supplies are being exhausted, while both surface and ground water sources are deteriorating in many areas;

(h) Water is often needlessly used inefficiently and wastefully, calling for the application of scientific knowledge and setting of appropriate service level;

(i) Application of expensive technology to water development projects in some places is eating away the development fund earmarked for some other projects;

(j) Conflict about rights and priorities among different users intensifies as the increasing demands for available water grows.

These problems affect different centers countries in different ways over the local, regional and national levels. The problems of semi-arid and arid regions vary from the tropical and temperate climate regions. The situations have not materially improved since the eighties.

Fresh Water is not Plentiful

In any case, the total amount of world water is theoretically constant and can neither be increased nor diminished. The ocean water constitutes almost the entire quantity of global water, but for a small portion of less than 3%, which is fresh water. Of these, the most of the fresh water is locked frozen in the Arctic ice-caps, glaciers etc.; a further very small part of less than 1% in lakes, swamps, rivers and streams is available for human consumption. Hence, fresh water is not, in reality, plentiful (see Tables 2.4 and 8.2).

Estimates of the volumes of fresh supply can be derived from observation and measurement of precipitation, evaporation and ground water. The shortcomings in the appraisal methods and general lack of basic hydrological data do affect the accuracy of the estimate but the overall estimates for individual countries, continents and the world do serve very useful services. Such data reveal tremendous variations in local supplies among rocky terrain, semi-arid tracts, tropical rivers with surplus perennial water and temperate developed countries. Water quality by minerals, chemical or bacteriological contaminants also vary equally.

II
Global Water Stocks

Hydrologic Cycle

The origin of water on our planet is not clearly known. The current thinking is that the primordial earth had no oceans and perhaps very little atmosphere. It is surmised that volatile constituents, trapped inside the earth, leaked to the surface through various natural processes including volcanic eruptions, rock movements and hot springs and condensed to form oceans and the atmosphere.

The seawater, which constitutes 97 per cent of all water, contains high concentration of salts making them unusable for human, agricultural and industrial needs. It is possible to separate water from salts by desalination distillation technologies, but that is, by and large, highly energy-intensive and generally uneconomic. Fortunately, the sun performs that service for us when it evaporates seawater and leaves the salt behind. Sea surface constitutes 70 per cent of earth's surface and the major part of the solar heat energy reaching the earth's surface gets converted into latent heat, removing seawater and water from the wet soil by evaporation and the water from the plant leaves by transpiration.

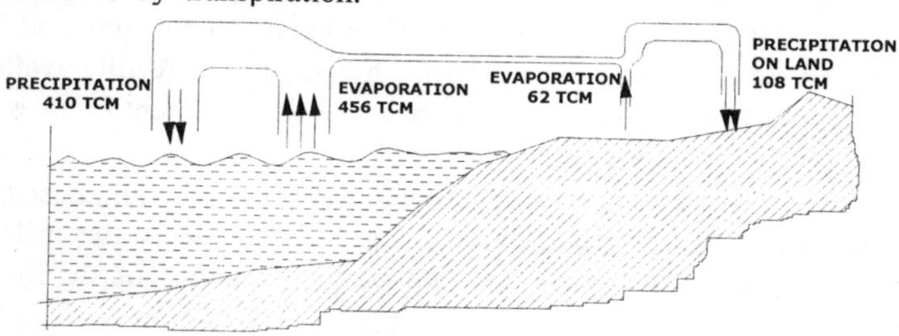

Fig. 8.1. Global Hydrological Cycle.

The resulting water vapour moves to the upper atmosphere and eventually condenses into water and returns to the earth as precipitation. Over the oceans, there is more evaporation than precipitation and over the land, it is the reverse, that is more precipitation than evapotranspiration. The difference between the precipitation and the evapotranspiration on land is the surplus water available for feeding the surface water comprising the rivers and the streams, lakes, springs and the ground water and the balance is the run-off back to the sea.

According to Budyko (Masters 2000), various quantities of evaporation and precipitation on land and sea are given in Table 8.1.

The above phenomenon of Evaporation/Evapotranspiration and Precipitation as a source of potable water is illustrated in Fig. 8.1 and the same is known as the Hydrologic Cycle.

Table 8.1: Annual Evaporation and Precipitation in thousand cu. km.

	Sea Area	*Land Area*	*Total*
Evaporation/Evapotranspiration	456	62	518
Precipitation	410	108	518
Surplus Water on Land	-	46	-
Run-off to Sea	46	-	-

Through the hydrologic cycle, water circulates from the earth to atmosphere to earth, driven by the sun's energy. Water moves endlessly from oceans to the atmosphere to the continents. But the precipitation that falls on the oceans does not contribute to the water resource. Similarly, of the precipitation on land, the amount that is evapotranspirated to the atmosphere is again lost to the water resource. As per the Table 8.1, only 46 thousand cubic kilometers of precipitation on land form the renewable water resource, which is renewed annually feeding the mother earth.

Global Measures of Water

Recent estimates show that the total volume of water on earth is 1400 million cubic kilometers. The Table 8.2 shows that 97.3 per cent is ocean water. Of the 2.7 per cent fresh water, 77.2 per cent is frozen and stored in the icecaps and glaciers, 22.4 per cent is ground and soil moisture and 0.04 per cent in the atmospheric vapour. The fresh water that is readily available for the human and allied consumption consists of 0.35 per cent in lakes and swamps and 0.01 per cent in streams.

While icecaps and glaciers and ground water account for most fresh water resources, for all practical purposes, it is the surface water in lakes, swamps and streams that constitute the basic available supply for the humans.

Another important feature to note that contrary to public perception, the stock of water in lakes and swamps is 35 times the total water in all rivers and streams of the world.

Although surface water provides most of man's supplies, groundwater use has been heavily developed in many parts of the world, especially for agricultural irrigation. This has resulted in the overexploitation and depletion of the water bearing structure lowering the water level. There is inadequate knowledge about the groundwater and its recharging and even about the quantum of water in a specific aquifer. With more studies, it should be possible to plan higher utilisation of groundwater in the overall management and development.

Table 8.2. Water Stock of the World

	Percent	*Quantity (Million Cu.km)*
Ocean Water	97.3	1362.2
Fresh Water	2.7	37.8
Total	100	1400
Location of Fresh Water:		
Icecaps and Glaciers	77.2	29.18
Ground water and Soil Moisture	22.4	8.47
Lakes and Swamps	0.35	0.13
Atmosphere	0.04	0.02
Streams	0.01	0.004
Total	100	37.804

Seawater

Seawater amounts to about 1360 million cu. km. At 97.3 per cent, as shown in Table T 4.2, Sea covers about 72 per cent of earth's surface. But the average depth of oceans is 3800 m, which is only 1/1600 of the earth's radius and the mass of the seawater amounts to only 1/4500 of the mass of the earth. Though to a human observer the oceans appear immense and vast, put to an appropriate scale, it is merely a shallow lake.

The origin of seawater has long been contested. Some maintain that seawater was originally fresh water to which salt were carried by rivers. This view is contested by the fact that seawater contains 89% chloride, 10% sulphates and 0.3% carbonates, whereas fresh water contain 80% carbonates, 13% sulphates and 7% chlorides. The water of the great lakes inland sea (like Caspian sea) is not seawater; it is much less salty and contains three times more calcium carbonates. It seems reasonable to infer that seawater was salty from the beginning (Furan 1967).

II A
Assessing Water Resources

Appraising Surface Water

Within a river basin, all water draining into the streams constitute the potential surface stock within the basin. The extent of availability varies with the flow of the stream at the specific location, its flow velocity and the river channel section.

Whether estimated or directly measured, the mean annual flow of a stream over a period of many years will thus represent the gross surface water supply. Full use of this supply would require that demand coincide with the flow at any time. But the actual supply that is used will depend upon the technology of water collection and utilisation, actual distribution and flow available at the particular time and place. Consequently, any appraisal of water supply must take economic and environmental as well as hydrologic data in order to harmonize plans of demand and supply within acceptable technological and economic limits.

Hydrologists and other water engineers have many scientific tools and methods for assessing water supply in absence of direct measurements. The improvement of these methods and the dissemination were the main objectives of the International Hydrological Decade (1965-1974) and of the International Hydrological and Operational Hydrology Programme of UNESCO and WMO, including the water balance.

Fortunately, the water balance is derived from the equation of continuity. That is, for any hydrological unit and specific time period,

$$\text{Outflow} = \text{Inflow} +/- \text{Storage} \qquad 8.1$$

This principle is applicable to a small catchment, a major river basin, a reservoir, a continent or the world as a whole. Since changes in storage become negligible when projected over a long period of time, the mean water balance for a river basin can be stated as,

$$\text{Run-off} = \text{precipitation} - \text{Evapotranspiration} \qquad 8.2$$

(Evapotranspiration, defined earlier, as transpiration from plant leaves and evaporation from soil, water and all other surfaces).

Maps of average annual precipitation, evaporation and run-off, are best derived from systematic water balance analysis. The isoclines can be constructed so that a balance of three terms in the equation 8.2 are available at all locations, thereby enabling quick, approximate estimate of the run-off on basin-wide geographical basis.

The water balance concept is basic to water resource analysis. It provides the necessary hydrological data for all developmental plans. The water balance equation 8.1 is applicable for any period of time. For short period of time, changes in storage become a predominant consideration and cannot be neglected. For short period, modeling technique can be applied with computer-generated data for the daily values of components in the cycle.

Appraising Precipitation and Evapotranspiration

Precipitation is the primary source of water supply for world food production, though irrigation for agriculture does still make additional demand on stream flow and ground water. Irrigation has to supplement precipitation in basic requirements of crop, where evapotranspiration is high. Precipitation data therefore play a vital role in water resource development and management.

Precipitation data, though are more numerous than stream flow; because, precipitation can be easily measured with simple instruments. What is perhaps more important is to draw precipitation isolines, correlate them to physiological factors and derive long-term trends, if possible, taking into consideration any significant climate change. For example, it is important to know if widespread and prolonged draught can be expected within the natural year-to-year cycle, of the Sahelian zone of West Africa type.

Evapotranspiration is the most difficult term to evaluate in the water balance equation. Evaporation can be obtained by measuring free-water evaporation from water, knowing its weight loss or its draft in a graduated tank. Evapotranspiration can be measured by knowing weight loss from a potted plant and a similar pot with soil but without a plant. But it is important to study the influence of the methodological conditions, e.g. radiation, wind, Humidity and ambient temperature by employing suitable techniques.

It is interesting to note that evaporation from Lake Nasser represents about quarter of the normal flow of Nile at Aswan.

Appraising Ground Water

Ground water is stored in aquifers (rock or soil containing and transmitting water). Groundwater and surface water generally form a somewhat linked water system. Flow can be in either direction, and the rate of flow varies geographically and chronologically.

The interchange is significant for some aquifers. But it has been estimated about 30 per cent of total flow in a surface stream is supplied by the groundwater, and seepage from streams is known to be the principal source of flow to some aquifers. For example, water withdrawn from riparian wells along an alluvial stream can cause an appreciable reduction in surface flow and on the other hand, the diversion of surface flow can reduce ground water recharge. In an other instance, withdrawal from either source has no apparent effect on supply from the other.

According to UN Document (Mordel Plata 1977), there are two ways to the exploitation of water. In the first instance, the withdrawal should match the average rate of recharge. In the second approach, the underground water is mined for a specific task for a specific period. For example, the aquifer, which supports 1.5 million hectares of irrigated land in Texas Panhandle, will be exhausted in a few decades at the current pumping rate.

There is again another instance; the rate of recharge is very insignificant like the artesian basin of the intercalary of the northern Sahara, which have immense stock of fresh water built over 40,000 years. The withdrawal of water from this type of aquifer to any significant degree will lead to gradual depletion of the aquifer.

It will be the concern of the development and management of the groundwater supplies on the regional or national basins to take an overview over the strategy of groundwater exploitation. It could be on the safe yield basis or fixed-period basis with alternative provision for time beyond the period. It is again a fact that supply recharge in many cases tends to accelerate with draw-down.

The appraisal of aquifer storage in particular requires extensive on-site exploration. Although the methods of appraisal are different for ground and surface water, supplies from either source cannot be evaluated separately unless it is established that the interflow and seepages are minimal. Flow rates inside the aquifers may be extremely slow. For example, a recharge at one area may take many decades to reach the other area of the extensive artisan aquifer where the withdrawal is planned. This time lag is to be taken into consideration in the ground water management.

The quality of water available under specific conditions may be estimated from the water basin equation. The equation can be written down for the surface and sub-surface components of a basin as a whole or for the groundwater reservoir only. The later approach is required for adequate understanding of the groundwater regime and its management, while the former allows a macro-level evaluation of the total surface and ground water supplies. Either way, a sound estimate, requiring reliable hydrological data and aquifer characteristics study, is quite complex and computer-aided modeling and solutions may be called for.

Assessing Water Quality

The physical, chemical and biological properties of water determine its suitability of its quality vis-à-vis various uses.

Water is considered polluted when it is altered in composition, directly or indirectly, to the degree that it is less suitable or even unsuitable for a particular use or function. Pollutants reach water bodies from 'point source' such as power station, factories, mines and city sewer outlets and from non-point source such eroding basins and fields.

Pollution of water results mainly from human activities but there are instances of natural degradation, e.g. underground water getting toxically contaminated with arsenic, fluoride or other heavy metal traces, water in arid and semi-arid zones getting saline contamination.

It is not possible to infer water quality of a long stream by local measurement at a single station. Accordingly, observations on water quality must be based by collecting and analysing samples periodically at a network of stations along streams, lakes and at bore holes. Typically, for a long stream like Ohio and Rhine, measurements are taken for temperature, sediment load, dissolved solids, major nutrients such as phosphorous and nitrogen, organic material by biochemical oxygen demand, dissolved oxygen, salts, acidity (or pH values), bacteriological count of coliform, heavy metals etc.

Reliability of Appraisals vis-à-vis Quality and Quantity

The reliability of appraisals is per se a function of sample size and its accuracy. The organisation of comprehensive national network and data-processing services remain one of the most urgent tasks in water appraisal, development and management and framing a national water policy. At the international level, this is the core of the WMO Operational Hydrological Programme. The operation of unified network is critical, whenever basins or aquifers are shared by two or more countries that require using commonly agreed data for all operational management and sharing of water resources.

III
Global Water Supplies

Surface Run-off Water

In Table 8.1, we presented global stock of water as estimated by U N. We indicated there that all of the fresh water stock at 37.8 million cu. km. is not available to human use. As a matter of fact, the mediam quality of fresh water that can be converted to human use is the annual run-off water to the sea. In the Table 8.3, the total global run-off amounts to only about 42,700 cu km; this is about 0.1 per cent of the global stock of fresh water. The more dependable value of fresh water run-off is 46,000 cu km as given in Table 8.1.

Precipitation on land represents the total renewable water availability. The evaporation represents the loss of the major part of that precipitation, which is more or less beyond control. The Balance is the run-off. The Table 8.3 is built up on the data of Baumgartner (1975), USSR Monograph (1074), and Luvovich (1974), which have all been averaged. The values of precipitation, Evaporatiion and run-off in Table 8.3 are average values for further of the sourcc.

Table 8.3: Average Annual Water balances, 1970

Region	*Global Average Norms 1000 cu. km.*		
	Precipitation	*Evaporation*	*Run-off*
Europe	7.4	4.4	3.0
Asia	31.9	18.7	13.2
Africa	21.3	17.2	4.1
Australia	6.9	4.6	2.3
N. America	15.9	9.2	6.7
S. America	28.6	17.4	11.2
Antarctica	24	0.3	2.1
Total Land	114.4	71.4	42.7
Total Ocean	418.3	461.0	-42.7
World	523.7	523.7	0

Source: Developed from USSR Water Balance Report 1974 (Luvovitch and Baumgartner)

It may be seen that over the Asian land, there is larger proportion of run-off, that is, 13,200 cu. km. out of total precipitation of 31,900 cu. km. The run-off over the African and is only 4,100 cu. km. out of total precipitation of 21,300 cu km whereas over Antarctica there is hardly any evapotranspiration. Finally, from Global water balance, the run-off is zero over the combined land and ocean areas, precipitation being equal to evapotranspiration.

Lakes and Reservoir Water

The total volume of fresh water stored in lakes amount to 130,000 cu km, equal to about 3 times the global annual run-off from all land areas. In comparison, the amount of water in man-made reservoirs totals to only 5,000 cu km, that is, about 11 per cent of the yearly run-off.

Lakes and reservoirs help to raise the usability of run-off. While the run-off through the streams is seasonal, lakes and reservoirs store the water for use while the run-off might have nearly dried up. But the lakes are not the primary source of fresh water resource, which are to be renewed every year from the run-off.

Ground Water Resource

For the most part of the world, the ground water resource is not scientifically mapped nor the process of recharge properly understood.

From UN sources, it is understood that the quantity of groundwater up to 4000 metres, is quite large at about 10 million cu km, but the amount of groundwater which forms a part of the hydrological cycle is only13,000 cu km, about 1/3 rd of the total run-off. If deeper groundwater is mined, it will bring down their level without recharge.

However, most countries have drawn up an estimate of their shallow ground water reserves, which are mostly used for agricultural irrigation.

Water Quality

The availability of suitable water is influenced by its quality. Surface water may contain suspended soil and many other organic and inorganic contaminants. On the other hand, ground water contains far more dissolved salts than the surface water. Only water, that is non-toxic and free of bacterial contamination, should be included in the availability of water. However, much of this water can be made potable after due processing. We will deal further more comprehensively with the quality of water in later section.

IV

Global Water Demands

Categories of Water Demand

In Table 8.4, it has been proposed to classify water uses by withdrawal, on-site and in-stream and take into effect the nature of use i.e. consumptive or non-consumptive, so that it will be possible to have a better understanding of the effect of water use on the human environment.

Table 8.4: Characteristics of Different Water Uses

Categories of Demand	*Substitutes*	*Quality Effect*	*Percentage Consumed*
Drinking	None		1-15
Other Domestic			1-15
Public Urban			1-15
Livestock	None	Organic	1-15
Soil Moisture	None	Sediment	0-100
Irrigation	None	Salt	10-80
Drainage		Salt	0
Wetland Habitat	None		0-10
Aquatic Habitat	None		0
Navigation	Land transport		0-10
Hydropower	Other energy		0
Mining			1-5
Manufacturing			
Cooling	Air	Thermal	0-3
Processing	Mechanical	Organic/toxic	0-10
Waste Disposal	Air/mechanical	Organic	0
Recreation			0
Aesthetic			0
Flood Loss	Land use		0

Source: Gilbert White 1976.

Usages, which require to actually withdraw water from waterbody – drinking water, irrigation, industry, for example and thus deplete the source, unless water is returned to the source by drainage or otherwise after treatment.

In-stream uses – navigation, hydropower, fisheries, for instance, generally leave the source volume of water unchanged.

On-site uses – for rain-fed crops or wildlife habitat including wetlands, for example – may or may not affect the volume of flow.

Eventually, all the above uses, whether consumptive or non-consumptive, will generally cause a change in quality. Besides, the above uses will also involve the potential losses of supply through evaporation, transpiration or diversion, so that they may not deplete a source stream or aquifer. Admittedly, while some uses, such as irrigation or wider withdrawal for potable water can wholly deplete a stream, other uses like repeated use of thermal cooling will reduce flow only marginally.

The demand for water has also an economic underpinning. It may mean the amount of water supply related services that would be demanded and used at a given price. But, in contrast to most other economic commodities, water, on the concept of plentifulness, has been regarded as a free natural commodity. Thus, it is not easy to establish reasonable levels of water demand and supply through the pricing policies. However, in the coming decades, water might have to be priced, preferably publicly in some measure, for its conservation and improvement of its quality and equitable availability to all.

In general, demand for water can be projected in aggregate terms. In Hungary, for example the range of demand was projected till the year 2050 on the basis of historical relationship between per capita aggregate use and gross national product.

More detailed studies require examination of each category of demands outlined in Table 8.4, projected at the national level or even for smaller areas such as basins and sub-basins. To do this, the most rudimentary method is extrapolation of rate of change observed in the past time period. This method is of course used very widely. Unfortunately, the influence of socio-economic factors on demand or even the influences of technologies and policy framework on demand are not adequately included. Hence the long-term projection on basis is unlikely to be reliable. But the short-time projection will remain valid.

Domestic Demands

To a person living in a semi-arid land, domestic water use may mean a few litres a day, mainly for drinking and cooking and another few litres for washing and bathing. In such small communities in non-industrialised countries, likewise the water demand is generally low, say 10 to 20 litres per day, namely because of distant access of water. In higher income societies, watering of lawns and kitchen gardens account for some considerable water. In middle and upper income sections of large cities, household appliances, swimming pool and the washing of automobiles considerably increase water demands. A farming household may also use its supply to water livestock and irrigate a small garden.

Therefore, the quantity of water required domestically will vary with local circumstances including the method of excreta disposals etc; it is finally income-elastic. It is common to design new community supplies for delivery of an average 100 to 350 litres per capita per day for house connections and 10 to 25 litres per capita daily for people using public stand pipes. In rural areas, the design figures used by government agencies differ according to the livelihood and the availability of natural water sources. The above general consumption ranges are illustrated in Fig. 8.2.

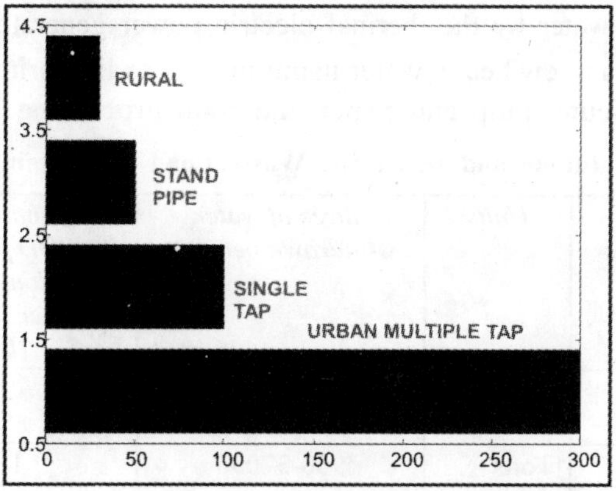

Fig. 8.2. Daily Consumption of Domestic Water Uses per capita.

According to the 29th World Health Assembly, May 1976, the minimum targets of water supply and the human excreta disposal in the developing countries were to be achieved by 1980 on the following order:

In urban areas:

Safe Water supply for 91% population, consisting of

68% population by house connections,

23% population by standpipes.

Safe sanitation for 94% population, consisting of,

38% population by public sewers,

56% population by household systems.

In rural areas,

36% population to have reasonable access to safe water,

24% population to have excreta disposal facilities.

In the above targets, there are wider sections of rural population uncovered even by the minimum health provisions. In this perspective, UN Millennium Development Goals adopted in 2000 exhorted the earliest remedial measures for eradication of poverty and provision of water and sanitation to all.

Industrial Demands

The economic development of a country and its application of technology influence the use and demand of water for industry to even greater degree than they influence the use of water for domestic purposes.

The major uses of water in industry are for cooling, processing, boiler water and other general purposes including air conditioning, cleaning etc. Cooling is the principal purpose of withdrawal and commonly accounts for as much as 60 to 80 per cent of industrial withdrawal. This is the reason for the heavy use of water by the thermal electric power generating plants.

Table 8.5 lists a few heavy water using industries like primary metals, chemicals, petroleum, pulp and paper and food processing.

Table 8.5: Water Intake and Discharge Waste Load in Certain Industries

Industry and Selected Products	*Units*	*Litres of water withdrawn per unit*	*Pounds of 5-day BOD per 1000 gallons Process Water discharge (USA)*
Primary Metals: Steel	Tons	8000-61000	-
Chemical: Soap	Tons	960-37000	16.70
Petroleum: Gasoline	Kilolitres	7,000-34,000	2.50
Pulp and Paper: paperboard	Tons	62,000-376,000	2.21
Food Processing: Sugar beets	Tons	1800-20,000	9.16

Sources: Wollman and Bonem, 1971 (UN Country Reports 1957-68)

The critical considerations in the industrial use are how much water is taken in, how much is consumed and how much and what kind of waste is discharged, since this can gravely affect human and environmental health.

Consumptive use rarely exceeds 20 per cent of withdrawal. In case of steam power generation, consumptive use is less than 2 per cent, which economy is still being improved in recent times.

The waste discharge affects the quality of receiving water, precise effect depending upon flow and quality of the receiving water, its temperature and its current characteristics pH, BOD load etc.

Today, inside the industrial plant, it is necessary to audit the water engineering as to its use, processing, and economy. Some less engineered plants may use several times more volume of water than a more modern processing unit. The waste may be treated within the plant or disposed off directly without treatment at great social cost. Thus, it is imperative that the volume and the quality of industrial waste be improved through process design, in-plant recycling and improved home keeping procedures.

Agricultural Demands

As the world population grows, so does the need for increased food production. The augmented agriculture, in turn, demands greater supply of water. The supply of water will need more effective use of rainfall and greater conservation and management of water. To extend the agriculture to still higher level of productivity may call for irrigation by diverting and withdrawing lake, stream or ground water.

Agriculture is the largest user of water, accounting for some 80 per cent world consumption, mainly through irrigation for higher production and also possibly for multi-cropping. In addition, vast areas use rainfall as the principal source of water for crops. Approximately 200 million hectares of land were irrigated in 1970; almost half of these in the developing countries where food needs are especially acute.

While it is difficult to define irrigation per se, any planned agro-development still requires an ensured controlled water supply to the crop. The quantities and acceptable standards of reliability of supply are determined by the local climate, agronomic and economic conditions and water requirements are established at the local project level. Table 8.6 provides water estimates of gross water estimates for land irrigation and harvesting based on UN sources 1970. To arrive at the estimates of gross water demands of Table 8.6, FAO assumed the demand to 700 mm of water per season, when the staple food is a dry cereal and 1500 mm where the it is rice. These values are presumed to be the total quantity needed by the crop, whether supplied from rainfall or irrigation, with the typical losses associated with storage, conveyance and application taken into account. Still, part of the water lost in application in one area may become a source of supply elsewhere through the seepage, flow back and ground water recharge.

Other essential agricultural needs, including rural domestic uses, make relatively small demands on overall water resources. Water demand for livestock, for example, is estimated at about 62000 million cubic metres per year for 1700 million animals, and this allows 100 litres per day per animal to cover all animal needs, Fisheries represent a small volume use, which is of course highly remunerative in yield per unit of water consumed.

Table 8.6 : Estimated Irrigated harvested Areas and Gross Water Demand, 1970

Regions	*Irrigated Harvested Areas (Million Hectares)*	*Gross Water Demands (Billion Cu. Metres)*
Africa (excluding Egypt, Libya, Sudan)	2	14
Latin America and Caribbean	8	59
Near East (including Egypt, Libya, Sudan)	16	109
Asia (excluding China and USSR)	75	899
Asia (China and other communist countries)	78	1167
USSR and Eastern Europe	14	95
North America, Australia and Oceania, Western Europe	33	228
Total	**226**	**2571**

The turbo-plants to capture hydro-energy are concentrated for the most part in Europe (notably in Russia), North America, Japan, China, India and Brazil. In some countries e.g. Brazil, Canada, Morocco, Norway and Uganda, hydropower accounts for as much as 80 to 90 per cents of electric power generated.

Often the hydroelectric water reservoirs are also used for irrigation programmes, navigation, domestic water supply or aquatic habitat. Thus, the development of hydroelectric power is often linked to multi-purpose programmes.

Apart from the fact that hydroelectric power is highly economic, it is one of the best source of environment-friendly energy, not emitting any polluting gases, GHG or otherwise.

Waterborne Transportation

Inland waterways make use of natural streams, lakes or artificial canals. Some of the great inland water systems are the Rhine, Seine, Danube, Elbe, Volga-Don, Ganges-Brahmaputra and the Mississippi. But the maintenance of the inland navigation ways involves quite elaborate management of flows, siltation problems and operation of various lock gates.

However, barges carrying as much as 3000 tons of cargoes with drafts of 2.5 metres are found on European waterways. Inland vessels in North America could have drafts of 3 to 4 metres.

Inland water transport faces stiff competition with the roadway and railway transport. But one thing is sure, given the infrastructure of inland port terminal with many road connections, maintenance of water navigation depths by dredging and water lock gates, water transport is the most energy economic for every ton-kilometre compared with the Road and Rail. By this count, the inland transport will also be comparatively more environment friendly

Flooding

A natural function of all stream valleys is to carry the peak discharges that occur seasonally or at rear intervals. All forms of water channels must try to provide for all these flows as much as possible. But, as we know, it is not possible to prevent flooding of plains and valleys by peak discharges.

To mitigate the effect of flooding, there are several actions, which are advised as follows:

(a) Trying to control flood flows by engineering works as embankments etc.;

(b) Reducing flows by upstream land treatment and inter-basin water transfer;

(c) Modifying building and land uses to reduce vulnerability to loss;

(d) Sharing the burden of flooding through relief and insurance scheme;

(e) Finally accepting rare and really heavy peak flows leading to limited loss of property and lives, if any.

Unfortunately, on a world scale, the average annual per capita property damage from floods is estimated to be increasing, as is vulnerability to catastrophic damage caused by exceedingly rare floods with a probability of recurrence, say 100 years or more. Urbanisation generally increases peak flows. And poor use of land has effects of even greater magnitude throughout the developing world.

Inland Fishing

It is understood that about 13 per cent of the total world catch at 63, 000 million metric tons is obtained from inland fresh water sources. Unlike marine fisheries, which have shown marked decline in catch resulting from overfishing and inadequate harvesting since 1970, fresh water fish stocks are stable and increasing. In India, inland fish catch and matching harvesting have grown several times over the decades.

The demand for fish as a food supply is intensified by the shortages of other animal proteins, minerals and vitamins. Fresh water fish production thus plays a small, yet significant role in helping meet increasing food needs. In multiple purpose, water planning, provision for fish production is of high priority.

V

Global Water Security and Vision 2025

Global Water Supply-Demand Management

In the world today, the essential problem in the management of water is how to reconcile increasing demand of the growing human society with the available fixed supply of fresh water. There are three typical scenarios in the global mismatch in the dynamic demand and constant supply of water as obtained below:

(i) At the one extreme are the regions, countries or parts of countries with a large natural supply and a relatively low level of consumptive demand, which can be met without the requirement of an overall regulatory regime. Here water management implies management of various water projects on basically local considerations only.

(ii) At the other extreme are areas with scarce supply and relatively high levels of demand, which are satisfied only through overall regulation and rationing of water resources. The regulatory regime mainly concerns itself with the demand management including relocation of demands of humans and industries.

(iii) In the intermediate situations, where there is overall balance of water resource in supply and demand, the water management calls for both increased availability of supply at some locations and regulated demand by making more efficient use of water at some other locations

Demand-Supply Management in Arid Areas

Demand-Supply management presents special difficulties in arid areas and semi-arid lands. The population in these regions is faced with overgrased, perched or sterile pasture lands; a degraded or vanishing forest cover; very poor agriculture aggravated by uncertain rainfalls and river flooding. The surface water is very scanty or non-available. In many of these areas, there is considerable ground water potential, which needs to be developed as an integrated water work as in the artesian basins of sub-Sahara. Forestation, land development, dry farming and long-term integrated water sourcing planning are other measures. The last alternative is relocation of population to more benign neighbourhood.

Long-term Water Development Planning

Solving problems of water supply and development need long years of infrastructure development works. From the UN report we know that Tisza River basin (a 156,000 sq. km. sub-basin of the Danube shared by five countries), water management began 150 years ago and likely to continue for another fifty years or so. The Salt River basin (32,000 sq. km. in the United States) developed complete regulation over 60 years period. Similarly, Netherlands undertook reclamation of sea over many decades to develop both new lands and water sources. In India too, river valley projects like Damodar Valley, Bhakra Nangal and Hirakund took more than one decade each for completing the core projects, the auxiliary developments having still continued.

To conclude our global overview of water resources, it should be emphasised that water development work is often undertaken on the consideration of its social benefits and economic costs. This is unavoidable, considering that there is always a demand for funds for other developments in the areas of alleviation of poverty, and improvement of education etc.

Freaks of Water Resources

We round up the total global annual available fresh water at 40,000 cu km But of this amount, only about 3,800 cu km (Cosgrove 1998), roughly 10 per cent were withdrawn (diverted or pumped) for human uses in 1995. Further, of the amount withdrawn, only 2100 cu. km. were consumed. The remainder was returned to the streams and aquifers, usually with significant reduction in quality. Now if we are withdrawing only 10 per cent of renewable water resources, and consuming only about 5 per cent, where is the water scarcity? This is the rub of the riddle.

The problem is that world's fresh water is very unevenly distributed in space and time. For example, in India's Brahmaputra-Barak basin, out of the total annual flow of 586 cu. km. only 24 cu km, amounting to 4 per cent is utilizable, because of short duration of the rainy season with flood flow, geographical non-feasibility of large storage and thinner population. Similarly, human demands are very small in the Amazon basin, Northern Canada and Alaska. On the other hand, in the arid and semi-arid river basin, the withdrawal and use amount to 80 to 90 per cent. In fact, the withdrawal of water is so high that some of the world's largest rivers do not reach the sea. In the wake of the multifarious developmental activities, total or near total of river run-off is either withdrawn or diverted. The Huang Ho (Yellow) River in China did not reach the sea on some days in 1972 and for seven months in 1997. The Colorado River in South-Western United States and the Indus River adjoining India and Pakistan are two of the many prominent rivers in the similar predicament.

Unsustainable Scale of Water Use

The general pattern of sectoral consumption of fresh water is of the following order:

Agriculture / Irrigation	70%
Industry	20%
Domestic (Drinking/Cooking/Bathing)	10%

As seen above, agriculture claims by far the largest share of water use. Much of that water withdrawn for agriculture is lost through evaporation from reservoirs, canals and soil and transpiration from crop plants, which also further absorb water.

A widespread revolution has come about in drawing groundwater for irrigation with hundreds of diesel engine driven pumps, privately owned, particularly in Asia. In Pakistan, it is believed there were 360,000 private pumps in 1993 alone drawing groundwater heralding new growth phenomena. Half of India's massive irrigated land measuring about 50 million hectares sources water from the groundwater. In China more than 2 million pumps irrigate 9 million hectares. In the United States, in the very large Ogallala aquifers, a privately funded groundwater well system feeds irrigation sprinklers is in operation. It is estimated that 150-200 cubic km more ground water is pumped out each year than is recharged in overexploited aquifers. As a result, ground water tables are falling significantly with risk of collapse of agriculture system in North China, US high plains and some major aquifers in India and Mexico. Naturally, a well-coordinated long-term water management plan is called for to regulate and to remedy this unsustainable and wasteful agricultural irrigation practice.

Industrial uses at 20 % withdrawal of water for power plants and processing appear quite profligate and wasteful. Some industries, for the same process, withdraw many times more water than a similar process in an efficient industry, practicing economy of water resources. Today, as the water is becoming scarce and more costly, some industries have introduced nearly 100 per cent recycling. Alternatively, reprocessed water is released for other multiple uses.

No doubt, municipal/domestic use accounts for only 10 per cent of world consumption of fresh water. Here again, there is much profligacy of water use for lawns, clothes wash, car wash etc. The following per capita consumption data speak for themselves:

Per capita Consumption of water

	Liters per day
USA	400
EU	200
Asia/Latin America	200-600
Sub-Sahara	10-20

In Asian countries, of the water consumed, 70 per cent is lost through leaks. Improvement in maintenance and service will immensely increase the availability of the much-needed potable water. Similarly, OECD countries can easily reduce their consumption substantially by redesigning and rationing their various washing, flushing, and lawn water processing.

Let us remember that Israel is efficiently managing their scarce water with annual consumption of water at 380 cu m per capita-year compared with the 1220 cu m per capita-year of the United Kingdom (UN Report).

Hydro-energy

Of the total global run-off of 40,000 cu km, about 6,000 cu km is held dammed in reservoirs offering water for hydropower, drinking water, flood control and recreation, all highly benefiting the mankind. But large dams and hydropower projects lead to large scale displacements of human habitation and also cause many other adverse environmental impacts.

Majority of rivers of the developed countries have their discharges regulated by flow control, diversion for agricultural and even regulation of river channels by dams. Dams lead to accelerated evaporation also. Socio-economic planning for hydro-energy projects is highly desirable, so that there is net socio-economic gain and no specific section of the deprived and poor societies are made to suffer.

World Water Vision 2025

The World Water Council under the distinguished Co-Chairmanship of Ismail Segeldin and Frank Rijsberman drew up a world water scenario for 2025 under business as usual (BAU) model, given the natural development of growing population, continuing economic progress and improving technology application. The gist of the findings of the model in respect of categorywise water withdrawal and consumption is presented in Table 8.7. It will be seen that the global withdrawal of water will increase from 3800 to 4200 cu km at 10 per cent growth annually. Similarly, water consumption will increase from 2100 to 2300 cu km again at 10 per cent growth annually

Table 8.7: Categorywise Water Visions – Withdrawal and Consumption 1995 and 2025

World Water Uses	*World Water Use 1995 (Cu. km.)*	*World Water Vision 2025 (Cu. km.)*	*Percentage Increase 1995-2025*
Agriculture			
Withdrawal	2500	2650	6
Consumption	1750	1900	9
Industrial			
Withdrawal	750	800	7
Consumption	75	800	7
Municipal			
Withdrawal	350	500	43
Consumption	50	100	100
Reservoirs (Evaporation)	200	220	10
Total			
Withdrawal	3800	4200	10
Consumption	2100	2300	10
Groundwater Consumption	200	0	-

Source: Cosgrove and Rijsberman (2000)

The scenario further highlights that the per capita availability of renewable water resources is likely to fall from 6,600 to 4800 cu.m. per capita-year. Given the highly uneven distribution of water resources, some 3 billion people will live in countries, wholly or partly arid or semi-arid and will have water less than 1700 cu.m. per capita-year, quantity below which a human is believed to suffer from water stress. Examined in an alternative perspective, 4 billion people will live in countries, where by 2025, about 40 per cent of renewable water resources will be withdrawn, which is another indicator of apparent water stress.

In further elaboration of Table 8.7, regionwise water withdrawals and consumptions are given for Africa, Asia and Europe in Table 8.8.

Table 8.8 : Regional Water Withdrawal (cu. km./year)

Categories	*Europe*		*Africa (Sub-Sahara)*		*Asia (East & South)*		*Total*	
	1995	*2025*	*1995*	*2025*	*1995*	*2025*	*1995*	*2025*
Agriculture	140	130	42	40	1359	1266	1541	1436
Industry	130	130	6	20	153	263	289	413
Domestic	80	80	12	40	114	471	206	591

The portion of land areas and the people who will become conspicuously water stressed by 2025 is shown in Table 8.9. The Table 8.9 shows that in

Table 8.9: Regional Water Stress: Global Perspectives

	Europe		*Africa (Sub-Sahara)*		*Asia (East & South)*		*Total*	
	1995	*2025*	*1995*	*2025*	*1995*	*2025*	*1995*	*2025*
Areas of Water Stress (million sq. km)	1.5	13.5	2.6	4	6.5	7.9	10.6	25.4
Population under Water Stress (million)	160	180	110	300	1.1	2.4	271.1	482.4

Europe, the total withdrawal of water in 2025 will, in fact, marginally decrease from 350 to 340 cu km because of higher water efficiency in Agriculture.

The withdrawal levels in Domestic and industrial uses will remain unchanged, again because of technology and improving operational efficiency.

In Africa, the domestic water withdrawal/consumption will jump up conspicuously from 12 to 40 Cu.km. per annum because of expected substantial improvement in living standards and partly demographic increase. Consequently, overall withdrawal will increase from 60 to 100 cu kms.

The Asian scene is similar to the African scene, but increase in withdrawal is less pronounced. Domestic use will increase from 114 to 471 cu km, but agriculture will withdraw less, town to 1266 from 1359 cu.km, because of superior agricultural practice and water management.

Towards the Water Crisis Resolution

The United Nation Millennium Development Goals (MDG) enunciated that in 15 years nearly all people in the world in the cities, towns and villages enjoy safe and adequate drinking water and sanitation and have enough food to meet their hunger and nutritional needs. The participants of the above 2025 vision exercises, having similar goals for water even before the MDG, had proposed five major actions to realize the goals:

(i) Involve all stakeholders in integrated water resources management;

(ii) Move towards full cost pricing of water resources of water services for all human uses;

(iii) Increase public funding for research and innovation in broad based public-private water resource management;

(iv) Seek international cooperation to improve integrated water resource development in multinational basins;

(v) Invest generously in all water-related infrastructure;

Among the specific measures, it must be recognised that agriculture is by far the largest consumer of water at the global average of about 65 per cent. Hence emphasis should shift to avoiding waste and saving as much water in agriculture. Dry farming and rain-fed farming including green farming should be increased. Even where irrigated farming is essential, excess water withdrawal shall be avoided and water should be utilised prudently as in sprinkler farming practiced in Israel. In any case, excess water deteriorates soil fertility by increase of salinity and possible soil erosion.

There is scope of economy in domestic use particularly in washing and gardening and flush-sanitation. New methods of eco-sanitation, not dependent on water as carrier, could be evolved, which may reduce pollution and offer human waste as agriculture fertilizer.

Natural and artificial wetland might be used to improve pollution water and treat domestic effluents. There are manifold ways to improve water resource availability in the coming decades to raise quality of life, provided adequate effort and funds are directed to water resource development.

VI
Water Quality and Environment

Water quality is very important consideration for all water development projects, as it affects all aspects of water use – for humans, for animals, for crops and even for industry. With respect to water quality in lakes and reservoirs, it is necessary to understand the problem of thermal stratification and its effect leading to the loss of dissolved oxygen. Further stratification also prevents the nutrients in the lower strata not getting distributed and mixed in the upper layers.

Among several other undesirable effects in quality, it may be mentioned that water under the spillway may become supersaturated with nitrogen due to high turbulence. The situation may also affect the mineral and organic content leading them to settle out. In the arid zone, the water often develops high salt concentration. With high evaporation and low precipitation regime, the impounded water gets high salinity.

Flooding of tropical forest may also cause hydrogen sulphide in impounded water, which causes aesthetic problem for human consumption, toxicity for fish and corrosion for metal parts. Decaying of plants in floodwater may cause high rise in nutrient contents, leading to eutrophication and loss of oxygen in water.

Impacts on Ground Water

Irrigation water may have impacted on ground water in various ways. In arid zone, if more irrigation water is supplied than required for evapotranspiration, it may lead to accumulation of salt in root zones. High irrigation water may also lead to seepage of salt, fertilizer and pesticides to the groundwater and cause harmful contamination.

Considering that there are 225 million hectares of irrigated land in the world, ground water contamination by irrigation projects could indeed be extensive. Irrigation schemes in Punjab, Pakistan have raised the ground tables by several metres in their first 10 years of operation.

On the other hand, excessive pumping of ground water in India and USA has led to depletion of ground water levels conspicuously.

Water Logging and Salinity

Degradation of soil fertility by water logging and salinity development is common, but one of the most serious impact of irrigation. Increase in irrigation without proper drainage facilities usually leads to increase in soil salinisation and alkalisation as water evaporates, leaving behind salt particles on top soil. This occurred in Pakistan during the 1960s, where 24,300 hectares of fertile cropland in the country was lost every year (Biswas 1980). Soil salinisation and soil erosion due to wrong irrigation practices are becoming serious problems throughout Canada's agriculture areas, particularly in the dry prairie environment.

As much as half of world's irrigation schemes suffer in some degree from water logging, Salinisation and alkalisation affecting 1 to 1.5 million hectares annually.

Seepage from irrigation canals is a common problem along all unlined canals. This is evidenced in Pakistan, where the ground water level, as mentioned earlier rose by 7 to 9 metres in last 10 years.

Occurrence of Earthquakes

It has been proven that earthquake can be induced by large reservoirs. The main earthquake in Greece in 1966 had its epicenter under the reservoir of the Kremastha Dam. Likewise, the 1967 Koyna Dam disaster in Indian peninsula was caused by an earthquake whose epicenter was located at the dam itself (Biswas 1978).

Effects on Climate

Some small changes take place around a new water reservoir. For example, some change in rainfall pattern has been observed around Lake Volta in Ghana, the peak having shifted from October to July/August. Aswan High Dam, by more than doubling the surface area of Lake Nasser has increased the total evaporation from 6 cu km to 10 cu km. The total loss by evaporation amounts to 11 per cent. The point is made that such change will always have some effects on humans.

Biological Impacts

- Direct killing of animals and plants in the area flooded by the dam;
- Opening of new roads and navigational channel and migration of new inhabitants may affect the eco-systems and bio-diversity;

- Development of fisheries in the new reservoir may be affected by water stratification, eutrification or other toxic attack;
- Fisheries downstream might get adversely affected due to loss of nutrients;
- Water resources development projects are often related to waterborne diseases.

Socio-economic Impacts

- Loss of timber forest and low-lying farmland due to flooding, clearings, construction or salinity;
- Deposition of silt upstream reduces the agricultural and/or fishing yield downstream;
- Water development projects attract migration of people from other areas for better job or commercial opportunities;
- A water development project may also cause considerable upheaval on the socio-cultural structure and some disruption of the local people's livelihood and resettlement in new areas.

Lessons Learnt

Generally all water resources processing and development projects have social and environmental implications, which are usually long-term, and widespread, but often reversible. Perhaps, the most well-known example is the Aswan High Dam, which created considerable adverse environmental impacts like excessive evaporation, decrease of fertile land downstream, and the killing of the sardine fishing industry in the Mediterranean and many other ill-effects.

But at the same time, Aswan Dam enabled Egypt's electricity generation capacity to double, helped prevent disastrous floods, improve water navigation and created a vast potential fishery in the reservoir. Hence, while weighing the negative impacts, we should not ignore greater positive benefits, the real raison d'être for development project. This calls for a very balanced evaluation of the new water projects in respect of development vis-à-vis environment.

VII
Sustainable Management of Water Resources

Water resource is not so plentiful as could be splurged and wasted. Besides, water is very precious for life and living. From the UN Report, we already know that global annually renewable fresh water, available as run-off, is only 42,000 cu. km and if all the water could be withdrawn and apportioned among the present 6 billion people, the share per person would have been 7000 cu.m., which is apparently sufficiently plentiful for human needs. Unfortunately only 10 to 12 per cent water can be harnessed today, such that only the actual globally available water is 700 to 1000 cu. m per person per year, which indicate high level water stress. Thus, there is great need to raise the recovery level to say15 to 25 per cent of the global run-off depending upon the local water demand. Such upgradation of water resource could only be achieved through long-term planning and investment in waterworks and other related infrastructure.

Secondly, to make matters worse, the global run-off of 42,000 cu km of renewable fresh water is very unevenly distributed over the continents and regions. In the Amazon basins, Northern Canada and Alaska, vast amount of water flows away to the sea untapped and in India's Brahmaputra-Barak basins, only 24 cu km water is withdrawn of the gross flow of 1000 cu km annually. On the other hand, in the Indus river basin of India-Pakistan and Amu Darya and Syr Darya basin in the Central Asia, Huang Ho (Yellow) river basin in China, water is so heavily withdrawn away that sometimes no river water reaches the sea. This grave situation underlines the urgent needs of long-term planning for redistributing the water resources by possible inter-basin transfer of water, tapping ground water or corrective water shortage and conservation scheme.

The third dimension of the water problem is the local management of water resources, regulating supply, demand and consumption, upgrading water quality, minimising leakage and wastage. Such measures will help to mitigate and improve the glaringly discriminatory annual water consumption levels—2000 cu. m. in Europe, 1000 cu.m. in India, 1000 cu. m. in East Africa, 400 cu. m. in Israel, 100 cu. m. in Kuwait or less than 100 cu. m. in Sub-Sahara.

Let us now shift our attention to the developing countries. There is no doubt that for developing countries, which are all located in the tropical and sub-tropical regions, with erratic rainfall patterns, efficient control and management of water is an essential requirement for their continued development. Without proper water management, self-sufficiency in food and energy will suffer. Scarcity of water and reliability of its supply are major constraints for agricultural development of arid and semi-arid countries. Given the above scarcity-prone water scenario of developing countries, the demand for water is likely to increase continuously with the improvement of the standard of living of the people. No doubt, these countries need to undertake ambitious programmes of water resource development adopting principles of modern management and economic marketing with social forces for enduring viability of projects.

Thus, we are given to infer that the principles of modern management are needed to be applied to alleviate the water problem, adopting tools of long-range planning, modern engineering and project management for the following broad objectives.

(i) To increase gross withdrawal of run-off water from 10 to 20 per cent by building dam and reservoir, distributional canals and any other means inside a basin;

(ii) To redistribute the increased available water to more water-stressed and arid regions through a network of storage facilities and distribution systems using measures of conservation and waste-minimisation, soil conservation, forestry and if possible by inter-basin water transfer.

(iii) To modernize all local water distribution systems in the urban and rural habitation with the aim of high utilisation, least wastage, recycle and reuse.

(iv) All programmes of development and management of water resources should ensure as not to adversely impact the environment.

Outlining the above modern management of water resources, we enlist some of the prime considerations in the managerial assessments and executions enunciated by Prof. A.K. Biswas, a long experienced international water specialist of high repute (Biswas 1980):

(i) Comprehensive project Review including searches of all possible management options;

(ii) Recognition of water as an economic good;

(iii) Adoption of the river basin as a unit of water resource development and management;

(iv) The consideration of environmental impacts;

(v) The assessment of social impacts includes the involvement of public in policy formulation, planning, execution and financing.

Water as an Economic Good

In recent times, there has been some loud thinking to levy a charge on the use of water with two principal objectives. Charged water will be judiciously used avoiding wastages, preserving water for greater and wider use. Secondly, charges realised can be further deployed to enhance water infrastructure.

Other natural resources like forests, fisheries and minerals are chargeable natural resources,; but there is a basic objective to treat water as economic good as water is as essential for life as air and shall be equitably available to all, irrespective of their economic status. Every society should therefore carry social cost-benefit analysis to determine if water should be chargeable even at subsidised rate.

River Basin as a Unit of Water Management

The River basin has increasingly become accepted as an integral unit for water resource management. The rationale follows from the concept of a river as an organic whole, so that interference with one part of it will directly influence its operation at another point of the river system. Engineers and geographers dearly espoused the idea. Economists have strongly recognised it as a viable way of internalising externalities.

In India, United Kingdom and France, river basin is taken for all phases of water management. Elsewhere, such concept is first catching up.

Environmental Impacts

In the previous section, we have observed how water development projects can have some serious impacts on other facets of the ecosystems. Under UN initiatives, water management agencies are obliged to identify and evaluate the impact of water projects on the environment and the overall ecosystems. In fact, the environmental impacts are now an elaborate process and in India, it is statutory to undertake environmental Impact study and due clearance before a Project can be implemented.

Social Impacts

Over the recent decades, the project planning for the major water projects has extended its study from the purely technical evaluation to wider social, environmental and other public concerns. In fact, the public and other social groups are being involved with the formulation of the project to maximize social and economic benefits.

In India, we have seen how much social turmoil was created by the very large Narmada Valley project, causing widespread displacement of population. Similar situation was also experienced in China. In New Zealand and Australia, the right of the aborigines of the affected valleys had to be duly recognised. Box 8.1 shows schematically the mechanism of Public Involvement.

Box 8.1 Mechanism for Public Involvement	
1. Public opinion polls and other surveys; 2. Referendum; 3. Ballot Box; 4. Letter to editors or public authority; 5. Public hearing of Inquiries; 6. Representation by Pressure groups; 7. Public Meeting /Demonstration; 8. Organising Workshops; 9. Formal Protest and Application; 10. Court Action	Increasing Effectiveness of Public ↓

B. INDIAN WATER RESOURCES

VIII.

Challenging Water Security Issues

Issues in Ancient and Medieval India (Prasad et al *1987)*

Some of the great ancient river valley civilisations are the Egyptian civilisation on the Nile valley, the Mesopotamian civilisation in the Euphrates and Tigris valley, Indian Harappa civilisation in the Indus valley and the Chinese civilisation in the Huang Ho river valley. All these civilisations, which flourished from 3000 to 2000 BC, had the common feature that their basic economic activity was agriculture supported by the water resources of the respective rivers.

The later Aryan civilisation, during which the prime religious anthology 'Rig Veda' was written, dating around 1500 BC, also flourished along river Ganga. The Rig Veda and the other Vedas refer to the damming and control of river flows, water wheels and canals (Prasad et al 1987) for purposes of irrigation.

From the 6th century for about six centuries, there was considerable political instability and turmoil. Even then, dug-well irrigation and prosperous agriculture along Ganga continued. In the 14th century, two reservoirs and a network of canals were constructed in the Gangetic plain by the Tuglaq rulers.

In the16th century, with the establishment of the Mogul empire, there was considerable political stability for the next three centuries. Irrigation by means of tanks, canals and wells was widely practiced by the farmers to lessen the vagaries of the monsoon and to boost the agricultural production. There was nevertheless no major irrigation scheme undertaken.

India again enjoyed two hundred years of peace and stability during the British rule. In the beginning, not much attention was paid to the irrigation or agriculture by the trade-minded British rulers. But in the first half of the 19th century, various parts of the country came in the grip of famine and which forced the British Government to constitute two commissions :

- the Famine Commission 1880 followed by,
- the Irrigation Commission 1991,

which recommended renovation of several defunct or dilapidated irrigation works and also proposed new irrigation schemes.

However, modern irrigation development came to India through the British along with the relevant technology and the organisation. The Thomson College of Engineering at Roorkee was established around 1850 and a research organisation named Central Water and Power Research Station at Pune was set up in 1916.

Issues in India Today (W.B. Report 1991)

India is blessed with a large number of rivers and also gets reasonably good amount of rainfall of 1100 mm on the national average. The problem is of space and time distribution of that water resource. In some river basins like Brahmaputra-Barak, there is high excess of water unutilised and flowing out to the sea. In some parts of India, typically in parts of Rajasthan in Western India, the annual rainfall is as low as 100mm. Further, the major rainfall occurs for only three to four months in a year leading to low utilisation of the available water. To make the matter worse, the Indian population is steadily increasing from about 400 millions in 1950 to about 1000 millions in 2000.

The World Bank in conjunction with the Ministry of Water Resource of the Indian Government undertook studies of Indian Water Resources (World Bank Report 1999). It observed that the water availability was as high as 5000 cu. m. per person annually about the time of Indian independence in 1950, which was plentiful compared with UN standards of water stress at 1000 cu. m. per capita. It was apprehended by them that the population will grow to 1.4 billion by 2025 and the water availability will come down to about 1500 cu. m. per annum. The adversarial relationship between population and water availability is illustrated in Fig. 8.3 of World Bank, which indicates that the water availability may come down to as low as 1000 cu. m. per capita in 2050, unless issues of development and management are tackled rather seriously and urgently.

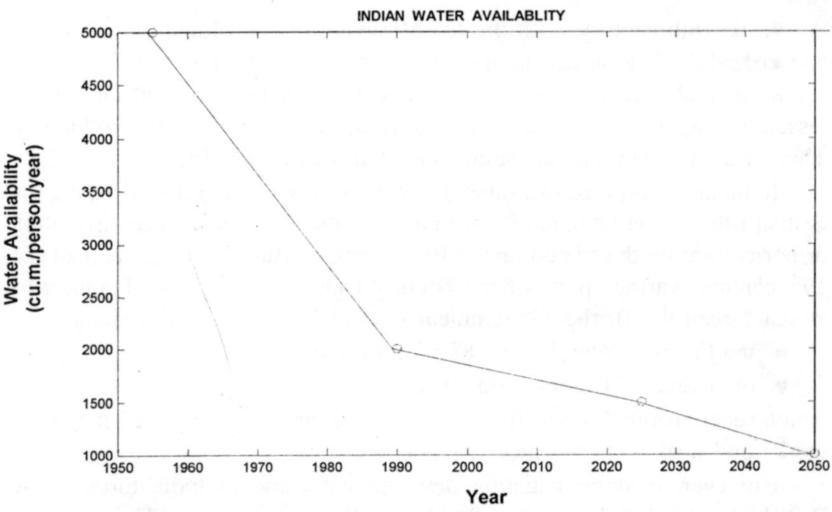

Fig. 8.3: Indian per Capita Water Availability

The most important framework of issues that should be addressed on all India vision:

- Finding solutions for competing inter-sectoral demands within a region and/or State keeping in view efficient water utilisation minimising wastage;
- Planning, developing, and managing water on a river basin basis applying the latest technology, economic and social justice;
- Equitably and prudently allocate water of a river basin under mutual or judicial awards;

To assist the above process of efficient and responsible management of water resources, following infrastructure and institutions are required to be strengthened and developed.

(i) National Water Policy for the whole of India;

(ii) Effective Legislation and Regulations for handling live issues of water;

(iii) Institutions for planning, development and dispute resolution if any;

(iv) Economic management including pricing and incentives in the use of water;

(v) Update data generation, analysis and public dissemination.

Some of the enabling factors and initiatives for augmenting water resources include:

- Private-Public Partnership in water project management and maintenance;
- Decentralisation of water project management;
- Stakeholder participation and involvement;
- Involvement of grassroot organisation;
- Water demand management to minimize wastage and maximise utilisation;
- Enabling environment for change and improvement of all initiatives;
- Economic Incentive framework.

Universal Access to Water

Though much work is being undertaken to make available water resources over all inhabited space and over the whole year, there are still many pockets where water does not reach in the developing countries. As a result, many humans do not still have ready and easy access to water.

The Millennium Development Goals (Target 10 of MDG) of the United Nations had ordained in 2000 that the percentage of population not having access to potable water should be halved by 2015. In Table 8.10 are given the trends of water accessibility at the global levels and India:

Table 8.10

Region	*Year*	*Population % (with access to water)*
World	1990	**77**
	2002	**83**
India	1990	**68**
	2000	**86**

The Table 8.10 shows a high jump in India's progress, beyond the average global progress. Of course, in the developed countries, almost everyone has satisfactory access to water and sanitation.

Two more important water issues in India which need to be duly identified and resolved for the long-term development of Indian water resources pertain to inter-state cooperation in sharing of water and inter-basin transfer of water.

The catchment areas of all major river basins in India are located in more than one administrative States of India and many rivers have a number riparian States. It is therefore in the fitness of the situations that all planning and development of a river basin must be conceived on the basis as whole and the states share the water on mutually agreed and equitable basis.

The inter-basin transfer of water from the surplus basins to the deficient basins to raise the levels of all India utilisation of water is a live issue which calls for finalisation of technical solutions awaiting the political consensus. This is however dealt with in a later Chapter.

IX

Supply and Demand Situations of Water Resources

We have already mentioned that India has a large number of rivers and also gets reasonably good monsoon rains. But at the same time, the country has a big population, large area of irrigated agriculture land and considerable industrial operation, which generates high demand for water. Thus, the water resource remains critical for India and her development plans. We present here a short outline of India's water resources.

Water Supply Perspectives

Rainfall

India's water resources are predominantly based on rainfall. The average rainfall is 1170 mm annually. But there is considerable variation in rainfalls across the country and again from year to year. The rain occurs mostly in the southwest monsoon season over three to four months, June to September in the year. Thus, there arises the necessity of creating large dam storage capacity for maximum utilisation of the surface run-off.

Regional rainfall variations are also high, ranging from the low 100 mm to high 11,000 mm in Meghalaya in the Northeast. Generally, there is adequate rainfall in the Eastern and Northeastern India including West Bengal, Eastern Uttar Pradesh, Bihar, Northern Orissa and the Cauvery delta – typically from 1000 mm to 2000 mm annually. In Western India – West Uttar Pradesh, Punjab, Haryana, Rajasthan and the rain-shadow of Deccan plateau, rainfall ranges from 300 to 800 mm.

Surface Water

India's average annual surface water run-off generated by the rainfall and snowmelt is established to 1869 bcm (billion cubic metres). However, it is estimated only about 690 bcm or 37 per cent of the potential surface water resources can actually be mobilised to effective use, because of the uncertain rainfall during the short period of few months in the year.

Consequently, the water availability is highly variable. While the northern rivers are more or less perennial fed from the rainfall and supplanted by the snowmelt in the summer, the southern rivers get dry during the major part of the year. By contrast, the Brahmaputra-Barak river system due to flood flow nature, is generally utilizable to a low of 24 bcm out of its average annual flow of 586 bcm.

Groundwater

India's rechargeable annual groundwater potential has been estimated to about 452 bcm in the aggregate terms. About half of the amount is located in the Ganga-Brahmaputra-Barak basins system. On an all India basis, using aggregate estimate, only about 30 per cent of the groundwater potential could be tapped for irrigation and domestic purposes.

However, the above aggregate use has been ballasted by the comparatively higher utilisation in the surplus Eastern and the Northeastern India. The situation is adverse in other parts of India, where underground reserve has fallen below the recharge level by several metres. In particular, Punjab and Haryana have already exploited most of the shallow water rechargeable reserve. In other areas like Rajasthan, Gujarat, most of the western UP and the Deccan States, the water tables are fast depleting at an unsustainable rate.

In fact, in the Tenth Five Year Plan (2002-07) document (GOI, 2002) notes that in 310 out of 5711 blocks in the country, groundwater was considered to be overexploited i.e. average withdrawal is in excess of replenishment. Consequently, the groundwater level has gone down by as much as 4 metres in the affected districts between 1981 and 2000.

Per Capita Water Availability – India vs. World

India has about 16 per cent of world's population but has only about 4 per cent of world's total renewable water resources (Indian 10th Plan, GOI, 2002). This highlights the pressure of population on available water resource, which would have been otherwise considered adequate at 1100 mm average annual rainfall.

We have already shown how Fig. 8.3 prophesizes the future gross water availability scenario from 5000 cu. m. per capita in 1950 to 1000 cu m per capita in 2050, which is the UNDP threshold of water scarcity. However, many nations are making good living with much lower level of water availability. Table 8.11 shows Israel managing with per capita water of 382 cum and Singapore with 211 cum, two fairly developed countries. These examples underline the importance of improvement in water use in India.

India ranks at the top of world in drawing 558 bcm per year for irrigation (FAO) of 72 million hectares. In fact, the irrigation water amounts to 84 % of total water usage. This underpins the necessity of improving the irrigation technology and makes do with much less water without adversely affecting agricultural productivity.

Assessment of Water Potentiality by River Basins

This topic is covered in fuller details in the Appendix . It is now only to be observed here that in as many river basins as Cauvery, Pennar,

Sabarmati, West-flowing rivers in Kutch and Sourashtra (including Luni), east-flowing rivers in between Mahanadi and Godavari, and Pennar to Kanya Kumari, the per capita water availability is below 1000 cum annually even today. This is a forewarning of general water stress and deserves special attention of water developers and managers. On the other hand, Ganga and Brahmaputra river systems with 1110.62 billion cu m water account for 59.4 percentage of total Indian surface water for 44.7 per cent of total land area. This also indicates an overall water scarcity in the rest of India outside the Ganga-Brahmaputra basin areas.

Water Demand Perspectives

Irrigation has played a major role in converting India from a highly food deficient country to a surplus country. However, Indian agriculture uses water most extravagantly. India ranks at the top of all countries in withdrawing 558 bcm of water for irrigation (FAO 2003). In fact, the irrigation water constitutes nearly 84 per cent of total water usage. That underpins the necessity of improving the irrigation technology to higher efficiency and economising water consumption without adversely affecting agricultural productivity.

Table 8.11: Global per capita Water Availability

Country	*Per capita Water Availabilit Cum per annum)*
Africa	
Algeria	528
Egypt	923
Kenya	1069
Libya	111
Morocco	1110
South Africa	1206
Asia	
Singapore	211
Europe	
Belgium	1236
United Kingdom	1219
Middle East	
Israel	382
Jordan	314
Kuwait	103
Lebanon	1854
Oman	892
United Arab Emirate	1047
Yemen Republic	359
South America	
Peru	1682

Source: World Resource Institute, 1997.

In any case, this historical situation of very high agricultural water consumption, while the industrial and domestic consumptions are minor players is definitely to change. As the country gets increasingly industrialised, demand for industry and power will multiply to much higher share of water and the domestic demand will conspicuously go up for higher quantity and quality of water. In 1995 in Europe, the demand for agriculture, industry and domestic use was respectively in percentages 38, 38 and 24. However, compared with Europe, much areas of Indian agriculture are semi-arid necessitating irrigation and hence requiring higher allocation of water.

X
National Water Policies 2002

Basically, the air a man breathes and the water he drinks are the inalienable rights of the humankind. In this sense, water is private goods. However, because of the very wide influence of water on life, climate, transportation, industry and agriculture, the water resource is coming under the purview of national laws and regulations. Here we present a very brief review of the National Water Policies.

Forty years after the Indian independence came the first National Water Policy in 1987 during the Prime Ministership of Rajiv Gandhi of the Congress Party, enunciating inter alia (Anand 2007):

(a) Water resources available should be turned into 'utilizable' resources to the maximum extent possible;

(b) The drinking water should be given the first priority, followed by irrigation and hydropower;

(c) Planning should be done on the basis of hydrological basins and appropriate organisations should be set up for the planned development and management of such basins;

(d) Water should be made available to water-short areas including inter-basin transfers;

(e) Water rates should be used to convey the scarcity value and encourage the economical use of water;

However, the main specific commitment made in the Policy document that "adequate drinking water facilities should be provided to the entire population both in the urban and the rural areas by 1991" remains unfulfilled even today, albeit considerable improvements achieved in the water resource development.

The Second National Water Policy was announced in 2002 during the Prime Ministership of Atal Behari Vajpayee of BJP. The 2002 NWP retained the basis of the earlier NWP with some additional elaborations (Anand 2007):

(a) The need to manage carefully the resettlement and rehabilitation of persons affected by the projects;

(b) Considerations of financial and physical sustainability replacing the water rate concept (which was perhaps considered inappropriate for the large poor population);

(c) Welcoming private sector participation;

(d) Equitable water sharing and distribution among the States and the need to amend the Inter-State Water Disputes Act 1956 (seeking better implementation imperatives),

(e) The need for a paradigm shift in the management ethos and to improving the performance of existing water resources infrastructure.

However, because of change of Government back to the Congress party, NWP 2002 was not put into effect. In any case, the policy statements are not legally binding but symbolic of good intention. However, as broad guidelines, they help to focus the approach, planning and budgeting prior to implementation.

XI
Water Institutional Framework

The Institutional frameworks, both formal and informal, for the development and management of water resources have grown considerably in India today. But there are structural weaknesses and lack of clarity in defining roles, responsibilities, and authority and finally the preponderance of bureaucracy, which all combine seriously hampering the operational efficiency. But, hopefully the wind of change is already blowing.

Now that the State has the primary responsibility for water, we start institutional survey with the State organ.

State Level Organisation

The water is primarily a State responsibility; still the State level institutions, governing multi-sectoral water use and allocation, are not strongly developed in all States. Because the irrigation is the largest user of water, State Irrigation Departments have till recently been responsible for construction, maintenance and management of water resource development schemes. Thus, irrigation enjoyed the priority of water use in violation of the National Water policy.

Recognising the need to rectify such untenable arrangement, States like Tamilnadu and Orissa, under the inspiration of the World Bank, were the first States to create independent management for water, e g State Water Council in Tamilnadu and State Water Resource Board in Orissa. Other States have since taken to the model as well.

Local Level

Under the State level Water Board or Council, there is need for the local specialist institutions to actually run a particular water project or river valley. Some of the authority of the State Board should devolve on these specialist institutions.

Because the local Governments, like the Zilla Parishad or a Panchayat, developed after the 73rd and 74th Constitutional amendments (1992), that build, operate and maintain the local water supply should be further trained and strengthened technically, professionally and managerially for upgradation of all water services.

Grassroot Level

It might be necessary to identify specialist organisation below the local level or Zilla Parishad but possibly above or at Panchayet level for the grassroot management of water issues and problems. For many, water management issues have their origin and solution at the very location level, where the grassroot team can identify and solve the issue mere consensually and effectively. The availability of water is becoming a critical issue for rural water supply system throughout India, especially in the Deccan plateau and arid western regions. Further, because of overuse by the irrigation, local ground water levels are depleting; contamination of drinking water is also a problem in cities with inadequate effluent treatment in large villages, which typically do not have sewage disposal network. Very often, these issues are entirely resolvable if the local community got together and took pragmatic ameliorative actions.

Inter-State Institutions

All major Indian rivers and their tributaries flow over several States. Inter-State Institutions, under the aegis of the Union Government, are a necessity for allocation, planning and scientific management of water of such basins for the optimum and equitable development.

Among the successful inter-state institutions, mention may be made of the Damodar Valley Corporation, Bhakra-Beas management Board, Upper Yamuna River Board and flood control institutions like Ganga Flood Control Commission and Brahmaputra Flood Control Board, etc.

All India-Level Institutions

At the Central level, the Ministry of Water Resources (MOWR) is recognised as the apex nodal Ministry responsible for water. Under the MOR, there are a number of technical agencies, e.g.,

- Central Water Commission (CWC), concerned with surface water;
- Central Ground Water Board (CGWB), concerned with groundwater matters;
- National Water Development Agency (NWDA), primarily assessing inter-basin water transfer options.

Other agencies and Ministries are also involved with water, e.g., Ministry of Environment and Forest (MOEF), the Central Pollution Control Board (CPCB), Rajiv Gandhi National Drinking Water Mission in the Ministry of Rural Affairs dealing with rural water supply and sanitation.

Civil Society and the Public-Private-Partnership

No large water utility has yet been fully privatised. The NWP 2002 did conceive of the Public-Private-partnership (P-P-P) model of developing water projects. A matrix of such cooperative models are recommended in Table 8.12 (Anand 2007). But with the development of the Capital Market, on the one hand and the enormous need to mobilize capital investment, the Governments today are seeking and experimenting with idea of private participation in water infrastructure in some form. There is also realisation that the grassroot waterworks or even larger water projects will be more effective with the community involvement or by partnership with private owner. For such water utilities, following models are being tried also (Anand 2007).

(i) **Performance criteria :** These contracts are for medium terms of Service Contract: where the ownership and basic operations remain with the State sector, various peripheral services are being contracted out to private parties on competitive tender basis, commonly on yearly contract.

(ii) **Management Control:** In Navi Mumbai, four private sector contracts have been awarded to manage the water supplies in 15 areas specified for 5 years.

(iii) Leases and Concessions: These are longer-term contracts. A Water treatment plant in Delhi has been given to a multi-national company to run. In Tripur, a private sector has been engaged to build, operate, and maintain a water supply project of Tripur Water Development Authority through a special purpose vehicle (SPV) supported by the U S Agency for the International development.

Reform Initiatives

The first task is to improve and modernize the water infrastructure and Institutions including inter alia the following:

(a) **Policy:** Improving Policy framework, for example, National Water Policy, 1987 or 2002 version should also clearly name implementing agencies for various policies enunciated therein;

(b) **Legislation:** Strengthening the legislative and regulative framework providing for explicit empowerment of the Central Government explained in later parts;

(c) **Institutions:** Institutions are very vital for water development, particularly the public-private-participatory models for multiplying manifold investment, execution and operation management;

(c) **Economic Return:** Time is soon coming to think about some economic pricing or return to provide, if not for investment, for efficient maintenance and preventive wastage;

(d) **Technology:** Upgrading technology to enhance water productivity, particularly for agriculture and domestic distribution;

(e) **Information and Data:** Collection of data and gathering of information require to be strengthened to develop project models, to improve performance and to draw up long term plan.

Public vs. Private Water Rights

With the goal of more equitable and effective use of the precious water

Table 8.12: Recommended Matrix of Implementation Agencies (Anand 2007)

	Formulating Water Policies, Laws etc	*Water Resource Management, Inter-state Rivers, Aquifers, Large-scale Water Projects*	*Water Resource Management, Inter-State Rivers, Aquifers, Small-scale Water Projects*	*Drinking Water Supply Infrastructure Investments (Treatment Plants, Mains)*	*Operation and Maintenance of Drinking Water Supply*	*Finances, Accounts, Billing*
Federal Government	*	*				
State Government	*	*	*	*	*	*
Local Government			*	*	*	*
Private Sector- Large Firms				*	*	*
Private Sector- Small Scale				*	*	*
Individual Citizens				*	*	

resources, there is a strong logic of regulating the extravagant and wasteful withdrawal of water by individuals and agencies. Water in future should be treated as public goods and the concept of usufructuary or absolute rights to available water needs to be revisited and limited. Local Governments, Panchayats or the community institutions should be empowered to set limits to water use through quantified permits with some legal safeguards to users. For surface water, automatic riparian rights of unlimited withdrawal of water should be derecognised. Similarly, for ground water, the presently practiced absolute discretion of the user to pump out water should be subject to user's quantity to be issued by the local empowered Institution with some judicial protection to the landowners.

References

Anand, P.B. (2007), Semantic of Success or pragmatics of Progress? (An Assessment of India's Progress with Drinking Water Supply), *The Journal of Environment and Development*, Vol. 16 Number 1, March 2007, pp. 32-57, Sage Publications.

Biswas, Asit K., (2007 January), Lecture on Water Management in the 21st Century for the Developing World at the Indian Institute of Technology, Kharagpur, India, c/o. Third World Centre for Water Management, Atizapan, Mexico.

Biswas, Asit K., (1990), *Systems Approach to Water Management.*

Biswas, Asit K., (1990), "Objectives and Concepts of Environmentally-sound Water Management", *Environmentally Sound Management*, (Ed. A.K. Biswas).

Cosgrove William (1987), "Water Resource Sector in the Coming Decade: Global Perspective, Water Resource and Sustainable Development in 21st Century", (Ed. Kamala Prasad.)

Chitale, M.A., (2000), South Asian Perspectives for Water Resource Development, *Water Resource and Sustainable Development in 21st Century.*

Deb, Arun K., (1987), "Dual Water Supply for Future Urban Water Management, Proceedings of the International Symposium on Water for the Future", Rome.

Furan, R., (1967), "The Problem of Water – A World Study".

Masters Gilbert M., (2000/1991), *Introduction to Environmental Engineering and Science*, Prentice Hall, New Delhi.

Prasad T., Kumar, B.S., (1987), "Water Resource Development in India – its Central Role in the Past and Crucial Significance for the Future, Rome.

Reddy, V. Ratna, *Water Security and Management.*

Rao, K.L., (1979), *India's Water Wealth*, Orient Longman, New Delhi.

Than, N.C. and Tam, D.M., (1990), "Water Systems and the Environment", Environmentally Sound Management, Asit K. Biswas.

UN Water Conference (1977), Mardel Plata, Argentina, Part I, *Water Development and Management*, Ed. Asit K Biswas.

World Bank Report (1999), "India: Water Resource Management, Inter-Sectoral Water Allocation, Planning and Management" in collaboration with the Government of India, Ministry of Water Resources.

9

Global Warming and Ecological Sustainability

I

A Short Overview of Global Warming

The earth is receiving heat energy from the solar radiation all the time. At the same time the earth is radiating away heat to the open universe. So assuming heat balance, the earth should remain at the same temperature.

But the earth is covered with a deep blanket of atmosphere, which is interfering with the transfer of heat radiation. The basic atmosphere consists principally of Nitrogen (79%) and oxygen (21%) approximately, which do not affect heat transfer. But there are some polluting gases, which are mainly generated as a result of human activities; these gases pass the high frequencies solar radiation to reach the earth surface but absorb and obstruct the low-frequency heat radiation from the earth to the atmosphere. These gases – carbon dioxide (CO_2), nitrous oxide (N_2O), methane (CH_4), and CFC (chlorofluorocarbons) are called Greenhouse gases (GHG). Because, they work in the similar way to the plant greenhouse i.e. they allow short wave heat radiation in but stop long wave heat out.

It has been estimated that because of green house effect, the earth surface has reached today an average temperature of 288 K (15°C), instead of 255 K (–18°C), a positive rise of 33°C, which is welcome making life more comfortable.

Earth's Basic Atmosphere

The earth is covered with normal atmospheric gases consisting of mainly nitrogen and oxygen but also many other polluting gases in small quantities including the GHGs. Above the earth surface, the temperature varies. Following upon the temperature profile, the atmosphere seems to be divided into four horizontal layers:

Troposphere: temperature range 288 K to 210 K at top;

Stratosphere: temperature range 210 K to 260 K at top;

Mesosphere: temperature range 260 K to 160 K;

Thermosphere: temperature range 160K to very high value.

Fig. 9.1 shows the temperature map of the atmosphere, which also shows the above four layers. In the Troposphere, most of the pollution accumulates. But here again precipitation from the clouds takes place and rains washes away much of the pollution back to the earth.

Stratosphere is filled with the atmospheric ozone, which absorbs the carcinogenic ultra-violet rays, thus saving the earth and the humans.

Thermosphere (Ionosphere) has the unique property of reflecting the radio waves from the earth back to the earth making possible global wireless communication till the advent of satellite communication.

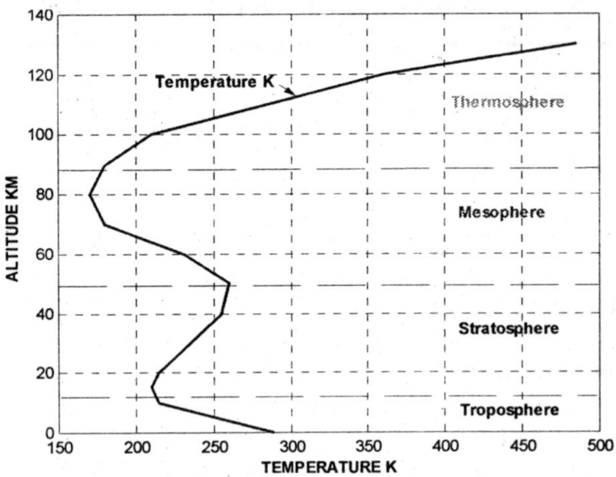

Fig. 9.1: Temperature Profile of Atmosphere

Generation of Greenhouse Gases

Carbon dioxide is generated with the burning of any fossil fuels in industrial furnaces and domestic kitchens; deforestation also causes decrease in capturing of carbon from the atmosphere and hence increases of carbon dioxide in the atmosphere.

Methane is produced from anaerobic decomposition of solid waste, coal mining, oil and gas production, livestock excreta, agriculture paddy fields, marshlands and also substantially from auto-emission.

Nitrous oxide is produced primarily from use of fertilizer, automotive combustion and stationary fossil combustion at the powerhouses and process industries.

CFCs are generated from the refrigeration industries and plastic-process industries. CFC generation has been controlled very effectively under the international agreements. Of the GHGs, HFC/CFCs trap 15000 times as much heat as an equal volume of carbon dioxide, while methane traps about 25 times and nitrous oxide traps 200 times more than that CO_2 does.

But because of higher concentration of carbon dioxide, it contributes about 70% to the global warming compared to 7% by CFCs, and 14% by N_2O and 8% by methane (Table 9.1). These are tentative and average estimates.

Table 9.1: Types of GHG and their Effects on Global Warming

GHG	*Atmospheric concentration (ppm)*	*Relative Greenhouse Effect*	*Respective Effect on Global Warming (%)*	*Principal Sources of GHG Emission*
Carbon Dioxide	351	1	70.5	Fossil fuels, deforestation
Methane	1.7	25	8.4	Wetlands, rice, livestock, fossil fuels
Nitrous Oxide	0.3	230	14.3	Fuels, fertilizers, deforestation
CFCs	0.002	15,000	6.8	Foams, aerosols, refrigeration, solvents
Total			100	

Sources: Masters.

Table 9.2 shows the global concentrations of the four main GHGs in the years 1982, 1986 1990, and 1994. Please note gradual increases in their concentrations.

Table 9.2: Atmospheric Concentration of Principal Green House Gases

Year	CO_2 (ppm)	CFC (ppt)	NO_2 (ppb)	CH_4 (ppb)
1982	341	500	301	NA
1986	347	603	305	1600
1990	354	718	307	1645
1994	359	760	309	1666

ppm = parts per million; ppb=parts per billion; ppt=parts per trillion.
Source: Sandler 1997.

Table 9.3 shows annual emissions of GHG (CO_2 equivalent) of top ten countries in 2000. United States have the ignominy of highest contribution at 6928 million tonnes annually amounting to 20.6% of total pollution of GHG (CO_2 equivalent).

Table 9.3: Top Ten GHG Emitting Countries, 2000

Countries	*CO_2 Emissions (million tons)*	*Per Cent Global Emissions*
United States	6928	20.6
U S S R	4938	14.7
China	4725	14.0
Japan	1915	5.7
Germany.	1884	5.6
India	1317	3.9
United Kingdom	1009	3.0
Iraq	851	2.5
Canada	680	2.0
Italy	654	1.9
Global	**33,665**	**100**

Source: WRI 2005.

On the other hand, it is seen from Table 9.4 that China and India contribute 16% and 14% respectively of total global emissions of methane, though USA and USSR follow closely at 11.6 and 11.2 percentages respectively.

Table 9.4: Major Methane Emission Countries

Countries	*Methane Emissions (Million Tons)*	*Per cent Global Emissions*
China	40	16
India	35	14
United States	29	11.6
U S S R	28	11.2
Brazil	10	4
Bulgaria	9	3.6
Indonesia	7	2.8
Bangladesh	6.2	2.5
Thailand	6	2.4
Australia	4.5	1.8
Global	**250**	**100**

Source: Sandler 1997.

IA

Basic Physics of Global Warming

The earth gets heated up receiving solar radiation, the sun being the primal source of heat energy. We determine the temperature of the earth's surface by applying Stephan-Boltzmann equation to two simplified energy transference models. The first model considers direct heating of the earth as a black body without the atmosphere and the second model includes the existence of the earth's atmosphere and also treats the earth as a gray body with 'albedo' effect.

Greenhouse Effect

Now one of the important reasons of the Greenhouse effect is that the solar radiation is of higher frequency (i.e. lower wave length) energy wave, while the thermal energy wave emitted by the earth is of lower frequency (i.e. comparatively of higher wave length).

Stephan-Boltzmann Law gives the rate of energy radiated from a blackbody. But the radiated energy is spread over a range of wavelengths or frequencies forming a spectrum similar to the two spectra shown in Fig. 9.4, one for the sun radiation and the other for the earth radiation. But using the temperature of the radiating body, the Wien's Law (E9.1) defines the wavelength at the mode of the spectrum i.e. at the maximum of the spectrum intensity.

Box 9.1A: Determination of Earth's Temperature (No Atmosphere)

Earth without Atmosphere:

The earth is assumed as a black body. It therefore absorbs all the incident radiations from the sun and emits an equal amount of heat energy. Referring to Fig. 9.2A, the solar radiation hits the earth within a radius R equal to earth's radius, over an area (Pi).R^2. But the intensity of solar radiation at earth's surface is known as I=1372 W/m2. The heat energy absorbed Ea is therefore:

$$Ea = I.(Pi).R^2, \quad \text{... E9.2A}$$

But the earth emits energy over all its global surface of area 4.(Pi).R^2. If we now assume that the earth's surface is isothermal, having temperature T, the energy radiated Er is given by Stephan-Boltzmann equation:

$$Er = S.4.(Pi).R^2.T^4, \quad \text{...E9.2B}$$

Where S = the Stephan-Boltzmann constant

= $5.67 \times 10^{(-8)}$ $w/m^2.K^4$ for the black body.

Since the earth's temperature is assumed constant, there is no change of internal energy and hence equating energy absorbed (E9.2a) with the energy emitted (E9.2b), we get:

$$I.(Pi)R2 = S.4.(Pi)R^2.T^4 \quad \text{...E9.2C}$$

Or, T = 279 K,

This temperature of 279°K is quite close to the actual average temperature of 288K (15°C) of the earth. But the above proximity of result is accidental. For the above model is not real, being simplistic without atmosphere and the earth being assumed a black body.

Lemax = 2898/T micrometre ...E9.1

Where L = wavelength in micrometer at emax= maximum energy intensity; T = temperature of the radiating body in degree K.

Assuming the sun as a blackbody and having surface temperature of 6000 K, we get from E9.1

[Lemax] sun = 0.48 micrometre ...E9.1A

Similarly, taking the earth's surface temperature at 288 K, we get from E9.1,

[Lemax] earth = 10.1 micrometre ...E9.1B

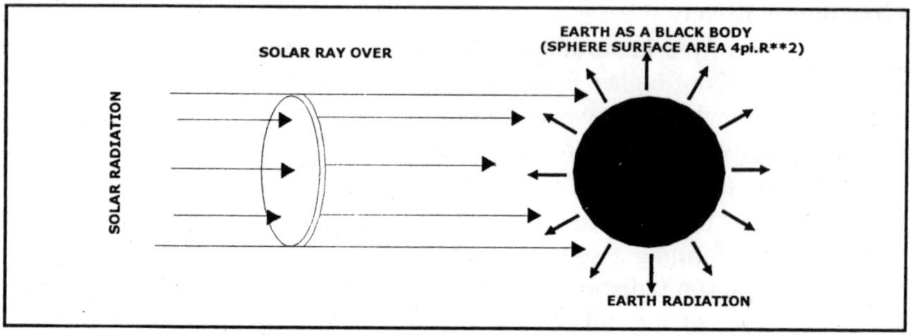

Fig. 9.2. Earth as a Black Body without Atmosphere

Box 9.1B: Determination of Earth's Temperature (with atmosphere) Earth as the modified Blackbody (with Modified Atmosphere)

Here the assumptions are –

(i) The earth and the atmosphere being not perfect blackbody, part of the incident radiations are reflected back from the earth and the atmosphere.

(ii) Though the atmosphere takes part in reflecting the incident radiation, there is no greenhouse gas effect from the atmosphere i.e. secondary warming of the earth by the atmosphere.

The reflected solar energy is called 'Albedo', denoted by 'A' and amount to 30 % approximately on annual average basis. The albedo is shown in Fig. 9.3, which diagrammatically represents the second earth model.

With the above assumptions, the equality equation E9.2C of the incoming and emitting energy stands modified as;

$I.(1-A).(Pi).R^2 = S.4.(Pi).R^2.Tm^4$ ...E9.3

Where 'A' is the albedo and the Tm is the modified temperature of the earth and from E9.2,

$$Tm = 255°K$$

The above revised temperature of 255°K (–18C) is still further away from the actual temperature of 288°K (+15C) of the earth's surface. But this difference of 33°C is actually a correct evaluation, for even the second model has included the albedo loss of radiation but did not include the secondary warming of the earth by the polluted atmosphere of the earth, otherwise known as the Greenhouse Effect. We will deal with the Greenhouse effect next.

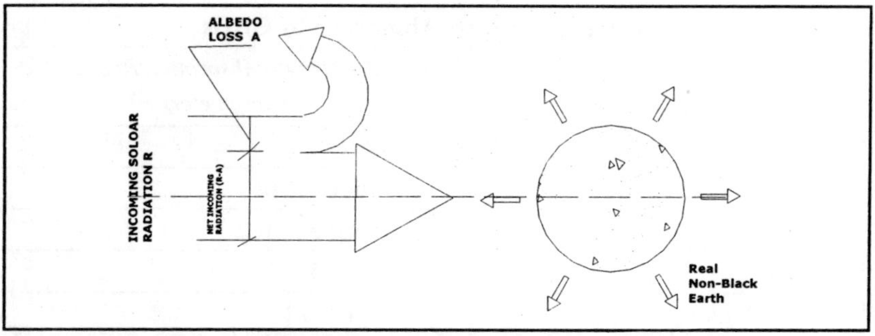

Fig. 9.3. Solar Radiation on Real Earth with Albedo Loss A

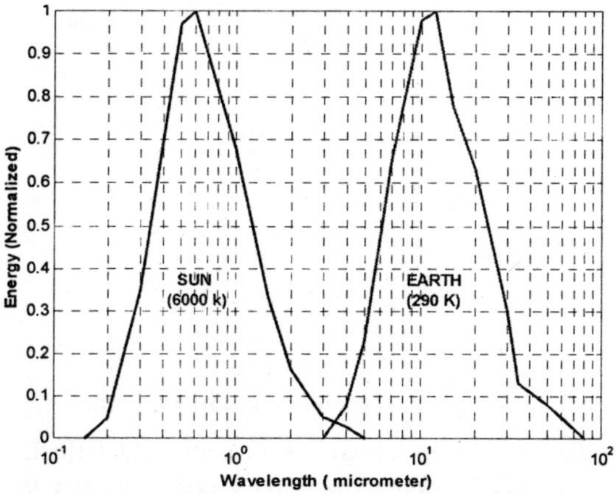

Fig. 9.4. Spectra of Solar and Earth Radiation

From the Fig. 9.4, it will be seen that the solar radiation spectrum is distributed over the wavelength from 0.15 micrometers to 4.5 micrometer, having its modal maximum at 0.5 micrometers. Similarly, the earth's radiation spectrum is distributed from 3 to 90 micro-m wavelengths, having its modal maximum at 10.1 micro-m wavelength.

Now the earth is covered with atmospheric air of oxygen, nitrogen etc., which freely allows solar radiation to reach the earth and the radiation from the heated earth to pass out to the space. But the earth's atmosphere is polluted with anthropogenic gases and other gases. In particular, the immediate atmospheric Troposphere layer is polluted with water vapour, clouds, carbon dioxide, Nitrous oxide, CFC. These pollutant gases have the property that they allow the low wavelength solar radiation below 4 micrometer to pass through to the earth but absorb the earth's radiation of relatively higher wavelength above 4 micrometer (see Table 9.5 and Fig. 5). This phenomenon leads to the gradual warming up of the earth. This is the reason why the earth is warmer by 33°C than what was the derived temperature arrived at from the second global warming up model above.

Table 9.5 : Energy Absorption by GHGs

GHGs	*Wavelength Absorption Range (micrometer)*
Water Vapour	above 18
Carbon Dioxide	from 13 to 18
Ozone	from 9.5 to 10.5
Methane	from 3.5 to 7.5
Nitrous Oxide	from 8 to 13

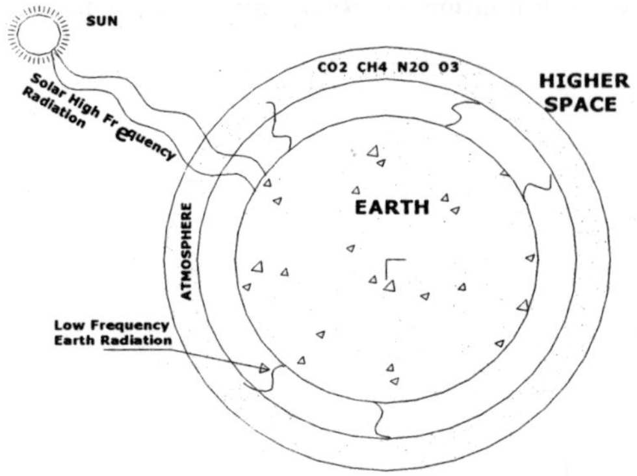

Fig. 9.5. Greenhouse Phenomenon

The secondary effect is known as Greenhouse Effect, similar to the greenhouse effect, which saves the plants and saplings from the severe winter of the western countries.

Generally, all gases, which absorb energy radiation of wavelength above 4 micrometer, are called Greenhouse gases. The following are the principal greenhouse gases (often abbreviated as GHGs) and they are shown with energy spectral ranges absorbed by them:

II
Anthropogenic GHG Emission and Climate Change Crisis

The pre-industrial level of natural CO_2 emission stock in the environment around 1750 remained moderate at 280 ppm (parts per million by weight). This level of carbon dioxide has remained steady for many millenniums of years (*Economist* March 20-26, 2010). Even during next two hundred years till 1950, the pace of industrialisation was slow and steady, and the fossil fuel was just getting established as the main energy driver. The rise in the carbon dioxide emission level was still quite moderate at about 310 ppm by1950 (WRI Report 2004). But after 1950, the automobile and consumer goods revolution swept all over the industrial world, energised by the fossil fuel to an unprecedented scale and by the year 2000, the anthropogenic carbon dioxide level zoomed up to 370 ppm, and the spectre of high global warming and climatic change created high disquiet among the world's scientific community and the United Nations (see Fig. 9.6).

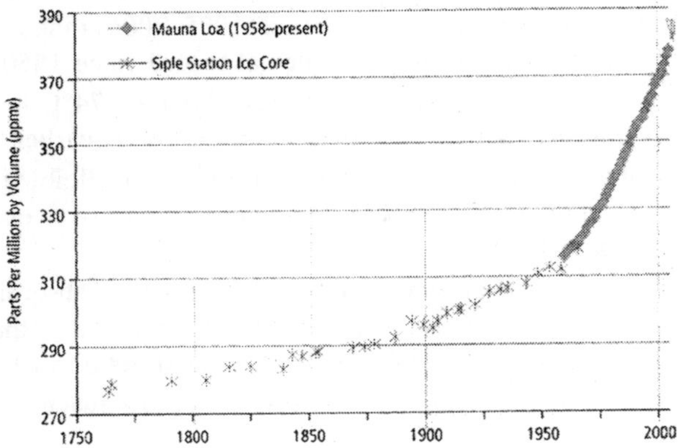

Fig. 9.6: Atmospheric Carbon Dioxide Concentration (1750 - 2000) (*Source:* WRI; A. Neftel et al 1994; Keeling and Wharf 2005).

The combined effect of all the GHG gases (carbon dioxide, methane, nitrous oxide and CFC) in the atmosphere was measured by several scientific organisations and is believed to have caused rise in the global temperature to 0.74 degree Celsius from the pre-industrial level. The quantum jump in the cumulative accumulation of CO_2 from 5 billion tons in 1950 to 25 billion tons in 2000 was alarming and it is now apprehended that the rise in average temperature could range from 1.5 to 3 degrees Celsius in the present century. The rise of temperature of this scale could cause catastrophic effect on the climate, increased precipitation in lower latitude, extended draught in crop growing regions, shift of seasonal rain affecting cropping practice, frequent and severe storm and typhoon, degradation of soil by heavy evaporation, melting of glacial ice and rise of sea level and coastal inundation and manifold other miseries especially to the poorer underdeveloped countries.

According to the IPCC Fourth Assessment Report, the accelerated anthropogenic GHG contamination of the atmosphere has taken place over the pre-industrial millennia as shown in Table 9.6:

Table 9.6: Increase in Anthropogenic GHG Contamination

Types of GHGs	*Pre-industrial Level (Natural Increases over 650,000 years)*	*2005 level (High Anthropogenic Growth)*
Carbon Dioxide	180 -300 ppm	379 ppm
Methane	320 -790 ppb	2774 ppb
Nitrous Oxide	270 ppb	319 ppb

IPCC Observations on Anthropogenic Warming

Cold days, cold nights, and frost events have become less frequent. Hot days, hot nights, and heat waves have become more frequent. Additionally:

- Eleven of the twelve years in the period (1995–2006) rank among the top 12 warmest years in the instrumental record since 1850.
- Warming in the last 100 years has caused about a 0.74°C increase in global average temperature. This is indicated by us earlier.
- Urban heat island effects were determined to have negligible influence (less than 0.0006 °C per decade over land and zero over oceans) on these measurements.
- Observations since 1961 show that the ocean has been absorbing more than 80% of the heat added to the climate system, and that ocean temperatures have increased to depths of at least 3000 m (9800 ft).
- "Average Arctic temperatures increased at almost twice the global average rate in the past 100 years."
- It is likely that greenhouse gases would have caused more warming than we have observed if not for the cooling effects of volcanic and human-caused aerosols.
- Average Northern Hemisphere temperatures during the second half of the 20th century were very likely higher than during any other 50-year period in the last 500 years and likely the highest in at least the past 1300 years.

Warming of Polar Ice, Glaciers and the Oceans

The SPM documents increases in wind intensity, decline of permafrost coverage, and increases of both drought and heavy precipitation events. Additionally:

- "Mountain glaciers and snow cover have declined on average in both hemispheres."
- Losses from the land-based ice sheets of Greenland and Antarctica have very likely (>90%) contributed to sea level rise between 1993 and 2003.
- Ocean warming causes seawater to expand, contributing to sea level rising.
- Sea level rose at an average rate of about 1.8 mm/year during the years 1961-2003. The rise in sea level during 1993-2003 was at an average rate of 3.1 mm/year. It is not clear whether this is a long-term trend or just variability.
- Antarctic sea ice shows no significant overall trend, consistent with a lack of warming in that region.

Hurricanes

- There has been an increase in hurricane intensity in the North Atlantic since the 1970s, and that increase correlates with increases in sea surface temperature.

- The observed increase in hurricane intensity is larger than climate models predict for the sea surface temperature changes we have experienced.
- There is no clear trend in the number of hurricanes.
- Other regions appear to have experienced increased hurricane intensity as well, but there are concerns about the quality of data in these other regions.
- It is more likely than not (>50%) that there has been some human contribution to the increases in hurricane intensity.
- It is likely (>66%) that we will see increases in hurricane intensity during the 21st century.

Apart from the greenhouse effect of the CFC, it was observed for the first time in 1974 by Molina Rowland that the anthropogenic CFC was causing destruction of the thin ozone layer in the stratosphere creating ozone holes. The ozone holes allowed the carcinogenic UV-B radiation bombarding the earth. The British Antarctic survey of 1985 reconfirmed such serious health hazard from CFC-induced ozone holes. Coinciding with the discovery of the ozone hole over the Antarctic, the Convention for the Protection of the Ozone Layer came into being in 1985 at Vienna under the aegis of the United Nations. Extending the Vienna Convention, the Montreal Protocol was signed in September 1987, setting specific limits to the emission of ozone destroying gases, particularly CFC-11 and CFC-12.

Global Warming has now come to be known as anthropogenic in character. Nevertheless, it has become intractable to control the problem through international agreements among nations because of varying interest. The cooperation among nations remains mainly elusive. The global warming arises, because earth's atmosphere has been getting polluted with certain gases, which allow the sunlight to pass through; but absorb the returning infrared radiation from the earth. The captured heat, between the atmosphere and the earth, warms up the atmosphere and the earth. The warming effect is similar to the Green House effect and hence, these gases are called Green House Gas (GHG). The principal GHG gases are carbon dioxide, methane, CFC (chlorofluorocarbon) and nitrous oxide (N_2O).

The carbon dioxide is formed and emitted by the combustion of fossil fuel in the automobiles, power plants, in residences etc. Carbon dioxide also escapes to the atmosphere because of extensive deforestation. Methane is generated from solid waste, coal mining, oil and gas production, livestock and rice agriculture. Nitrous oxide is generated from use of fertilizer and fuel consumption at the power plant. CFCs are released from refrigeration and certain chemical reactions.

Thus, it is seen from the above that the greenhouse gases are generated mainly as a result of human, economic, industrial, and agricultural activities.

Adverse Effect of Global Warming

The world community has now come to realize that unabated and unregulated accumulation of GHGs can raise average global temperature by about 1.5° to 3°C by the turn of the century. We will also see an empirical relationship between the GHG concentration and global temperature, that has been tried by many analysts but the format of the relationship is yet to be fully evidenced and proven. Further, the contributions of other atmospheric pollutants need also to be fully assessed scientifically. Thus, there is considerable doubt about the estimate of global warming.

Given a certain level of temperature rise, there is no global model of climate change beyond a computer weather modeling. Even such computer meteorological forecast sometimes does not hold good at all. However, generally speaking and as indicated by IPCC studies, global temperature rise should affect quality, duration and timing of rainfall, affecting agricultural production schedule and even productivity, glacial melting and rise of sea level, occurrence of storm and possibly other geological and meteorological phenomena. But the effect of the global warming will not be uniform, some countries will have more rainfall and more agriculture and other countries will suffer from intense aridity and long-term fear of desertification.

It is then likely that the potential gainer countries might prefer global warming, while the potential loser nations will wish to prevent global warming, which can be achieved only by reduction of emission of GHG. The reduction of GHG will be our next topic of consideration.

Some Reduction Measures of GHG Emission

Carbon dioxide contributes about 70 per cent or more to global warming among all principal GHGs. The main source of CO_2 emission is due to combustion of fossil fuels (petroleum oils, Petroleum gases and coal) in automobiles, transportation in general (aviation and railways), power plants and process industries. It is very difficult to change all these plants and automobiles to change over to newer developing carbon-free energy sources, if any. Thus, what can be tried is to undertake optimisation, efficiency-enhancement and preventing system wastage through energy audit. These measures might help to reduce GHG emissions marginally, say by an order of 10% to 20%. The cost of these marginal reductions could be about, say less than 1% of global GDP and may amount to a figure of many billion dollars as is apprehended in some quarters.

Other measures, which are feasible, are to switch over more and more to nuclear power and hydroelectric power, increasing their proportion in the national grid. Firstly, though the nuclear energy sources are generally GHG-free, there is still strong public opposition to nuclear energy. Secondly, the disposal of radioactive nuclear waste is a big problem technically, because of long half-life of radioactive wastes. Thirdly, both these energies are highly capital-intensive; and large scale change-over may cost to the order of trillion dollars annually. Thirdly, the availability of hydropower is comparatively limited.

Fortunately, there is a recent development of Bio-diesel, which has about 50 per cent less carbon content than the conventional fossil petroleum diesel. The bio-diesel is produced from vegetable oils, which can be extracted from the seeds of the non-edible shrubs like Jatropha circus grown profusely on arid cultivable land and fallow hilly terrain. The bio-diesel is presently useable up to 20% with petroleum diesel without any change of the plant or engine. Hence the change-over can be implemented economically. Germany, USA and Brazil have already adopted bio-ethanol, another bio-fuel but recently are also introducing bio-diesel energy quite significantly. India has a big plan for developing bio-diesel technology in a big way.

Now, all these measures call for large capital investment today, but the results in the form of carbon dioxide reduction will take place after some years. It will be seen from the Table 9.3 that three developed countries United States, old Soviet Union and Japan together account for 49.3 per cent of global GHG emissions; and naturally they were to control the emission first, by deploying necessary large investments, which they were not till recently very willing to incur. Under the 1992 UN Convention on climate Change, Kyoto Protocols, 1997 placed the developed countries, producing heavy emission, under Annex I and LDC (the Less Developed Countries) under the Non-Annex Countries. Under the Kyoto Protocols, Annex I countries are required to lower their CO_2 emission by an average of 5.2 per cent below the 1990 levels by the period 2008-12, while the non-Annex countries, having quite a low per capita carbon emission, are not required to take any action. USA, the largest producer of CO_2 emission, has not signed the Kyoto Protocols, insisting that the large LDC countries like China and India must also give commitment of controlling CO_2 emission, as they are likely to become large economies in the next 25 to 50 years.

Again fortunately, the situation in respect of CFC emission and consequent problem of ozone hole is much better controlled. Most countries including India have acceded to the 1985 Convention and the 1985/1987 Montreal Protocol on substances that deplete the ozone layer. The production and uses of CFC have now almost gone out of practice since the Montreal Protocols. One reason of this success is due to the fact that only few countries were producing CFCs to start with. The number of production units was also limited.

IIA

Relation between GHG Emission Level and Global Temperature

Relationship of Atmospheric GHG and Temperature

There is considerable uncertainty in modeling the atmosphere admixed with GHGs vis-à-vis temperature change. The complexities arise, inter alia, from some of the following factors:

- GHGs raise global temperature by greenhouse effects, but they may also increase the rate of albedo, which may create opposite effect;
- Global warming may intensify evaporation and cause cloud formation, which has also greenhouse effect, thus further enhancing temperature increase;
- As the ocean is a large heat sink, the long-run equilibrium temperature takes a long time, perhaps many years, to stabilize much after the GHG concentration leads to some short-run temperature rise.
- There are also many other effects of atmospheric dynamics – adiabatic lapse rate effect, which change temperature profile continuity.

Temperature Change due to

CO_2 Concentration:

The temperature model estimates temperature change over every doubling of CO_2 concentration. The National Research Council on Climate Change recommends the following model formulation:

$$\Delta T = \left[\frac{\Delta T_d}{In_2}\right] \times In\left[\frac{(CO_2)_t}{(CO_2)_0}\right] \qquad \text{...E9.4}$$

Where, ΔT = average global equilibrium temperature increase at time t;

$(CO_2)_t$ = concentration of CO_2 at time t;

$(CO_2)_o$= concentration of CO_2 at initial time;

ΔT_d = equilibrium temperature change predicted for the doubling of CO_2 concentration.

Typically, ΔT_d is considered between 1.5°C and 4.5°C. Even if the lower limit of 1.5°C is taken as the quantum of temperature change, such change is believed to have not taken place in the last 10,000 years. Hence our view is to accept ΔT_d = 1.5°C-2.0°C.

Fig. 9.7 gives the average surface temperature changes over the years as recorded by the Goddard Institute for Space sciences. From the regression analysis curve, it is seen that from1885 to 1985, the surface temperature increased by about 0.6 that corresponds approximately with the calculated result using the equation E9.4 as shown in Box 9.3 for the period 1850 to 1984. Firstly, the actual temperature rise in Goddard Institute data of Fig. 9.7 might not have been done due to increase in CO_2 concentration level only, for other factors like situational location of the Institute, etc.

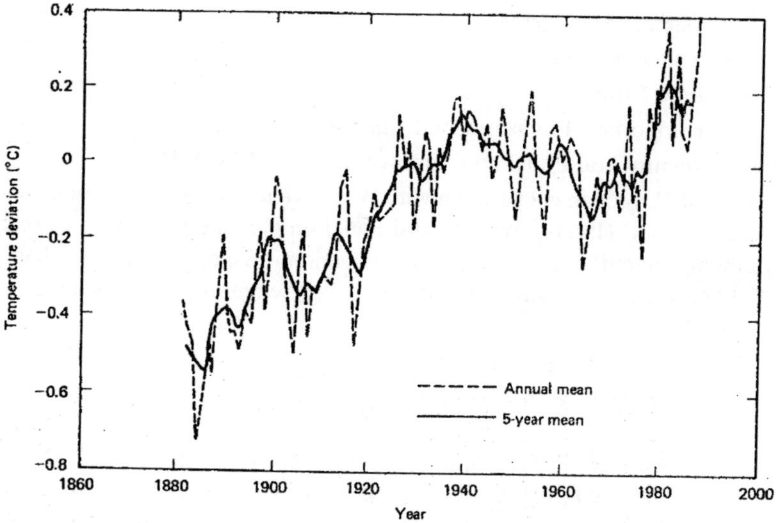

Fig. 9.7 Global Surface Temperature Change (Goddard Institute)

> **Box 9.3: Calculated Temperature Change by Equation E9.4**
> Predicted Temperature Change in 1985
> Assume,
> CO_2 concentration level in 1850 = 280 ppm
> CO_2 concentration level in 1985 = 345 ppm
> Assume ΔT_d =1.5 C
> Using equation E9.4,
> $\Delta T = [\Delta T_d / \ln 2] \times \ln[345/280] = 0.602C$

III

Global Climate Change Movement and UN Initiatives

Realising the critically adverse impact of climate change due to anthropogenic global emission of carbon dioxide and other greenhouse gases, the United Nations had established important institutions like UNEP and IPCC and initiated a series of international conferences and Conventions like UNFCCC. We will describe them briefly now.

United Nations Environment Programme (UNEP), 1972

The need for urgent and concerted global action to meet the challenges of climate change was recognised by UN and the world. The United Nations created a permanent secretariat UNEP (United Nations Environment Programme) in 1972 to coordinate and promote all environment-related work of UN through the different forums. The UNEP today have blossomed into an important UN Institution with following six major operational areas:

Climate Change;

Disaster and Conflicts;

Ecosystems Management;

Harmful substances;

Resource Efficiency;

Other important Environmental Functions.

Inter-Governmental Panel on Climate Change (IPCC), 1988

In 1988, UNEP in cooperation with World Meteorology Office established an Agency IPCC (Inter-Governmental Panel on Climate Change) to directly work on all scientific, technical and socioeconomic aspects of climate change. IPCC have since done comprehensive work on the anthropogenic interference with the global climate.

IPCC have since produced four Assessment Reports, the 1st Report in 1990, 2nd Report in 1995, 3rd Report in 2001 and the 4th Assessment Report in 2007.

The 7000-page IPCC Fourth Assessment Report, 2009 earned it a joint Nobel Prize with the former US Vice-President Al Gore. The Fourth Assessment Summary concludes:

- Warming of the Climate is unequivocal;
- Most of the observed increase in global average temperatures since the mid-twentieth century is very likely due to the observed increases in anthropogenic greenhouse gas concentration.

United Nations Framework Convention on Climate Change (UNFCCC), Rio de Janeiro, 1992

The United Nations Conference on Environment and Development (UNCED), informally known as the Earth Summit, took place in Rio de Janeiro from 3 to 14 June 1992. At the UNCED, was adopted the United Nations Framework Convention on Climate Change (UNFCCC). The objective of the treaty is to stabilize greenhouse gas concentrations in the atmosphere at a level that would prevent dangerous anthropogenic interference with the climate system.

The treaty itself sets no mandatory limits on greenhouse gas emissions for individual countries and contains no enforcement mechanisms. In that sense, the treaty is considered legally non-binding. Instead, the treaty provides for updates (called "protocols") that would set mandatory emission limits. The principal update is the Kyoto Protocol, which has become much better known than the UNFCCC itself. It entered into force on March 21, 1994. As of December 2009, UNFCCC had 192 parties. The UNFCCC is also the name of the United Nations Secretariat charged with supporting the operation of the Convention, with offices in Bonn, Germany.

The parties to the convention have met annually from 1995 in Conferences of the Parties (COP) to assess progress in dealing with climate change. Since then, a series of COPs took place till the recent COP16 at

Cancun, Mexico from 29 November to 10 December 2010; all COPs are listed in Appendix I for ready reference. In 1997, the Kyoto Protocol was concluded and established legally binding obligations for developed countries to reduce their greenhouse gas emissions.

One of its first tasks was to establish national greenhouse gas (GHG) inventories of GHG emissions and removals, which were used to create the 1990 benchmark levels for emission limits of OECD or Developed Countries (so-called Annex I countries) to the Kyoto Protocol and for the commitment of those countries to GHG reductions. Updated inventories must be regularly submitted by Annex I countries.

The UNFCCC did recognize that the historical responsibility of the industrialised countries for the accumulation of the carbon stock in the space and the primacy of the poverty alleviation and development of the poor countries; and hence the responsibility for containing and overcoming Climate Change falls primarily on the OECD countries. The developing countries will have also a common responsibility, but that responsibility is "common but differentiated" by the respective capabilities of the developing countries. The highly enlightened objectives and principles of the UNFCCC may be summarised as follows:

(i) The need to protect the climate system shall be based on equity so that States should have common but differentiated responsibilities based on their respective capabilities;

(ii) While the developed countries are required to mitigate or reduce their green house gas emissions, developing countries should be pursuing environmentally sustainable strategies of development and in this task they shall be supported by financial assistance and by technology transfer.

(iii) The international cooperation shall be organised and made available to undertake adaptation measures to overcome the adverse impacts of the climate change. This objective is very important to the resource-starved developing countries.

(iv) The UNFCCC also mandated, as already mentioned, to build up a national GHG inventory for the base year 1990 and to enhance scientific knowledge for cost-effective measures.

Taking cognizance of the higher carbon emission of the developed countries and at the same time their higher income and capabilities, the countries were identified into two groups:

- Annex I Countries including OECD countries, Eastern European countries (Transitional Economies) and former Soviet Union;
- Non-Annex I Countries (all other countries deemed developing).

Some Important Conferences of Parties (COPs) of UNFCCC

The complete list of all the Conferences of Parties from 1995 to 2011 and the relevant Meetings of the Parties MOPs are presented in the Appendix I. Brief proceedings of some important COPs (COP1 Berlin Mandate 1995, COP6 Bonn 2001, COP 7 Marrakech 2001 and COP 11 Montreal 2005) are also included in Appendix I.

But the Kyoto Protocol which took place under the auspices of UNFCCC in Japan in 1995 and which outlined the various obligations under the Convention for the first time and have probably become more important than the UNFCC Convention itself are discussed here next.

The latest UNFCCC Conference on Climate Change which took place at Cancun in Mexico just a few months ago are included here, though the summarised proceedings of the two immediately preceding Conferences which took place in 2007 at Bali, Indonesia and in 2009 in Copenhagen, Denmark are presented in Appendices II and III.

Kyoto Protocol on Climate Change, 1997 (COP3)

COP 3 took place in December 1997 in Kyoto, Japan. After intensive negotiations, it adopted the Kyoto Protocol. Most industrialised nations and some central European economies in transition agreed to legally binding reductions in greenhouse gas emissions of an average of 6 to 8 per cent below 1990 levels between the years 2008-2012, defined as the first emissions budget period. The United States would be required to reduce its total emissions to an average of 7 per cent below 1990 levels, however neither the Clinton administration nor the Bush administration sent the protocol to Congress for ratification. The Bush administration explicitly rejected the Kyoto Protocol in March 2001.

The signatories to the Framework Convention on Climate change met at a conference from December 1 to 11, 1997 at Kyoto to draw up a Protocol of implementation of the recommendations of FCCC 1992 within an international Climate Change regime. The key elements of the Kyoto Protocols are:

(i) Agreements of quantified emission limitation or reduction commitments by Annex I countries of FCCC, which will reduce the aggregate GHG emission by an average of 5.2 per cent below 1990 carbon dioxide level by the period 2008 – 12 with the leading Industrialised countries reducing by the following percentages by 2012:

European Union ...	... 8%
USA ...	... 7%
Japan ...	... 6%

(ii) Undertaking by the Annex I countries to demonstrable progress in this regard by 2005;

(iii) Undertaking by Annex I countries to the preparation of inventories of reduction of GHG emission and reporting national plans to the FCCC;

(iv) No requirements of any reduction of GHG emission by the developing countries of Non-Annex categories;

(v) Introduction of several instruments of carbon trading, which allows Annex I countries to fulfill their quota of GHG reduction by financing carbon reduction in other countries, e.g.:

- Clean Development Mechanism (CDM) in developing countries;
- Joint Implementation (JI) in the transitional economies of Eastern Europe;
- International Emission Trading (IET) among the OECD countries.

As mentioned, the main reason for project-based mechanism is to offer credit in the carbon reduction quota of Annex I countries. CDM allows a Annex I country to finance a valid GHG emission reduction project in non-Annex country and the actual reduction gets credited to the committed quota of the sponsoring-cum-financing country. This way, GHG emission reduction also gets accomplished in the developing country.

Similarly, a JI allows an Annex country to finance a GHG reduction project in another Annex country and to receive credit equal to part or all of emission reduction.

IET is a tradable quota regime, which establishes a national emission limit for each participating country and each country to hold quota equal to part or full of the emission reduction achieved.

Tradable carbon reduction is now a growing business in India, China and Philippines. But so far, USA and Australia did not sign the Kyoto Protocol.

Bali Action Plan, 2007 (COP 13/MOP 3), Indonesia

The Conference (COP 13/MOP 3) took place between December 3 and December 15, 2007, at Nusa Dua, in Bali, Indonesia. Agreement on a timeline and structured negotiation on the post 2012 framework (a successor to the Kyoto Protocol) was achieved with the adoption of the Bali Action Plan.

The Ad Hoc Working Group on Long-term Cooperative Action under the Convention (AWG-LCA) was established as a new subsidiary body to conduct the negotiations aimed at urgently enhancing the implementation of the Convention up to and beyond 2012.

These negotiations took place during 2008 (leading to COP 14/MOP 4 in Poznan, Poland) and 2009 (leading to COP 15/MOP 5 in Copenhagen).

In view of its contemporaneity and importance, the text of the Bali Action Plan is presented in the Appendix II.

Copenhagen Accord, 2009 (COP 15/MOP 5), Denmark

The United Nations Climate Change Conference (COP 15/MOP 5) took place in Copenhagen, Denmark from 7 December to 18 December 2009. The overall goal for Climate Change Conference in Denmark was to establish an ambitious global climate agreement for the period from 2012 when the first commitment period under the Kyoto Protocol expires.

However, on November 14, 2009, the *New York Times* announced that President Obama and other world leaders have decided to put off the difficult task of reaching a climate change agreement, but instead to make it the mission of the Copenhagen conference to reach a less specific "politically binding" agreement that would push the most difficult issues into the future. Ministers and officials from 192 countries took part in the Copenhagen meeting and in addition there were participants from a large number of civil society organisations.

The conference did not achieve a binding agreement for the post-Kyoto period. A 13-paragraph 'political accord' was negotiated by approximately 25 parties including US and China, but it was only 'noted' by the COP. But the UN Secretary General Ban Ki Moon hailed the accord. The accord came to be known as the Copenhagen Accord.

This important "Copenhagen Accord" is presented in the Appendix III (of this chapter) for perusal.

Comparative Evaluation of three Important Summits

Kyoto Protocol, Bali Action Plan and Copenhagen Accord

The United Nation Framework Convention on Climate Change laid down the philosophy – pursue Sustainable Development to simultaneously eradicate global poverty and preserve the ecology. This was the core message of the Brundtland Report 1987 of United Nation Commission on Environment and Development. This was also the major theme of the Declaration of Principles of the Earth Summit at Rio de Janeiro and the Agenda 21, which made history and was later followed by the Millennium Development Goals enunciated by the United Nations in 2000. It was at the Earth Summit, that the Framework Convention was presented and accepted by the representatives of 192 Parties to the Earth Summit.

The UNFCCC 1992 carried forward the concept of the sustainable development in its application to the climate change to new level and inferred:

(i) That the climate change affects all countries; that the causes of the climate change are anthropogenic in character; that it has been mainly caused by copious emission of carbon dioxide emission by the industries of the western countries and their energy-intensive lifestyle.

(ii) That the corrective actions, mitigation and adaptations, are to be undertaken by all, but with common but differential responsibilities and respective capabilities; that the developed countries have higher capabilities and higher responsibility than the underdeveloped.

(iii) The developed countries have been identified as the Annex I countries and the underdeveloped countries have been termed as Non-Annex I countries.

The Kyoto Protocol1997, representing the 3rd Conference of the Parties to the UNFCCC, laid down the quantitative bases to the various recommendations of the UNFCCC. That the Annex I countries are to commit to bring down their emissions below 1990 level by 5.2 per cent; they shall show demonstrable progress in this respect to FCCC by 2005; they shall also submit their inventories of GHG emissions and their plans of execution to FCCC; Annex countries could however finance CDM projects to reduce GHG emissions in developing countries and credit such reductions to their account; the non-Annex I were however not required to commit any quantitative reduction. The operational rules for the implementation of the flexible mechanism like CDM and JI were further refined at the 7th Marrakech COP. Today CDM market has become a many billion dollar business worldwide participated by India, China, Philippine, etc. In the process, Kyoto Protocol has kindled more public interest than the main Convention UNFCCC itself.

The Bali Action Plan 2007, representing the 13th COP, introduces for the first time the concept of developing countries undertaking "nationally appropriate mitigation actions". Otherwise it espouses the basic principle of the Kyoto Protocol in respect of "common but differential responsibilities based on respective capabilities" among the Annex I and the non-Annex. The Bali Action Plan also exhorts "enhanced action" in more explicit language on four principal arms of fighting climate change - mitigation, adaptation, technology development and transfer and provision of financial resources and investment. It recommends incentives for reduction of deforestation and degradation of forest (REDD). Even then it does not presume to have solved the problem of climate change but envisages at the 15th Conference of Parties at Copenhagen in December 2009 "full, effective and sustained implementation of the Convention through long-term cooperative action."

The language of the Bali Action Plan was definitely more precise and transparent than before. The language of the Copenhagen Accord was bolder, perhaps having been finalised at the level of the head of the delegation including that of the US. Regarding its general contents, we may summarise our views as under:

(i) The Copenhagen Accord did uphold the basic principles of the FCCC/ Kyoto Protocol in respect of the 'common but differential responsibilities and respective capabilities' and in stating in mind that social and economic development and poverty eradication are the first and overriding priorities of developing countries and in putting rightful emphasis on Adaptation and Mitigation tasks to combat climate change.

(ii) But what the Copenhagen Accord did go beyond Bali and Kyoto is to specifically mention the collective support of the Annex I countries of USD 30 billion for 2010 – 2012 and a further collective commitment of USD 100 billion per year by 2020.

(iii) Where the Copenhagen Accord departed from the UNFCCC/Kyoto was in respect of requiring the developing non-Annex I countries to submit mitigation action plan of GHG emission in the context of the sustainable development. This requirement is voluntary for least developed countries. The mitigation actions are to be monitored and measured domestically for international consultation. Further, in departure of the Kyoto Protocol, the Annex I countries are urged to undertake quantified mitigation of emission beyond the Kyoto but let them specify their commitment in the format of Appendix I of the Accord. There are really no such elaborate procedures of international monitoring. of the mitigation actions of the Annex I countries.

The new features of the Copenhagen Accord is that it recommends deep cuts in emission levels of all countries but does not prescribe any specific cut levels, even for the Annex I countries, which are to be projected by the respective countries, developed and developing, and submitted to the UNFCCC Secretariat in the prescribed formats of the Accord by 31st January 2010.The countries that submitted the proposals are shown in Table 9.7.

The latest UN Climate Conference at Cancun, Mexico, 2010

Coming as the latest such meet this year, the Cancun meet has put forward the tasks of the international community in reasonably clear language in a well-structured format regarding the main parameters of the climate change challenges:

I. Shared Vision for long-term Cooperative Actions;

II. Enhanced Action on Adaptation;

III. Enhanced Action on Mitigation;

(A) Nationally appropriate mitigation commitments or actions by developed country Parties;

(B) Nationally appropriate mitigation actions by developing country Parties;

(C) Policy actions on reducing emissions from deforestation and Forest degradation in developing countries;

(D) Using opportunities of Market to enhance the cost-effectiveness and promotion of Mitigation actions;

(E) Economic and Social consequences of Response Measures.

IV. Finance, Technology and Capacity-Building:

(A) Finance:

Fast-start Finance

Long-term Finance

Green Climate Fund

B. Technology Development and Transfer:

(a) A Technology Executive committee;

(b) A Climate Technology Centre and Network;

(c) Capacity Building.

V. Review

There is an obvious push towards the developing countries to participate more actively for meeting the obligations of mitigation and adaptations; Though the Agreement does still enumerate that the developing countries are to take nationally appropriate actions with common but differentiated responsibilities based on respective capabilities.

There is promise of technology transfer and financial help from the developed countries, presently of USD30 billion for the period 2010-2012 and later by 2020 of USD 100 billion annually.

There is also emphasis on Capacity building and adaptation measures with the technology being available from the developed countries.

There are no specific quota of emission reductions laid down but the Agreement did wish that developed countries and bigger developing countries like China, India, Brazil, etc to voluntarily declare their nationally appropriate target reductions as indicated in Table 9.7.

Table 9.7: CO_2 Mitigation Commitments by 2020

Countries	*Chief Negotiators*	*Details of commitments*
USA	Todd Stern	17% reduction 2005 base line
CHINA	Xia Zhenhua	40-45% reduction in emission intensity per unit GDP on 2005 base line
INDIA	Jairam Ramesh	20-25% reduction in emission intensity per unit GDP on 2005 base line
EU	Arther Runge-Nctzer	20% reduction on 1990 baseline (30% conditional offer)
JAPAN	Kazuya Endo	25% reduction on1990 baseline by 2020
BRAZIL	Sergio Serra	36-39% reduction on projected 2020 emission
Small Island States (AOSIS)	Dessima Williams (Grenada)	100 % on 2000 baseline by 2020

The trend of opinions and stands taken by some of the important countries are summarised in Table 9.8.

Table 9.8: Issues Raised by Major Countries / Parties

Countries/ Parties	*Issues*
USA	1. Key player behind Copenhagen Accord's bottom-up voluntary "pledge and review" approach; 2. Weakened by failure to deliver Legislation, resigned to incremental progress as a result; 3. Requires enhanced transparency from China on the measurement, reporting and verification (MRV) of emission.
CHINA	1. Desires to keep alive "dual truck" negotiation process with clear division between "developed" and "developing" countries; 2. Wants action on mitigation targets in developed countries, notably USA; 3. Wants delivery on climate finance commitments by developed countries; 4. Tolerates no concession on international MRV; 5. Refuses binding targets for developing countries.
INDIA	1. Desires delivery on climate finance commitments by developed countries; 2. Wants clear division between developing and developed countries maintained as under the Kyoto Protocol; 3. Wants mitigation targets for developing countries to reduce emission voluntary; 4. Requires US to come forward with firm commitments to reduce emission.
EU	1. Has a clear preference for a new top-down legally binding outcome to replace Kyoto Protocol; 2. Resigned to "step-down" approach at Cancun, with progress on deforestation, adaptation, climate financing and technology transfer; 3. Requires progress on MRV and development of carbon markets; 4. Will publish a low carbon 2050 road map in spring 2011; 5. 30 % target remains conditional – part of engaging debate.
JAPAN	1. Agree with EU in preference for a single framework covering all major emitters to replace Kyoto Protocol; 2. Selects strong commitments from large developing countries with focus on localy binding commitments from all players; 3. Wants uniform rules for MRV to record implementation.
BRAZIL	1. Progress on a mechanism for Reduced Emissions from Deforestation and Forest Degradation (REDD) is key. Has established an Amazon Fund to finance reduced deforestation which it hopes can be funded from financial commitments made by the developed countries.
AOSIS*	1. Desires a legally binding top-down agreement which would see global concentrations of CO_2 stabilize at 350 ppm; 2. Requires global emissions to peak by 2015 at the latest and decrease thereafter to at least 85% below 1990 level by 2050; 3. Wants immediate delivery of "fast-start" finance to fund adaptation to impacts of climate change.

A Summarised version of the Agreement is presented in the Appendix IV for perusal in more details.

Closing Observations: East and West, the twain should meet?

One fundamental task facing the world community in its endeavour to control global warming is to close the gaps in the perception of the developed and the developing countries regarding their roles in their obligations and responsibilities in their common endeavour. The extent of global warming depends on the net stock of the Green house gases in the space. In terms of the recommendation of the Fourth Assessment Report of IPCC, all efforts shall be made to restrict the global rise in average temperature to 2°C, beyond which the costs of mitigation and adaptation will rise sharply. This calls for limiting the GHG level in the space to 450-500 ppm against present concentration of GHG at below 400 ppm.

The industrialised countries have already overwhelmingly used up their share of the carbon space and they are now being required to undergo very deep cuts, of the order of 50 per cent below the 1990 level to be achieved by 2050. For the thin margin of carbon-equivalent space between 450 to 500 ppm has now to be equitably and fairly shared between the developed and the underdeveloped. The developed world has been asked upon to provide technological and financial support to the underdeveloped to enable them to take both nationally appropriate mitigation and adaptation. This is not being graciously taken by the developed. Hence, there is need for compromise and 'give and take' on both sides. History cannot be unwritten and then rewritten. The developing world should not take a doctrinaire or rigid approach at the past act of omission and commission by the developed countries, despite the polluters pay conundrum.

The fossil fuel, the prime culprit, was the only fuel available then to energize the industrial wheel and the long term ill effect was not known then. In the act of the commission, the developed nations have presented the world much of science and technological inventions, which are more or less world heritage today. Let us proceed with whatever best consensus we can arrive at, without rancour. As the global temperature scenario unfolds, we will be able to arrive at fresh consensus and more favourable commitment. As the developing countries get enriched, their 'respective capabilities' will strengthen and many of them will open up to their own initiatives and not always look up to the western grudging munificence. We particularly have India in view. For, China is already aspiring to the economic super-power tag!

APPENDIX I

United Nation Framework Convention on Climate Change

Different Conferences of the Parties

1995- COP 1, Berlin Mandate

1996 - COP 2, Geneva, Switzerland

1997 – COP3, Kyoto Protocol on Climate Change

1998 - COP 4, Buenos Aires

1999 COP 5, Bonn, Germany

2000 - COP 6, The Hague, Netherlands

2001 - COP 6, Bonn, Germany

2001 - COP 7, Marrakech, Morocco

2002 - COP 8, New Delhi, India

2003 - COP 9, Milan, Italy

2004 - COP 10, Buenos Aires, Argentina

2005 – COP11/MOP 1, Montreal, Canada

2006 – COP12/MOP 2, Nairobi, Kenya

2007 – COP 13/MOP 3, Bali, Indonesia

2008 – COP 14/MOP 14, Poznan, Poland

2009–COP15/MOP5, Copenhagen, Denmark

2010 - COP 16/MOP 6, Cancun, Mexico

2011 - COP 17/MOP 7, South Africa

2012 – COP 18/MOP8

The complete list of all the Conferences of Parties from 1995 to 2011 and the relevant Meetings of the Parties MOPs are presented in the Appendix I. Brief proceedings of some important COPs (COP1 Berlin Mandate 1995, COP6 Bonn 2001, COP 7 Marrakech 2001 and COP 11 Montreal 2005) are also included in Appendix I.

1995 – COP 1, The Berlin Mandate

The first UNFCCC Conference of Parties took place in the spring of 1995 in Berlin, Germany. It voiced concerns about the adequacy of countries' abilities to meet commitments under the Convention. These were expressed in a U.N. ministerial declaration known as the "Berlin Mandate", which established a two year Analytical and Assessment Phase (AAP), to negotiate a "comprehensive menu of actions" for countries to pick from and choose future options to address climate change which for them, individually, made the best economic and environmental sense. The Berlin Mandate exempted non-Annex I countries from additional binding obligations, in keeping with the principle of "common but differentiated responsibilities" established in the UNFCCC even though, collectively, the larger, newly industrialising countries were expected to be the world's largest emitters of greenhouse gas emissions 15 years hence.

2001 – COP 6, Bonn, Germany

COP 6 negotiations resumed July 17-27, 2001, in Bonn, Germany, with little progress have been made on resolving the differences that had produced an impasse in The Hague. However, this meeting took place after President George W. Bush had become the U.S. President, and had rejected the Kyoto Protocol in March; as a result the United States delegation to this meeting declined to participate in the negotiations related to the Protocol, and chose to act as observers at that meeting. As the other parties negotiated the key issues, agreement was reached on most of the major political issues, to the surprise of most observers given the low level of expectations that preceded the meeting. The agreements included:

1. **Flexible Mechanisms**: The "flexibility" mechanisms which the United States had strongly favoured as the Protocol was initially put together, including emissions trading; Joint Implementation (JI); and the Clean Development Mechanism (CDM) which allow industrialised countries to fund emissions reduction activities in developing countries as an alternative to domestic emission reductions. One of the key elements of this agreement was that there would be no quantitative limit on the credit a country could claim from use of these mechanisms, but that domestic action must constitute a significant element of the efforts of each country to meet their targets.
2. **Carbon sinks**: Credit was agreed to for broad activities that absorb carbon from the atmosphere or store it, including forest and cropland management, and re-vegetation, with no over-all cap on the amount of credit that a country could claim for sinks activities. In the case of forest management, an Appendix Z establishes country-specific caps for each Annex I country, for example, a cap of 13 million tons could be credited to Japan (which represents about 4% of its base-year emissions). For cropland management, countries could receive credit only for carbon sequestration increases above 1990 levels

2001 – COP 7, Marrakech, Morocco

At the COP 7 meeting in Marrakech, Morocco October 29-November 10, 2001, negotiators in effect completed the work of the Buenos Aires Plan of Action, finalising most of the operational details and setting the stage for nations to ratify the Kyoto Protocol. The complete package of decisions are known as the Marrakech Accords. The United States delegation continued to act as observers, declining to participate in active negotiations. Other parties continued to express their hope that the United States would re-engage in the process at some point, but indicated their intention to seek ratification of the requisite number of countries to bring the Protocol into force (55 countries representing 55% of developed country emissions of carbon dioxide in 1990). A target date for bringing the Protocol into force was put forward: the August-September 2002 World Summit on Sustainable Development (WSSD) to be held in Johannesburg, South Africa. The main decisions at COP 7 included:

- Operational rules for international emissions trading among parties to the Protocol and for the CDM and joint implementation;
- A compliance regime that outlines consequences for failure to meet emissions targets but defers to the parties to the Protocol after it is in force to decide whether these consequences are legally binding;
- Accounting procedures for the flexibility mechanisms;
- A decision to consider at COP 8 how to achieve a review of the adequacy of commitments that might move toward discussions of future developing country commitments.

2005 – COP 11/MOP 1, Montreal, Canada

COP 11 (or COP 11/MOP 1) took place between November 28 and December 9, 2005, in Montreal, Quebec, Canada. COP 11 was also the first Meeting of the Parties (MOP-1) to the Kyoto Protocol since their initial meeting in Kyoto in 1997. It was therefore one of the largest intergovernmental conferences on climate change ever. The event marked the entry into force of the Kyoto Protocol. Hosting more than 10,000 delegates, it was one of Canada's largest international events ever and the largest gathering in Montreal since Expo 67. The Montreal Action Plan is an agreement hammered out at the end of the conference to "extend the life of the Kyoto Protocol beyond its 2012 expiration date and negotiate deeper cuts in greenhouse-gas emissions."

Appendix II
Bali Action Plan, 2007

UNFCCC CONFERENCE OF THE PARTIES (COP 13)

Report of the Conference of the Parties on its thirteenth session, held in Bali, 3 to 15 December 2007

The Conference of the Parties,

1. *Decides* to launch a comprehensive process to enable the full, effective and sustained implementation of the Convention through long-term cooperative action, now, up to and beyond 2012, in order to reach an agreed outcome and adopt a decision at its fifteenth session, by addressing, inter alia:
   (a) A shared vision for long-term cooperative action, including a long-term global goal for emission reductions, to achieve the ultimate objective of the Convention, in accordance with the provisions and principles of the Convention, in particular the

principle of common but differentiated responsibilities and respective capabilities, and taking into account social and economic conditions and other relevant factors;

(b) Enhanced national/international action on mitigation of climate change, including, inter alia, consideration of:

(i) Measurable, reportable and verifiable nationally appropriate mitigation commitments or actions, including quantified emission limitation and reduction objectives, by all developed country Parties, while ensuring the comparability of efforts among them, taking into account differences in their national circumstances;

(ii) Nationally appropriate mitigation actions by developing country Parties in the context of sustainable development, supported and enabled by technology, financing and capacity-building, in a measurable, reportable and verifiable manner;

(iii) Policy approaches and positive incentives on issues relating to reducing emissions from deforestation and forest degradation in developing countries; and the role of conservation, sustainable management of forests and enhancement of forest carbon stocks in developing countries;

(iv) Cooperative sectoral approaches and sector-specific actions, in order to enhance implementation of Article 4, paragraph 1(c), of the Convention;

(v) Various approaches, including opportunities for using markets, to enhance the cost-effectiveness of, and to promote, mitigation actions, bearing in mind different circumstances of developed and developing countries;

(vi) Economic and social consequences of response measures;

(vii) Ways to strengthen the catalytic role of the Convention in encouraging multilateral bodies, the public and private sectors and civil society, building on synergies among activities and processes, as a means to support mitigation in a coherent and integrated manner;

(c) Enhanced action on adaptation, including, inter alia, consideration of:

(i) International cooperation to support urgent implementation of adaptation actions, including through vulnerability assessments, prioritisation of actions, financial needs assessments, capacity-building and response strategies, integration of adaptation actions into sectoral and national planning, specific projects and programmes, means to incentivize the implementation of adaptation actions, and other ways to enable climate-resilient development and reduce vulnerability of all Parties, taking into account the urgent and immediate needs of developing countries that are particularly vulnerable to the adverse effects of climate change, especially the least developed countries and small island developing States, and further taking into account the needs of countries in Africa affected by drought, desertification and floods;

(ii) Risk management and risk reduction strategies, including risk sharing and transfer mechanisms such as insurance;

(iii) Disaster reduction strategies and means to address loss and damage associated with climate change impacts in developing countries that are particularly vulnerable to the adverse effects of climate change;

(iv) Economic diversification to build resilience;

(v) Ways to strengthen the catalytic role of the Convention in encouraging multilateral bodies, the public and private sectors and civil society, building on synergies among activities and processes, as a means to support adaptation in a coherent and integrated manner;

(d) Enhanced action on technology development and transfer to support action on mitigation and adaptation, including, inter alia, consideration of:

(i) Effective mechanisms and enhanced means for the removal of obstacles to, and provision of financial and other incentives for, scaling up of the development and transfer of technology to developing country Parties in order to promote access to affordable environmentally sound technologies;

(ii) Ways to accelerate deployment, diffusion and transfer of affordable environmentally sound technologies;

(iii) Cooperation on research and development of current, new and innovative technology, including win-win solutions;

(iv) The effectiveness of mechanisms and tools for technology cooperation in specific sectors;

(e) Enhanced action on the provision of financial resources and investment to support action on mitigation and adaptation and technology cooperation, including, inter alia, consideration of:

(i) Improved access to adequate, predictable and sustainable financial resources and financial and technical support, and the provision of new and additional resources, including official and concessional funding for developing country Parties;

(ii) Positive incentives for developing country Parties for the enhanced implementation of national mitigation strategies and adaptation action;

(iii) Innovative means of funding to assist developing country Parties that are particularly vulnerable to the adverse impacts of climate change in meeting the cost of adaptation;

(iv) Means to incentivize the implementation of adaptation actions on the basis of sustainable development policies;

(v) Mobilisation of public- and private-sector funding and investment, including facilitation of climate-friendly investment choices;

(vi) Financial and technical support for capacity-building in the assessment of the costs of adaptation in developing countries, in particular the most vulnerable ones, to aid in determining their financial needs;

2. *Decides* that the process shall be conducted under a subsidiary body under the Convention, hereby established and known as the Ad Hoc Working Group on Long-term Cooperative Action under the Convention, that shall complete its work in 2009 and present the outcome of its work to the Conference of the Parties for adoption at its fifteenth session;

3. *Agrees* that the process shall begin without delay, that the sessions of the group will be scheduled as often as is feasible and necessary to complete the work of the group, where possible in conjunction with sessions of other bodies established under the Convention, and that its sessions may be complemented by workshops and other activities, as required;

4. *Decides* that the first session of the group shall be held as soon as is feasible and not later than April 2008;

5. *Decides* that the Chair and Vice-Chair of the group, with one being from a Party included in Annex I to the Convention (Annex I Party) and the other being from a Party not included in Annex I to the Convention (non-Annex I Party), shall alternate annually between an Annex I Party and a non-Annex I Party;

6. *Takes note* of the proposed schedule of meetings contained in the annex to this decision;
7. *Instructs* the group to develop its work programme at its first session in a coherent and integrated manner;
8. *Invites* Parties to submit to the secretariat, by 22 February 2008, their views regarding the work programme, taking into account the elements referred to in paragraph 1 above, to be compiled by the secretariat for consideration by the group at its first meeting;
9. *Requests* the group to report to the Conference of the Parties at its fourteenth session on progress made;
10. *Agrees* to take stock of the progress made, at its fourteenth session, on the basis of the report by the group;
11. *Agrees* that the process shall be informed by, inter alia, the best available scientific information, experience in implementation of the Convention and its Kyoto Protocol, and processes there under, outputs from other relevant intergovernmental processes and insights from the business and research communities and civil society;
12. *Notes* that the organisation of work of the group will require a significant amount of additional resources to provide for the participation of delegates from Parties eligible to be funded and to provide conference services and substantive support;
13. *Strongly urges* Parties in a position to do so, in order to facilitate the work of the group, to provide contributions to the Trust Fund for Participation in the UNFCCC Process and the Trust Fund for Supplementary Activities for the purposes referred to in paragraph 12 above and to provide other forms of in kind support such as hosting a session of the group.

Appendix III

The Copenhagen Accord, 2009

The Heads of State, Heads of Government, Ministers, and other heads of delegation present at the United Nations Climate Change Conference 2009 in Copenhagen, have agreed on this Copenhagen Accord which is operational immediately.

1. We underline that climate change is one of the greatest challenges of our time. We emphasise our strong political will to urgently combat climate change in accordance with the principle of common but differentiated responsibilities based on respective capabilities. To achieve the ultimate objective of the Convention to stabilize greenhouse gas concentration in the atmosphere at a level that would prevent dangerous anthropogenic interference with the climate system, we

shall, recognising the scientific view that the increase in global temperature should be below 2°C, on the basis of equity and in the context of sustainable development, enhance our long-term cooperative action to combat climate change. We recognize the critical impacts of climate change and the potential impacts of response measures on countries particularly vulnerable to its adverse effects and stress the need to establish a comprehensive adaptation programme including international support.

2. We agree that deep cuts in global emissions are required according to science, and as documented by the IPCC Fourth Assessment Report with a view to reduce global emissions so as to hold the increase in global temperature below 2 degrees Celsius, and take action to meet this objective consistent with science and on the basis of equity. We should cooperate in achieving the peaking of global and national emissions as soon as possible, recognising that the time frame for peaking will be longer in developing countries and bearing in mind that social and economic development and poverty eradication are the first and overriding priorities of developing countries and that a low-emission development strategy is indispensable to sustainable development.

3. Adaptation to the adverse effects of climate change and the potential impacts of response measures is a challenge faced by all countries. Enhanced action and international cooperation on adaptation is urgently required to ensure the implementation of the Convention by enabling and supporting the implementation of adaptation actions aimed at reducing vulnerability and building resilience in developing countries, especially in those that are particularly vulnerable, especially least developed countries, small island developing States and Africa. We agree that developed countries shall provide adequate, predictable and sustainable financial resources, technology and capacity-building to support the implementation of adaptation action in developing countries.

4. Annex I Parties commit to implement individually or jointly the quantified economy-wide emissions targets for 2020, to be submitted in the format given in Appendix I by Annex I Parties to the secretariat by 31 January 2010 for compilation in an INF document. Annex I Parties that are Party to the Kyoto Protocol will thereby further strengthen the emissions reductions initiated by the Kyoto Protocol. Delivery of reductions and financing by developed countries will be measured, reported and verified in accordance with existing and any further guidelines adopted by the Conference of the Parties, and will ensure that accounting of such targets and finance is rigorous, robust and transparent.5. Non-Annex I Parties to the Convention will

implement mitigation actions, including those to be submitted to the secretariat by non-Annex I Parties in the format given in Appendix II by 31 January 2010, for compilation in an INF document, consistent with Article 4.1 and Article 4.7 and in the context of sustainable development. Least developed countries and Small Island developing States may undertake actions voluntarily and on the basis of support. Mitigation actions subsequently taken and envisaged by Non-Annex I Parties, including national inventory reports, shall be communicated through national communications consistent with Article 12.1(b) every two years on the basis of guidelines to be adopted by the Conference of the Parties. Those mitigation actions in national communications or otherwise communicated to the Secretariat will be added to the list in appendix II. Mitigation actions taken by Non-Annex I Parties will be subject to their domestic measurement, reporting and verification, the result of which will be reported through their national communications every two years. Non-Annex I Parties will communicate information on the implementation of their actions through National Communications, with provisions for international consultations and analysis under clearly defined guidelines that will ensure that national sovereignty is respected. Nationally appropriate mitigation actions seeking international support will be recorded in a registry along with relevant technology, finance and capacity building support. Those actions supported will be added to the list in appendix II.

These supported nationally appropriate mitigation actions will be subject to international measurement, reporting and verification in accordance with guidelines adopted by the Conference of the Parties.

6. We recognize the crucial role of reducing emission from deforestation and forest degradation and the need to enhance removals of greenhouse gas emission by forests and agree on the need to provide positive incentives to such actions through the immediate establishment of a mechanism including REDD-plus, to enable the mobilisation of financial resources from developed countries.

7. We decide to pursue various approaches, including opportunities to use markets, to enhance the cost-effectiveness of, and to promote mitigation actions. Developing countries, especially those with low emitting economies should be provided incentives to continue to develop on a low emission pathway.

8. Scaled up, new and additional, predictable and adequate funding as well as improved access shall be provided to developing countries, in accordance with the relevant provisions of the Convention, to

enable and support enhanced action on mitigation, including substantial finance to reduce emissions from deforestation and forest degradation (REDD-plus), adaptation, technology development and transfer and capacity-building, for enhanced implementation of the Convention. The collective commitment by developed countries is to provide new and additional resources, including forestry and investments through international institutions, approaching USD 30 billion for the period 2010 - 2012 with balanced allocation between adaptation and mitigation. Funding for adaptation will be prioritised for the most vulnerable developing countries, such as the least developed countries, small- island developing States and Africa. In the context of meaningful mitigation actions and transparency on implementation, developed countries commit to a goal of mobilising jointly USD 100 billion dollars a year by 2020 to address the needs of developing countries. This funding will come from a wide variety of sources, public and private, bilateral and multilateral, including alternative sources of finance. New multilateral funding for adaptation will be delivered through effective and efficient fund arrangements, with a governance structure providing for equal representation of developed and developing countries. A significant portion of such funding should flow through the Copenhagen Green Climate Fund.

9. To this end, a High Level Panel will be established under the guidance of and accountable to the Conference of the Parties to study the contribution of the potential sources of revenue, including alternative sources of finance, towards meeting this goal.

10. We decide that the Copenhagen Green Climate Fund shall be established as an operating entity of the financial mechanism of the Convention to support projects, programme, policies and other activities in developing countries related to mitigation including REDD-plus, adaptation, capacity building, technology development and transfer.

11. In order to enhance action on development and transfer of technology we decide to establish a Technology Mechanism to accelerate technology development and transfer in support of action on adaptation and mitigation that will be guided by a country-driven approach and be based on national circumstances and priorities.

12. We call for an assessment of the implementation of this Accord to be completed by 2015, including in light of the Convention's ultimate objective. This would include consideration of strengthening the long-term goal referencing various matters presented by the science, including in relation to temperature rises of 1.5 degrees Celsius.

APPENDIX IV

Abbreviated Report
UNFCCC Climate Change Conference in Cancun, Mexico
11 December 2010

The Conference affirms a balanced package of decisions that set all governments more firmly on the path towards a low-emission developing world. The package was dubbed the 'Cancun Agreements' by the Parties in the final Plenary.

I. A Shared Vision for Long-term Cooperative Actions

The vision addresses mitigation, adaptation, finance, technology development and transfers, and capacity-building in a balanced, integrated and comprehensive manner to enhance and achieve the full, effective and sustained implementation of the Convention now, up to and beyond 2012.

II. Enhanced Action on Adaptation

Taking into account their common but differentiated responsibilities and respective capabilities, the parties agree adaptation is a challenge faced by all parties and that enhanced action and international cooperation on adaptation is urgently required aimed at reducing vulnerability and building resilience to climate change.

III. Enhanced action on Mitigation

A. Nationally Appropriate Mitigation Commitments or Actions by developed country Partners : It emphasizes the need for deep cuts in global greenhouse gas emissions and early and urgent undertaking to accelerate and enhance the implementation of the Convention, also with a view to reducing their aggregate emission of carbon dioxide consistent with recommended by the Fourth Assessment Report of the Intergovernmental Panel on Climate Change.

B. Nationally Appropriate Mitigation Commitments or Actions by developing country Partners : The Conference reaffirms that social and economic development and poverty eradication are the first and the overriding priorities of developing country Parties and recognizes the share of the global emissions originating in the developing countries will grow to meet the social and development needs.

It agrees that the developing country Parties will take nationally appropriate mitigation actions in the context of sustainable development, supported and enabled by technology, financing and capacity building provided by the developed country Parties and aim at achieving a deviation in emission relative to business as usual emissions in 2020.

C. Policy Actions on reducing Emissions from Deforestation and Forest Degradation in developing Countries : The Conference affirms that, in the context of the provision of adequate and predictable support to the developing country Parties, they should collectively aim to slow, halt and reverse forest cover and carbon loss.

The developing countries will thus contribute to mitigation actions in the forest sector in accordance to their respective capabilities and national circumstances –

(a) Reducing emissions from deforestation and forest degradation;

(b) Reducing emission from forest degradation;

(c) Conservation of forest;

(d) Sustainable management of forest;

(e) Enhancement of forest carbon stock.

D. Using opportunities of Market to enhance the cost-effectiveness and promotion of Mitigation Actions : Ensuring good governance and robust market functioning and regulation, it was recognised to provide incentives in support of low-emission development strategies by creating market-tradable carbon-reduction instruments under the broad outline of the Kyoto Protocol.

E. Economic and Social Consequences of Response Measures : The response measures to mitigate climate change should be coordinated with social and economic development in an integrated manner to avoid adverse impacts, especially for a developing country.

IV. Finance, Technology and Capacity-Building

A. Finance : *First-Start Finance:* The conference takes note of the collective commitment by developed countries to provide new and additional resources, including forestry and investments through international institutions, approaching USD 30 billion for period 2010-2012, with a balanced allocation between adaptation and mitigation.

Long-term Finance: The conference recognizes that developed country Parties commit in the context of meaningful actions and transparency on implementation to a goal of mobilising jointly USD 100 billion per year by 2020 to address the needs of the developing countries.

Green Climate Fund: The conference decides to establish a Green Climate Fund to be designated as an operating entity of the financial mechanism of the Convention under Article 11. The fund shall be governed by a Board of 24 members comprising an equal number of members from the developed and developing country Parties.

B. Technology Development and Transfer : The conference confirms the importance of promoting and enhancing national and international cooperative action on the development and transfer of environmentally sound technologies to developing country Parties to support actions on adaptation and mitigation now, up to and beyond 2012, in order to achieve the ultimate objective of the Convention.

The conference decides to establish a Technology mechanism to facilitate the implementation of actions to achieve the objectives of the Convention, which will consist of the following institutions:

(a) A *Technology Executive Committee* to undertake the many functions under Technology Development and Transfer as contained in paragraph 121 of this Agreement;

(b) A *Climate Technology Centre and Network* to undertake many functions for Technology Development and Transfer as contained in paragraph 123 of this Agreement.

C. Capacity-Building : The conference reaffirms that capacity-building is essential to enable developing country Parties to participate fully in addressing the Climate Change challenges and implement effectively their commitment under the Convention. Capacity-building is an integral part of enhanced action on mitigation, adaptation, technology development and transfer, and access to financial resources. The scope of capacity-building should be guided by decisions 2/CP7, and key factors identified in decisions 2/CP 10.

V. Review

The conference decides to periodically review the adequacy of the long-term global goal (referred to in paragraph 4 of this Agreement) in the light of the ultimate objective of the Convention and overall progress towards achieving it in accordance with the Convention.

(a) The review should be guided by the principles of equity, common but differentiated responsibilities and respective capabilities and take into account inter alia,

- The best available scientific knowledge, including the assessment reports of the Intergovernmental Panel on Climate Change;
- Observed impacts of climate change;
- An assessment of the overall aggregate effect of the steps taken by the Parties;

(b) The first review should start in 2013 and should be concluded by 2015.

(c) The Conference of the Parties shall take appropriate action based on the review.

References

Ahmed, Feroz 2006, Report: The Scary Story of Climate Change, *Business World*, 18 September 2006, Vol. 26, Issue 17, New Delhi.

Desai Nitin, Climate Change: The Challenge, pp. 298-307, "Environment and Climate Change", ed. Prof. S B Verma, Pentagon Press, 2007 *Economist*, March 20-26, 2010, pp. 69-75.

IPCC 4th Assessment Report, 2007.

Keeling, C.D. and Bacasta, R.B., 1977, *Impact of Industrial Gases on Climate Change, Energy and Climate*, Natural Research Council, Washington DC.

Krishnamurthy, Bala 2005, *Environment Management*, Prentice Hall India, New Delhi.

Masters, G.M., *Environmental Engineering and Science*, Prentice Hall of India, 2000.

McGraw-Hill Encyclopedia of Science and Technology, 10th Edition.

National Action Plan on Climate Change, Government of India; file:///E:/CLIMATE%20CHANGE/I%20climate%20Change %204AssReport KyotoProto/India's%20.htm

Saran Shyam, Climate Change: *Multilateral Negotiations*, pp. 294-297, "Environment and Climate Change", Ed. S.B. Verma, Pentagon Press, 2007.

Sincero, A.P. and Sincero Georgia A. 1996, Environmental Engineering – A Design Approach, Prentice Hall (USA 1996, India 1999).

Swamy, N.M. 2003, *Text Book on Environment Law*, Asia Publishing House, Hyderabad.

10

Green Energy for Security, Sustainability and Prosperity

I

FUTURE GLOBAL ENERGY POLICY OPTIONS

Modern life thrives on energy. When we drive a car on roads, ride on railways or fly in the sky, motive engine consumes energy all the time. From the kitchen to the drawing room to the office and factories, the modern civilisation is sustained on high energy consumption.

Diverse Contexts of Energy Policy

Although fuel and electricity amount for a relatively modest part of national economics and household budgets, but we noted above, the energy is absolutely critical to everyday life and the economy. Energy policy embraces diverse objectives and instruments and the mix changes over time. We examine these contexts (Scrase and Mackerron 2008).

- Before 1980, energy policy was Government led, embracing public ownership, and long range planning and direct control of markets.
- At the turn of the twenty-first century, it appears that the key challenges in energy policy was to reconcile some basic market failures in the environment domain, even as the energy becomes more market-driven for pricing and investment. How could the energy system be made more sustainable at the same time as market forces were prioritising short-term and low cost solutions?
- The context for energy policy is shifting again with the energy security as a high national priority, as crude oil prices increased from an average of USD 24 per barrel in 2001 to USD 130 per barrel in early 2008 and now in 2011 at USD 100 per barrel approximately, driven in particular by the rapid industrialisation of China and India.
- Oil and gas reserves are increasingly found concentrated in countries and regions that are seen as unstable or even as hostile. With the rising demand for energy, energy producers are trying to translate

their energy power for political influence and major energy users seeking to translate their political power into influence over energy.

- Most importantly, avoiding dangerous climate change is one of the defining challenges for humanity in the 21st century. Since, the energy system, by emitting carbon dioxide, is both the primary cause of climate change and again the primary means of mitigation, by reducing fossil fuel consumption or by switching to non-carbon fuel, the future evolution of energy policy and choice of fuel mix is of critical importance.
- Advocates of market liberalisation see unfettered competition as the best guarantee of market access and secure supplies. Bush administration was an advocate of this policy. But the British Chief Scientist Sir David King questions the role of the market in the energy sector vis-à-vis the climate change. In the context of expanding energy politics, countries are now reversing their liberalisation approach and retreating from the market mechanism.

Other Circumstantial Factors Moulding Oil Policy:

There are several broad circumstances which influence and tend to determine the world energy policy, namely:

- Energy security and oil price shock
- Pattern of oil peak production and reserve
- Climate Change Impact in global perspectives

In 1973-74, a combination of Middle East conflicts and the cartel action by the oil producing countries (OPEC) pushed the oil prices five-fold in two years. Prices doubled up in just one month at the end of 1973. Even before 1973, the oil producing nations had started to flex their collective oil power muscles and Western governments were reviewing their import dependencies on oil import; USA wanted to be oil independent by 1980 and European nations wanted to restrict the oil import to 40 % of energy needs, vowing to build 200 GWe of nuclear energy. France was able to sustain a nuclear programme reaching to nearly 80 per cent of their electrical power needs.

In spite of many expert projections, the actual peak oil production and subsequent decline had not occurred. The reserve estimates went up and up, the demand estimates trailing much below the forecasts and the oil producing countries distorting the status information to shore their margins. Nevertheless, two decades of investments and action plans are required to reduce demand and to develop new energy alternatives.

Stern Report (2006) states 'the scientific evidence is now overwhelming that climate change now presents very serious global risks' and it demands an urgent global response. IPCC 4th Assessment Reports (2007) that the global average temperatures are already 0.7°C above the pre-industrial level, another 0.5°C of warming is in pipeline as a consequence of the thermal inertia of the earth system. The Report concluded with 90% confidence. That the net effect of human activities since 1750 has been one leading to global warming.

Global primary energy consumption doubled between 1971 and 2004. But in 2004 OECD per capita consumption is nearly five times of the rest of the world and in CO_2 emission, OECD per capita emission figure is 4 times higher. In 2007 World Energy Outlook, the IEA projects that global primary energy demands could be 50 per cent higher in 2030, with the developing countries accounting for 73 per cent of this increase.

Primary Energy Choice Drivers: Energy for the Future

At the national level, the choices of energy are primarily determined by two policy drivers:

(i) Security of supplies;

(ii) Maintaining affordable prices for the citizen

At the global level, there are again two main dimensions influencing choice of energy policy:

(i) When petroleum and oil fail to meet the rising demands;

(ii) When the impact of the anthropogenic GHG debars the use of fossil fuels for energy.

The other secondary policy dimensions include

(i) Acceptability of Nuclear Power;

(ii) Acceptability of large scale CO_2 Capture and Storage system;

(iii) Sufficient development and effective use of renewable energy;

(iv) Acceptability of large sale land use changes (biomass, bio-fuels, crops etc).

To date it has proved extremely difficult to find a generally acceptable road to a safe and sustainable energy future, both within individual nations and for the world as a whole. There are varieties of technical solutions available, these unfortunately do not map onto an integrated power infrastructure. In OECD's commitment to privatisation and liberalisation the effort is concentrated on to maximising returns from existing and aging power plants and not to work for any energy resourcing. Energy is an issue with a uniquely long time scale and highly problematic political challenges.

IA. Some Important Energy Scenarios

In view of the difficulty expressed in designing a specific energy evolution route over the next few decades, we present a few energy evolution scenarios by well-known institutions of the world – IEA, Shell and IPCC.

IEA World Energy Outlook 2010

The International Energy Agency (IEA) has published the *World Energy Outlook* 2010 giving an overview of world energy demand and supply, projected up to 2035. We mention the salient points.

(i) The National Government's responses to the twin challenges of climate change and energy security that will shape future of energy in the long run.

(ii) In the New Policy Scenario – the Central Scenario this year - world energy demand increases by 36% between 2008 and 2035 or 1.2% per year on an average over the previous 27-year period.

(iii) In New Energy Policy Scenario, non-OECD countries account for 93% of the projected increase in global energy demand, reflecting mainly the faster rates of growth of economic activities. China's demand rises by 75% between 2008 and 2035. Aggregate demand in OECD countries rises very slowly. Nonetheless, by 2035, the United States become the world's second largest energy consumer behind China.

(iv) Global energy demand for each fuel source increases, with fossil fuels - coal, oil and gas – accounting for over 50% of the increase in total primary energy demand.

(v) Oil remains the dominant fuel in the primary energy mix to 2035. The share of nuclear power increased from 6% in 2008 to 8% in 2035. The use of modern renewable energy – hydro, wind, solar, geothermal, modern biomass and marine energy – triples between 2008 and 2035, its share in total energy demand increasing from 7 per cent to 14 per cent.

(vi) Natural gas is set to play a central role in meeting the world's energy needs for at least the next two-and-half decades. Despite its demand fall in 2009, its demand will increase by 44% between 2008 and 2035 – an average increase of 1.4% per annum. Growth in demand for gas surpasses other fuels because of its low carbon footprint; China's demand grows fastest, a fifth of global increases till 2035.

(vii) Oil demand (excluding biofuels) continues to grow steadily in the New Policy Scenario, reaching about 99 million barrels per day by 2035 – 15 million barrels per day up on 2009.

(viii) The eventual peak in oil will be determined by factors affecting both demand and supply. In the New Policy Scenario, the production does not peak by 2035.

Table 10.1, Table 10.2 and Table 10.3 show the world primary energy supplies, world generation of electricity and the world installed electric generation capacity respectively for 1971, 2002, 2010, 2020 and 2030 which indicate the trend of growths as indicated in the above overview.

Table 10.1 World Historic and Projected Primary Energy Supply
Million Tonne Oil Equivalent (Mtoe)

Energy Types	*1971*	*2002*	*2010*	*2020*	*2030*
Coal	1 407	2 389	2 763	3 193	3 601
Oil	2 413	3 676	4 308	5 074	5 766
Gas	892	2 190	2 703	3 451	4 130
Nuclear	29	692	778	776	764
Hydro	104	224	276	321	365
Biomass and Waste	687	1 119	264	1 428	1 605
Other Renewable	4	55	101	162	256
Total	5536	10 354	12 194	14 404	16 487

Table 10.2 World Historic and Projected Electricity Generation
Terawatt-Hour

Energy Types	*1971*	*2002*	*2010*	*2020*	*2030*
Coal	2 095	6 241	7 692	9 766	12 011
Oil	1 096	1 181	1 187	1 274	1 182
Gas	696	3 070	4 427	6 827	9 329
Nuclear	111	2 654	2 985	2 975	2 929
Hydro	1 206	2 610	3 212	3 738	4 248
Biomass and Waste	9	207	326	438	627
Other Renewable	5	111	356	733	1 250
Total	5 217	16 074	20 185	25 752	31 657

Table 10.3 World Historic and Projected Generation Capacity
(Million Tonne Oil Equivalent)

Energy Types	*1971*	*2002*	*2010*	*2020*	*2030*
Coal		1 135	1 337	1 691	2 156
Oil		454	499	524	453
Gas		893	211	1 824	2 564
Nuclear		359	385	382	376
Hydro		801	934	1 076	1 216
Biomass and Waste		34	53	71	101
Other Renewable		43	119	256	438
Total		3 719	4 539	5 822	7 308

Sources: IEA, World Energy Outlook 2004.

Shell International Scenarios (www.shell.com/scenario)

Humanity faces unprecedented challenges in searching for new paradigm of energy vis-à-vis a safe planet. This has been summed up by Shell in five words "more energy, less carbon dioxide". Shell has pioneered the systems of scenario building for seeking solution of energy and safe planet over the years.

In 1990s, Shell developed the scenario TINA – There Is No Alternative. The forces of market liberalisation, globalisation, and technology had created a global economic engine that was already beginning to engage vast populations in Asia. TINA helped people to apprise and resolve different faces of possible solution scenarios.

Then in 2005, Shell introduced scenarios that introduced the geopolitical crisis of security and trust that accompany TINA, as foreshadowed by unforeseen events as 9/11 or the Enron collapse. This led to the evolvement of new scenario – TINA's natural successor offering – TANIA – There Are No Ideal Answers.

The modern energy system is large, complex and demands major investment and development over a longer period. Hence its inherent inertia does not allow quick appraisal and redemption. It takes several years for effect of major changes to show up. The question is how to organize and grapple with changes.

Scenarios are used as tools to help identify such shifts and consider the plausible interactions between different perspectives and possibilities. Consequently new scenario tools like Scramble and Blueprints came to be developed by Shell in 2008.

To evaluate the possible scheme of energy development in future without endangering the planet by GHG emissions, Shell International has evolved, as indicated above, two new scenarios, described briefly as follows:

Scramble: In this scenario, policy makers pay little attention to more efficient energy use until supplies are tight and likewise green house gas emissions are not seriously addressed until there are major climate shocks.

Blueprints: In our second type of scenario, growing local actions begin to address the challenges of economic development, energy security and environmental pollution. A price is applied to critical mass of emissions giving a huge stimulus to the development of clean energy technologies such as carbon dioxide capture and storage, and energy efficiency measures. The result is far lower carbon dioxide emissions.

Shell submits that starting from "Scramble", outcomes of "Blueprints" offer best hope for a sustainable future with right combination of policy, technology and commitments from the Government, industry and society.

For better and fuller appreciation of various Shell Scenarios, readers should refer to www.shell.com/scenarios.

IPCC SRES Scenarios 2007: Future Energy Demand (mainly focused on Environment parameters)

The scenario uses four parameters – Economic versus Environmental and Global versus Regional, building following four scenarios as also indicated in the Fig. 10.1

A1 Future: Globalisation, market-based solutions and rapid economic development. Besides, reliance on carbon-intensive energy of the domestic coal, there are strong shifts to the unconventional, oil, natural gas, renewable energy and possibly nuclear future.

B1 Future: High level of environmental and social concern conveyed through international cooperation. High economic growth resource efficiency, clean technology; Gas, as the cleanest fossil resource is used as bridge development towards post-fossil energy future.

A2 Future: Lower level of international cooperation; high-income resource region move over to regime of renewable and nuclear energy. Low-income resource-rich countries fall back on fossil fuel-technologies and coal forming a significant share.

B2 Future: Regional economic regime with low level of international cooperation; would display increased concern over environmental issues compared with A2 but with a more insular, national perspective than B1. The emphasis is on developing local renewable resources.

The above four scenarios will be driven by forces of population, economic status, technology status, energy demand, and agricultural practices.

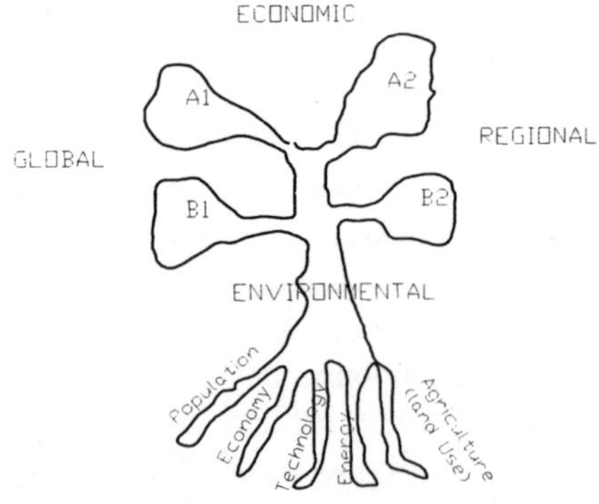

Fig. 10.1. IPCC SRES Scenarios.

IB. Future Indian Energy Policy Options

Some Innovations

Profile of Energy Crisis: Time has already arrived to seriously ponder what Indian energy options are. The energy scene is challenged by multiple complexities of climate change, energy security, depleting oil reserves and skyrocketing fuel prices, risk of nuclear power, etc. How should we order what options to choose and what directions to avoid? How to reconcile the various choices with the present day realities of economic competitiveness, equity and poverty alleviation and democratic consensus?

Whatever the set of options and challenges are addressed, we have to realise that the world is facing the prospect of a profound energy transition. Whether the ensuing energy transition is driven by climatic change, depleting stock of fossil energy, technological disruption, the energy economy is destined to undergo deep transformations in the coming decades.

Current Energy Scenario: Table 10.4 shows that in India (similarly in China), about 55 per cent population uses biomass fuel for most of their energy needs amounting to 200 million toe biomass, which works out to 30 per cent of the total primary energy consumption. Thus hydrocarbons and carbohydrate biomass together constitute about 96 per cent of total Indian energy, emitting carbon dioxide. In the Climate Change control regime such energy system is just not sustainable. Thus in spite of India's low per capita CO_2 emission, there is no option for India but to change over to a low-carbon energy regime.

Table 10.4A: Primary Energy Consumption (%) World & India 2010

	Coal	*Oil*	*Gas*	*Hydro*	*Nuc-lear*	*Bio-mass*	*other*	*Total (mtoe)*
World	26.3	31.2	21.3	5.9	4.9	10.4	0.9	12,460
India	36.8	22.2	7	3.5	.5	30	-	668

Table 10.4B: Electricity Energy Consumption (%) World & India 2010

	Coal	*Oil*	*Gas*	*Hydro*	*Nuclear*	*Bio-mass*	*Other*	*Total (TW-h)*
World	38	5.9	22	15.9	14.8	1.6	1.8	20,185
India	53.3	0.9	10.5	24.7	2.9	-	7.7	775

Table 10.4C: Electrical Generation Capacity (%) World & India 2010

	Coal	*Oil*	*Gas*	*Hydro*	*Nu-clear*	*Bio-mass*	*Other*	*Total (GWe)*
World	29.5	11	26.7	20.6	8.5	1.2	2.5	4539
India	53.3	0.9	10.5	24.7	2.9	7.7		152

With the economic well-being of the rural masses, there will be a steady shift away from the rural biomass chullas (domestic stakes), which are really very unhealthy, inefficient and inconvenient, raising the already high petroleum import. Such high import of petroleum will surely compromise our oil security.

Table 10.4B highlights that the thermal plants, mostly coal-based, generate 64 per cent of India's electrical energy. There is no doubt the situation will need to be mitigated with quicker switchover to hydro, nuclear and other renewable fuel driven power stations.

Table 10.4C shows that India, with a large population, has a very low electric power generation capacity at 152 GWe, which works out to about 700 KWh per capita compared to Chinese power consumption at 2000 KWh, while the world average is 2600 KWh per hour. Hence, India has to embark on a massive power build up to raise the economic well-being of the people. The present Integrated Power Planning Group of the Planning Commission projects a required power capacity of 800 GWe by 2032, which works out to 3000 KWh per capita on enhanced population base.

The carbon footprint of world primary energy and electrical energy scenarios are not much better than India's, but with their higher technological ability and larger basket of capital they are able to accelerate their transition to non-carbon energy regime. In the following paragraphs, we try to visualize how India could change to a new power economy with low-carbon or carbon-neutral energy paradigm, which could be produced from its own indigenous resources without compromising the national security.

Clean Coal Technology: In India, coal-based power at present constitutes more than 50 per cent of the generated electrical energy. Further Indian power generation for the large plants, will of necessity, will have to depend on coal, which is available indigenously or partially by import in absence of any other viable alternatives in the near future, say up to 2050. It is therefore most important that in order to contain the carbon emission all future coal-based power plants must switch over to clean coal technology.

The core of clean coal technology is the use of super-critical pulverised coal (SC-PC) system combustion or Integrated Gasification Combined Cycle (IGCC) to raise the plant efficiency to over 50 per cent. In IGCC, coal is first reformed with steam and oxygen to produce synthetic gas with provision for pre-combustion CO_2 removal (CCS) to first run gas turbine and in the second stage to run steam turbine. The IGCC is also a new technology as there are less than 100 working plants in the world.

Gas-based Power: More and more gas reserves are being identified in India and abroad. Newer sources of natural gas like shale gas, methane hydrate and coal-based methane (CBM) are getting discovered. The American Energy Information Agencies (EIA) estimates the present world reserve of gas at 400 trillion cubic metres, which will last 100 years at present consumption rate, outlasting oil, if not even coal. Gas indeed is a cleaner fossil fuel emitting only 0.45 gm/KWh compared to 0.87 gm/ KWh of coal. The plant efficiency can also be raised to about 60 per cent compared to 50 per cent of IGCC plant.

Nuclear Power: Next to coal and gas, nuclear energy offers the most viable electrical power supply system. It is also more or less carbon-neutral fuel and hence has generally been welcome in present CO_2-burdened global atmosphere. Nevertheless, there is still a section of people, who raises the question of nuclear power safety referring to two old nuclear reactor accidents at Chernobyl in former Soviet Union and Three-Mile in USA and the third recently at Fukushima, Japan. The majority of the scientific community today still believe that nuclear power is safe and will become safer with continuous improvement of technology both in operation and nuclear waste management.

No doubt, India is somewhat deficient in the availability of Uranium, the mother fuel of nuclear power reactor, so much so, that the Indian Atomic Energy Commission (AEC) has stated that only a power capacity of 10,000 Mw per annum can be sustained with the Indian Uranium. But India does posses a larger reserve of Thorium (U-232). For utilising this Thorium, India has developed a Three-stage Nuclear Power programme strategy employing three types of reactors: PHWR (Pressurised Heavy Water Reactor), FBR (Fast Breeder Reactor) and AHWR (Advanced Heavy Water Reactor). In the opinion of Dr. Anil Kakodkar, former AEC Chairman, a potential power capacity of even 600 GWe is sustainable following this route. For accelerating this strategy, the Indian Government has entered into new treaties with USA along the other members of the nuclear supply group, which has also been blessed by the International Nuclear Agency.

Hydro-Power: This is one of the best carbon-neutral power source. India has presently a hydropower capacity of about 40,000 Mwe, which can be raised to a 150,000 Mwe capacity using all the available hydel resources. Though the larger hydel power plant has raised the environment problem of rehabilitating large number of displaced persons; there is also adverse impact on flora and fauna.

Indian Long Term Energy Planning: With the above strategies, it appears that India, as a developing country, will be able to follow its own basket of technologically enhanced energy generation path with coal, gas, nuclear and hydro power and also keep its internationally committed limit of carbon emission. But India will have to improvise and import newer technologies in each area, perhaps other than hydropower. In the coal power, SC-PC technology is being introduced by the ten UMPP projects under the PPP (public-private partnership) schemes. India will now need to urgently import and embark upon the sophisticated IGCC technology with CCS. These twin technologies could make available coal power with very little GHG emission, within a decade. In gas power, we require large scale NGCC (natural gas combined cycle). Smaller NGCC units are already working in the Assam Electricity Board. Even without CCS, NGCC with high generation efficiency has low carbon footprint. Next comes the nuclear power, where India's famous three-stage – PHWR, FBR, AHWR – technologies are now steadily closing in to success. The imported LWRs under negotiation with the foreign NSG could provide the much needed plutonium-239 to jump start the FBR and AHWR stages.

In a study (see Table 10.5) presented in the inaugural issue of Journal 'Green Earth' edited by this author, it was observed that it was feasible to

Table 10.5: Projected Electrical Power Mwe India

Type of Power Sources	*2012*	*2022*	*2032*	*2052*
Coal/ Gas Thermal Power	135,000	260,000	470,000	660,000
Nuclear Power	7,300	20,000	100,000	600,000
Hydro Power	40,000	64,000	100,000	120,000
Solar Power	1,200	20,000	60,000	100,000
Wind Power	12,000	30,000	60,000	100,000
Other Renewable Energy	4,500	6,000	10,000	20,000
Total Power Required	**200,000**	**400,000**	**800,000**	**1,600,000**

provide a capita power of 3000 KWh per annum by 2032 by generating capacity of 800,000 Mwe, mostly from coal, gas, nuclear and hydropower, with the newer renewable energies like solar and wind contributing less than 10 per cent each.

The newer energy sources like wind and solar are natural renewable energies, for they continuously renew themselves and are inexhaustible over longer time; but they are limited over a time and a space. As long as these limits are not exceeded, there will not be any depletion of renewable energy.

Along with the development of the main energy sources – coal, gas, nuclear, and hydro, India has to also research and develop the newer energy sources like wind and solar, for they are totally carbon-free and their use will become mandatory globally in future. Wind energy is getting reasonably well-developed in India, being the fifth largest wind power country globally. However, there is need for further consolidation in technology, viability and a broad national policy is required to be enunciated. Solar power is really yet to be developed in India. West Bengal has taken some initiatives in a very small way. Fortunately, the Government of India has announced an ambitious Solar Energy Mission to produce 20,000 MWe solar voltaic energy by 2020 with an interim target of production of 1000 Mwe by 2012. There heightened activities in solar power sector are :

Hydrogen and Fuel Cells: A large part of the fossil fuel is burnt in the transportation sector, in automobile and railway locomotive, generating high volumes of sulphur oxides (SO_x) and nitrous oxides (NOx) and considerable carbon emission. Battery-driven electric automobiles have not yet been successful to replace the versatile fossil fuel. The current global research is now focused on hydrogen-fueled electrochemical cell, popularly known as Fuel Cell to generate electricity to electric motor. There is still considerable problem to produce hydrogen, which is not directly available in nature; further it is the known lightest element on earth; there is also problem of storing voluminous hydrogen in a vehicle.

This is then the story of the transition to new energies in this new century. It is a somewhat difficult transition, but the future energy evolution route through clean coal, gas, nuclear, hydro, wind, solar and finally hydrogen is slowly emerging. In the end the innovative spirit will triumph as always.

II. DEVELOPING CONCEPT OF ENERGY SECURITY- GLOBAL AND INDIAN

Energy Security – Some Concepts

The national energy policy, foreign policy and the national security are closely inter-linked. The failure to cultivating friendly foreign policy with oil producing countries may jeopardise the supply of energy fuel, which may greatly undermine the national security. Thus, a major component of the national security constitutes energy security.

The United Nations, as early as 1999, perceived energy security as requiring 'the continuous availability of energy in various forms and in sufficient quantities at reasonable prices.'

The energy security in terms of threat to supply at reasonable price may also include supply constraint or inadequacy, higher cost, import dependency, geopolitical conflict, absence of maritime safety of transport, natural calamities etc. In the context of India, the lack of energy security may affect our energy-driven growth and our quest remediating energy inequity i.e. poverty alleviation programme. Hence the Indian definition of energy security is more broad-based as enunciated by the 2006 Integrated Energy Policy Report of the Planning Commission:

'The country is energy secure when we can supply lifeline energy to all our citizens as well as meet their effective demand for safe and convenient energy to satisfy various needs at affordable costs at all times with a prescribed confidence level considering shocks and disruptions that can be reasonably expected.'

Multi-dimensional Challenges of Energy Security

Two important forces influence India's energy policy decisions and also pose challenges to the policy makers. The first dimension concerns *'energy and growth'*. The economy must be supplied with enough commercial energy to drive growth, to contain unsustainable consumption and to manage our ability to cope with high energy prices. The second dimension of the challenges concern India's energy needs to uplift the large Indian population afflicted with abject poverty. Basically the other face of the poverty is the large energy inequities and the need to manage the transition of the poor from their traditional biomass fuel sources to cleaner fuel; to provide lifeline energy required to eliminate poverty and provide a basic minimum standard of living to all citizens.

Arising out of the need to spur economic growth, there comes the necessity of increasing the domestic production of different fuels; this task also calls for the challenges of environmental challenges, both local and global. The secondary challenge arising out of this programme is the need to walk the political tight rope; the highly politicised and volatile nature of the international oil trade, particularly those of West Asia.

The second issue with energy security concerns the extreme 'energy poverty' of rural India. Energy access is a huge problem in Indian villages, where traditional biomass fuel dominates the energy mix. The rural energy deficit must be made good to pull out the masses from their chronic poverty.

Primary Energy Scenario

In the latest National Sample Survey of India (2000) it was found that 86 per cent of rural household continued to use biomass in the form of dung cake, firewood or wood chips for cooking. Even today only 5 per cent of rural households use LPG and 43 per cent are electrified (TERI 2005). While the penetration of modern energy sources was greater in urban households, 20 per cent still continue to rely on firewood and wood chips and fewer than half used LPG for cooking. The total contribution of traditional fuels at 6 EJ to total primary energy of 20 EJ remains very significant.

Table 10.6 Indian Primary Energy

Fuel Type	*Energy (EJ)*	*Energy (%)*
Oil	5.07	25.3
Coal	7.12	35.5
Gas	1.24	6.2
Nuclear	0.21	1.1
Hydro	0.30	1.5
Traditional	6.06	30.4
Total	20.00	100

Source: TERI 2005 (National Sample Survey 2002)

Energy for Growth

The supply of energy is the prerequisite of any developmental activities. The Indian Planning Commission has made a perspective plan of Indian growth envisaging a steady growth of Indian economy at 8+ per cent per annum from 2005 to 2030 for a period of 25 years. They have also worked out the incremental growth rate of primary energy to support this kind of growth of the real economy at 4.3 per cent p.a., if there is continuous improvement of energy efficiency. However, if energy efficiency is not assured and the developmental effort goes on business as usual mode, the annual energy demand will require to be grown at still higher rate of 5.1 per cent per annum.

In the above study, the total primary energy demand was estimated at 513 mtoe (i.e. million tonnes oil equivalent) or 21.86 EJ (1 ExaJoule is equal to 10^{18} Joules) in 2005. This amount of primary energy was required to be increased by about three times to 1536 mtoe by 2031. As the economic well-being of the rural people improves there will be shift from the traditional biomass fuel to the commercial fuel. Thus, the rate of growth of commercial energy will still be faster than the growth rate of the primary energy. This development has two implications, firstly the import dependency of energy will increase and secondly the fossil fuel consumption as a share of global consumption will have an upward trend from 3.7 per cent in 2005-06 to 7.6 per cent as shown in Table 10.7.

Table 10.7: Growth of Indian Energy Consumption

Year	*Primary Energy mtoe*	*Comm-ercial Energy mtoe*	*Import Dependency (%)*	*Fossil Fuel Consumption as % of Global Consumption*
2005-06	513	367		3.7
2030-31	1536	1351	33	7.6

Rural Energy Poverty

The Indian rural poverty has many dimensions. It is associated with many interrelated deprivations – lack of access to adequate levels of income, food, water, and clothing, shelter, sanitation, health care and education. Although, a definition will directly include energy as component, access to modern energy sources is a critical ingredient towards improving various dimensions of poverty we mentioned above. In this light, a number of studies have conceived 'energy poverty' as the absence of sufficient means of accessing adequate, affordable, reliable, high quality, safe and environmentally benign energy sources to support economic and human development (World Bank and UNDP 2005).

India is home to more than a quarter of the world's poor; it alone accounts for 50 per cent of the energy-impoverished people who have high dependence on traditional biomass fuel for cooking and heating, which causes more than 400,000 premature deaths (mostly women and children) in India annually.

To summarize, there is a huge deficit of modern energy in the rural sector, which is today mired in abject poverty. Electricity for households, other infrastructures and rural industries and LPG for domestic cooking need to be provided to eradicate poverty and modernize the rural sector. Thus, any National Energy Security programme shall include the large rural energy deficit.

Indian Energy Scenario – Contour of Change

The conspicuous Indian growth story during the last two decades has covered a wide range – from high economic growth rate, structural shifts from agriculture to services sector, population growth at 1.9 per cent p.a.,

Table 10.8: Growth of Indian Energy Demand

Growth Drivers of Energy Demand	*Extent of Energy Demand Growth*
Economic Growth	The rate of growth of Indian economy now hovers over 8 – 9 % p.a.
Structural Shift of Economy	Post-1991 reform, there is difficult structural shift away from Agriculture towards the Services sector
Demographic Trend	Growths of Population over last two decades have averaged 1.9 % p.a. Projected Growth in Urban population: 25.3% (1999); 40% (2030)
Expanding Transport Sector	In period 1980-2004: Petrol –driven 7.4% p.a. Diesel-driven 5.7% p.a.
Progressive Energy Transition	Biomass still major cooking fuel in rural India; Only 50% of rural households are electrified. Domestic sector consumes 25% of total electrical energy, whose share is growing.
Inadequate Indian Energy Reserves	India has 17% of world population, but only 0.8% of known oil reserves.

high growth of transport sector, slow but steady shift from biomass energy to commercial energy. A very brief contour of these changes is outlined in Table 10.8.

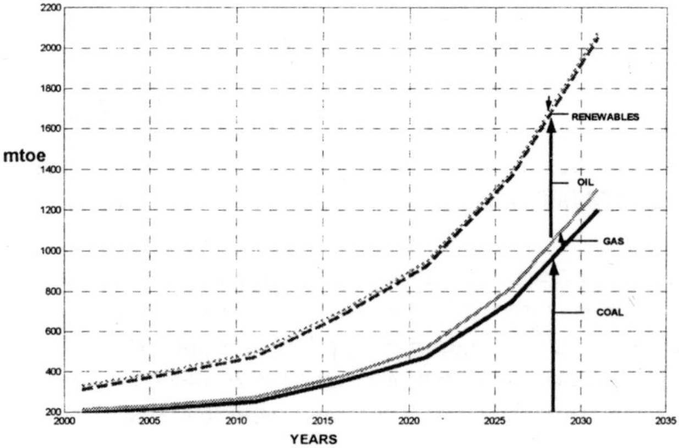

Fig. 10.2 Teri Projection of Indian Commercial Energy Consumption (BAU Situation)

Fig. 10.2 shows the Primary commercial energy mix is almost entirely composed of carbon-based fossil fuels coal, gas and petroleum oil. Of course, a significant part of energy supply is from the traditional biomass fuel (30.33%), which is not included in the commercial energy mix. However, in the electrical power sector, a growing proportion is from Hydropower (at about 25%) and nuclear power (4%). The total energy projections are on the high side, as the basis is on the business as usual. But all round today there is emphasis on energy efficiency.

Energy supply will be dominated by coal and oil in the foreseeable future in the next decade as presented by TERI 2006. However, significant development change is underway in nuclear power which is likely to grow with accelerated pace after India signed the Civil Nuclear Power Treaty with United States in 2009 in conjunction with the liberalisation of business with the Nuclear Supply Group of Countries. Developments in Renewable energy sector of solar, wind and solar power are taking place rather fast by large investments, spurred by the governmental incentive schemes.

Composition of Indian Energy Mix in Long Term Vis-à-vis Energy Security

Fig. 10.2 represents Indian energy growth projection per National Energy Map (TERI Report 2006) under the business as usual condition. There is another important energy growth model in the Integrated Energy Policy 2006 (Planning Commission 2006a). The Integrated Energy Policy (IEP) envisages a maximum growth of 4.5 per cent for renewable sources of energy, which may amount to about 100,000 MW of renewable energy over a span of about 25 years. Thus, both the TERI and IEP model projections show high dependency on fossil fuels of coal and oil, which will both be required to be imported to considerable extent. However, oil may be required to be imported to an extent of 90 per cent or more, jeopardising the energy security of the country. The other dimension of this catch-22 situation created by the near total dependency on fossil fuel is the worsening GHG emission scenario, which is very likely to be embargoed by the UNFCC Convention in the near future. In the following sections we will examine the nature of the coal and oil energies.

Indian Coal Power

The bedrock of Indian electrical power is the coal energy which constitutes over 53 per cent of the total energy capacity in 2009 as per Indian Ministry of Power. The coal consumption had risen to about 450 million tonnes by 2010. It is therefore not surprising that the IEP prognosticates 'coal shall remain India's primary energy source till 2031-32'. In fact, coal power may exceed 60 per cent by the end of this decade.

Possibly the single most important reason underlying India's increasing reliance on coal over the decades since independence has been that the indigenous reserves of coal are quite large – plentiful enough to sustain India's need for many decades. In fact, the proven coal reserves in India are estimated at about 101.8 billion tonnes as per Geological Survey Report (A K Sen 2010). The reliable *BP Statistical Review of World Energy* 2009 puts

India's proven reserve to production to 207 years. The coal industry in India stands nationalised since 1971-73; Coal India Ltd, the behemoth PSU is now the largest single coal production company in the world. However, in the recent years, the private thermal power plants have been allowed to own captive mines. Similarly till recently, the electrical power generation was solely monopolised by the State Electricity Boards and one major electricity generation Utility Company, National Thermal Power Corporation, is owned by the Central Government.

But since the enactment of the Central Electricity Act 2003, a major drive is on to privatise the power generation systems to boost the lagging power generation sector. Today a large number of power generation companies have come into existence. The new power generation companies have already surpassed the public undertakings in creating the new generation capacities. In the meantime the SEBs are gradually withdrawing or slowing down from the power generation business because of large losses they are incurring. One exception is the surging central PSUNTPC, which is dominating the power sector in competition with the private sector and is highly profitable.

At the time of independence in 1950, the total power generation capacity was about 1,500 mw. The power energy mix was 50 per cent thermal and 50 per cent hydropower. India was then vigorously pursuing rapid industrialisation which required more energy; coal based power technology, which was the oldest technology, simple and proven and was readily available, India opted for coal power in a big way. Now in 2009, with the total installed capacity of 1,52,360 mw, the coal thermal share of power plant stands at 53.3 per cent.

In power generation, modern technology is being introduced; all private sector UMPP s are to generate power by Super Critical Pulverised Coal (SCPC) technology. The adoption of this technology has been made compulsory by the Government of India for future power plants.

Till 1980 coal was mined underground employing large labour force. The labour unit cost was low; hence coal thermal power was still very competitive in price. As Indian mining switched over to open cast mining with mechanised shoveling and other labour saving devices, coal thermal power was still competitive in price with other power. Indian coal has generally high ash content and coal India is already supplying washed coal to the power plants; because of higher efficiency of washed coal, the power industry prefers washed coal only.

Oil–based Power for Transportation Energy

The vast road transport industry and partly Indian Railways are energised by petrol and diesel fuels and hence constitutes the largest consumers of petroleum oil-fuel. TERI 2006 Report estimates that the Petroleum product consumption of the transport sector which stood at 36 per cent in 2001 is likely to mount to 64 per cent by 2031-32.

The invention of the motor vehicles by Henry Ford has developed the way for the fastest means of transport by roads. Similarly, railways driven by steam engines and later powered by the diesel engine created the means of long-distance passengers and freight transport systems. People look upon them for convenience and necessity; today oil consumption for the transport sector has grown phenomenally. Now urban roads, cluttered with cars and buses, are named by the economists as the modern tragedy of the commons. For example, in Delhi, vehicle speed went down from 20-27 km per hour in 1997 to 15 kmph in 2002. In Chennai, the average speed in 2002 was just 13 kmph and in Kolkata 7 kmph.

Of course, Indian Railways are a monolith transport organisation but by and large it is serving the country well. However, trade and industry accuse railways of populism, raising goods freight rate to compensate the loss incurred due to artificially low passenger fares. On the overall, there is scope to improve efficiency of fuel energy and other capital resources.

But the real concern is felt as regards the urban motor transport. Indian cities have grown up unplanned with serious concentration of commerce and rich residence at the heart of the city and growth of moderately priced satellite townships in the outer periphery, some of which are situated quite far away. Regular and economically priced transport is an essential element of such urban life. In many of these cities such transport is provided by the government transport department or public sector companies, which in general are run inefficiently at a great loss to the exchequer, but guzzling valuable energy and causing wastages of resources in general. What is required is optimally designed private corporate enterprises which should be able to provide efficient regular services at economic price.

Fossil fuel is the prime energy provider of the transportation sector. There are two strong reasons to look for alternative transportation technologies, first to reduce the oil import which adversely affect the national energy security and secondly oil fuel emits highly undesirable green house gas (predominantly carbon dioxide) which is disastrously affecting the world climate. The search is on all over the globe for the green energy, which shall be carbon-neutral i.e. does not emit GHG. Renewable Green energy is required in any case as the fossil fuel shall be getting exhausted in not too distant future.

Among the possible alternatives are the electric-driven vehicles, where the electric power is provided by batteries. Where it is not practical to provide electric energy as the sole energy provider, hybrid car with electric drive and fossil fuel engine is the choice. A newer technology is under development which is hydrogen fueled Fuel Cell technology (FC). Some advanced technology car like Toyota Prius is already in use, but the capital cost is naturally high.

Indian Nuclear Power Perspective

That nuclear power is a viable source of non-carbon energy source other than the fossil fuel is well established. For France has 75.2 per cent of its electrical power supply from nuclear sources, while United States, Japan, United Kingdom and even Canada has sizable nuclear power as shown in Table 10.9. But the actual figure for Indian nuclear energy consumption is sill 4 per cent in 2010.

Table 10.9: Percentage Nuclear Power

Countries	*Percentage Nuclear Power*
France	75.2
Sweden	34.7
Germany	26.1
United States	20.2
United Kingdom	17.9
Japan	28.9
Canada	14.8
India	2.2

Source: IEA 2010

As per Ministry of Power, Government of India, India had nuclear capacity of 4120 MWe in 2009. In 2010 India has also signed the Civil Nuclear Treaty with USA and has now also become a member of Nuclear Supplier group. This will help India to implement the three stage nuclear power programme; it should now be theoretically possible to raise the indigenous capacity by 150,000 MWe by Fast Breeder (U-238) Reactor route and 300,000 MWe by Thorium (U-232) Reactor route by 2050. However, in spite of various options available, the civil nuclear power route with international cooperation has raised the most passionate debate on national sovereignty and safety. In the light of this debate, utmost transparency in the implementation of the nuclear programme will be welcome.

III. NATIONAL MISSION FOR ENHANCED ENERGY EFFICIENCY (NMEEE)

The National Action Plan on Climate Change was released by Honorable Prime Minister of India in June 2008. The Action Plan Outlines eight Missions, as indicated below, including National Mission on Enhanced Energy Efficiency (NMEEE).

1. National Solar Mission
2. National Mission for Enhanced Energy Efficiency
3. National Mission for sustainable Habitat
4. National Water Mission
5. National Mission for Sustaining Himalayan Eco-system
6. National Mission for a Green India
7. National Mission for Sustainable agriculture
8. National Mission for Strategic Knowledge for Climate Change

Enhanced Energy Efficiency is one of the most effective tools to very substantially reduce the electricity consumption and thereby reduce fuel consumption and bring down carbon dioxide emission. It will be mentioned later that the goal of the Indian energy efficiency programme is to save Rs. 74,000.00 crores or USD 25 billion in five years, which is a very large sum of money.

Four New Initiatives under NMEEE

The National Action Plan, para 4.2 on Climate Change enumerates following four initiatives under NMEEE:

- *Perform Achieve and Trade:* A market based mechanism to *perform* with cost effectiveness and to achieve energy efficiency in energy-intensive large industries and facilities, through certification of energy savings that could be traded.
- *Market Transformation for Energy Efficiency:* Accelerating the shift to energy efficient appliances in designated sectors through innovative measures to make the products more affordable.
- *Energy Efficiency Financing Platform:* Creation of mechanisms that would help finance demand side management programmes in all sectors by capturing future energy savings.
- *Framework for Energy Efficient Economic Development:* Developing fiscal instruments to promote energy efficiency.

Mission Goals: NMEEE

- Market-based approaches to unlock energy efficiency opportunities, estimated to the order of Rs. 74,000 crores.
- The Goals shall reach following goalposts by 2014-15:
  - Annual fuel savings in excess of 23 million toe
  - Cumulative avoidance of additional electricity capacity of 19,000 MW
  - CO_2 emission mitigation of 98 million tons per year

1. Perform, Achieve and Trade (PAT) Mechanism

(a) Specific Energy Consumption (SEC) Reduction Targets

- Targets would be percentage reduction of current SEC; Percentage reduction requirement based on current SEC as a ratio of the best in the sector
- Target setting for the power generation and fertilizer sectors through the existing tariff-setting processes
- SEC measurement and verification by BEE through accredited auditors.
- There are in all 685 Designated Consumers (DCs) in nine industry sectors as shown in Table 10.10.

Table 10.10: Number of Designated Consumers

Industry Sector	*No. of Identified DCs (Designated Consumers)*
Aluminum	11
Cement	92
Chlor alkali	21
Fertilizer	22
Pulp and paper	70
Power	154
Iron & Steel	110
Textile	197
Railways	8
Total No. of DCs	**685**

(b) PAT: Large Energy Usage Bandwidth : That there is vast scope of improving the industry specific energy efficiency is seen by the wide spectrum of energy consumption in Paper and Cement industry:

- Large bandwidth in specific energy consumption in all sectors (see Figs. 10.3A and 10.3B)
- In almost every sector, the most energy-efficient unit is also amongst the most efficient units in the world

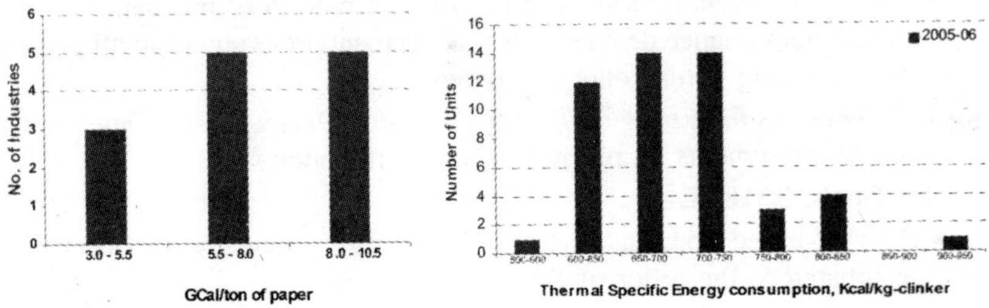

A. Cement **B. Paper**

Fig. 10.3 PAT–Huge Diversity in Specific Energy Consumption within sectors

The energy usage pattern varies widely in industries of a particular sector due to various diversities like

- Scale of Production (Installed Capacities)
- Use of Raw Material
- Process Technology
- Vintage of Technology
- O and M Practices
- Type of Product Output etc.

The factors of diversities are tabulated in Table 10.11

Table 10.11

Factors of Diversity	*Most Affected Sectors*
Raw Material Input	Pulp & Paper, Fertilizer, Power Plant, Textile
Quality of Raw Material / Fuel	All sectors
Process & Technology	Aluminium, Iron & steel, Chlor-Alkali, Paper
Final Product output	Textile, Iron & Steel, Aluminium
Vintage	All Sectors
Capacity Utilization	All sectors

(c) PAT - Energy Savings Certificates will incentivize actions.

- Issuance of Energy Savings Certificates (ESCerts) to units for energy-efficiency improvements in excess of their targets.

- EScerts can be traded, and used for compliance
- Trading can be carried out bilaterally, or on special platforms in the power exchanges
- Fungibility with Renewable Énergy Certificates can be provided.

(d) PAT – Current Status

- 685 units in 9 energy intensive sectors notified as Designated Consumers under EC Act. The sectors are: Aluminum, Cement, Iron and Steel, Chlor Alkali, Thermal Power Plants, Fertilizer, Pulp and Paper, Textiles and Railways.
- The studies for fixing of targets in all 9 industrial sectors are underway and will be completed by March 2010.
- The protocols for trading, monitoring and verification will be set up by Dec, 2010
- Amendments to EC Act passed by Parliament.

2. Market Transformation for Energy EfficiencyProducts

(a) MTEE - New Initiatives

- Leveraging international funds for promoting energy efficiency.
- Implementing a National Energy-Efficiency Road map.

(b) Market Transformation for Energy Efficiency – Current Status

- The Programme of Activities (PoA) to leverage CDM for agriculture and municipal sectors under development.
- The Bachat Lamp Yojana PoA got CDM registration in June 2010.
- Super Efficient Equipment Programme (SEEP) under development; would provide financial incentive to manufacturers for the early sale of "super efficient" products.

3. Energy Efficiency Financing Platform (EEFP)

(a) EEFP – New Initiatives

- Ensuring availability of finance at reasonable rates for energy efficiency project implementation- Expansion of EEFP to include other FIs and public and private sector banks.
- Create demand for energy efficiency products, goods and services- awareness, public policy, facilitation/ stimulation by preparation of bankable projects and markets.
- Credible monitoring and verification protocols to capture energy savings.
- Capacity building of banks and FIs

(b) EEFP – Current Status

- Energy Efficiency Services Ltd. (EESL) has been created as a corporate entity to provide market leadership.
- Investment-grade energy audits being prepared for government buildings, municipalities, and SMEs.
- SIDBI has initiated exercise for financing projects in SMEs.

4. Framework for Energy-Efficient Economic Development (FEED)

The scheme provides for the following:

- Providing comfort to lenders by provision of:
    - Risk guarantee for performance contract Partial Risk Guarantee Fund (PRGF).
    - Venture Capital Fund for Energy Efficiency (VCFEE).
    - Initial seed capital from Government budget-can be expanded by contributions from other agencies as well.
- Incentives to Central Public Sector Undertakings (CPSUs) to take up energy efficiency.
    - Policy guidance to CPSUs to take up energy efficiency project-special parameter (to be called Energy Efficiency Performance Index (EEPI) on the line of KPI may be added to the MoU guidelines from 2010-11.
    - Promoting Energy Efficient Public Procurement.
- Tax / Duty Exemptions for Promotion of Energy Efficiency.
    - Graded excise duty for STAR labelled equipments in favour of higher efficiencies.
    - Income and Corporate tax incentives for ESCOs/ Venture Capital funds, etc. in energy efficiency.
    - Providing infrastructure status to ESCO business.
- Guidelines for procurement of energy efficient appliances and services by public authorities under consideration of Government.

References

Diwan, P. and Sarkar, A.N. (2009), Energy Security, Pentagon Energy Press, New Delhi.

Dullon, G. and Page, M. (2008), 'The Prospect for Electricity and Transport Fuels to 2050 in Future Energy, ed. Trevor M Letcher, Elsevier.

Green Earth, Vol. 1, No. 2, July-Dec. 2010, Editorial Millennium Institute of Energy and Environment, Kolkata. www.mieem.com

Government of India, Indian Planning Commission: Summary of Integrated Energy Policy 2006

Ligia Noronha and Sudarshan (Ed.) 2009, *India's Energy Security*, Rutledge Contemporary South Asia.

National Mission for Enhanced Energy Efficiency, Dr. Ajay Mathur, Director General, BEE.

http://www.bee-india.nic.in

Scrase, I. and Mackerron, G., (Ed.) 2008, *Energy for the Future – A New Agenda*, Palgrave McMillan.

Singh, B., 2010, *India's Energy Security*, Pentagon Energy Press, New Delhi

www.shell.com/scenarios.

PART II: GLOBAL BASKET OF VIABLE GREEN ENERGY

11

An Outline of Proven Clean Fossil Energy

Clean Coal and Gas Power in New Century

Introduction: World Energy Scenario

Two major fossil fuel energy sources, coal and petroleum, dominate the world scenario today. The transportation sector - automobile, railways, aviation and shipping are all served by the petroleum-fuel. Coal serves mostly the world electrical power sector. Both these fossil fuels heavily pollute the atmosphere with GHG emissions. However, the power sector is also served by natural gas, which emits considerably less GHG than coal and oil. Secondly, the hydro-energy and nuclear energy, both of which are generally carbon-neutral, also serve the power sector. Biomass energy is serving the primary energy needs, particularly of the developing countries in the domestic sector. The common renewable energies – solar and wind power – though the basic technology is developed – have not reached the level of economic pricing and achieved supporting ancillary industry status of today's dominant energies coal, oil and natural gas and also hydropower and nuclear power. The much celebrated Hydrogen energy and its novel engine Fuel Cell will perhaps still require some decades more of development. Hence our prescription for the energy needs of the new 21st century is to continue to use coal with the improved clean coal technology and increase the proportion of Gas energy as much as possible. The coal and gas power can be further cleaned of GHG emissions by Carbon Capture and Sequestration (CCS) technology. Similarly develop more and more hydropower. These are all tested energy technologies. Further develop and add more and more nuclear power. In the medium term, develop more and more renewable energies – solar and wind and bio-fuels – bio-ethanol and bio-diesel. In the longer term let intensive research be directed to hydrogen energy.

IA
Coal Power

Most of the world power generating stations are coal-based Steam Turbines. In the Fig. 11.1 and Table 11.1A, we note that the coal power constitutes 37.5 per cent of the total generating capacities.

Table 11.1A: World Major Electrical Energy Sources

Power Source	*Energy (GW-h)*	*% Total*
Coal	3,925,766	37.5
Gas	2,181,239	20.8
Oil	3,181,239	3.7
Nuclear	2,262,217	21.6
Hydro	1,296,291	12.4
Biomass, Waste, Geothermal, Solar, wind, Tide	402,034	4
Total	**10,449,059**	**100**

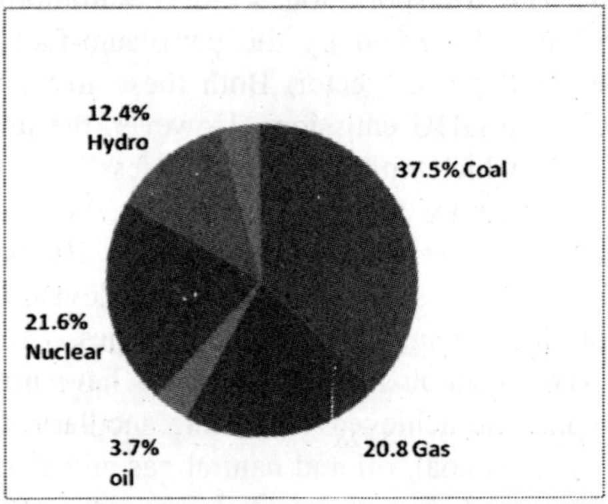

Fig. 11.1 Electrical Power Generation by Sources

Source: IEA/OECD, Paris, 2007.

But the world's major power generating nations are more heavily dependent on coal power, for example, US (51%), China (79%), India (53%), Australia (77%), South Africa (93%), Poland (95%) and Germany (51%) (Nersesian 2009). In Table 11.1B, it is seen that the renewable and sundry energies at about 4% have a long way to develop before the world can think of scaling down the coal power. Further, most of the Process Industries and chemical plants are operated on coal-based steam systems. Because of their

Table 11.1B: World Renewable Electrical Energy Sources

Power Source	*Energy (GW-h)*	*% Total*
Biomass	151,651	1.45
Waste	62,949	0.6
Geothermal	29,228	0.28
Solar PV	3,911	0.04
Solar Thermal	681	0.01
Wind	149,405	1.43
Tide	550	0.01
Other	3,659	0.03
Total	**402,034**	**3.85**

heavy carbon emissions, we may wish to replace them, but that appear well neigh impossible for the whole industrial – power complex will come to a grinding halt. There will be economic catastrophe. The obvious inference is that we have to choose one way or the other control the carbon emission and let the coal power definitely continue to serve the world economy, though in gradually decreasing proportional role, till at least 2050. The solution lies in adopting the Clean Coal Technology (CCT), which must include the Carbon Capture and Storage (CCS) Technology, wherever possible.

Coal – its Origin

Some 300 million years ago, Coal was transformed from biomass by the tectonic convulsion during the Devonian and carboniferous geologic epoch burying the biomass-rich swamps which then covered most of the earth. This resulted in withdrawal of huge amount of carbon from the atmosphere, creating a more benign environment for growth of animal life. In later periods also, coal continued to be formed, as some coal samples seemed to be only a million years old. Depending upon its age and maturity, the carbon content of coal varies and coals are classified in four basic grades globally into

Table 11.2A: General Grading of Coal

Type of Coal	*Carbon Content %*	*Approx. Heating Value Kcal/kg (KJ/kg)*
Lignite (soft)	25 – 35	3,000 (12600)
Sub-bituminous	35 – 45	5000 (20,000)
Bituminous (hard)	45 – 86	7000 (30000)
Anthracite	50 – 98	8500 (35000)

Source: Sen 2010.

Lignite (soft), Sub-bituminous, Bituminous and Anthracite (hard) as shown in Table 11.2A. The official grading of Indian non-coking coal is graded into A, B, C, D etc as shown in Table 11. 2B. The Grade A has the highest heating values but the least (Ash + moisture) contents.

Table 11.2B: Grading of Indian Non-coking Coal

Grades of Indian Coal	*Useful Heat Value (UHV) Kcal/kg (MJ/kg)*	*(Ash + Moisture) %*
A	Exceeding 6200 (26)	Not exceeding 19.5
B	6200-5600 (26-23.4)	28.8
C	5600-4900 (23.4-20.5)	34
D	4900-4200 (20.5-17.5)	40

Source: Sen 2010.

Coal Assisting Industrial Revolution

Though China seemed to have first used coal during 1000 BC, the use of the coal in the modern times came about during 1200 CE in England enabling it to emerge as the World's greatest seafaring and trading nation. By 1690, Britain was supplying Europe almost all of its coal needs. The invention of steam-powered pump for dewatering coal mines by Newcomen in 1725 and the development of a much improved steam engine by James Watt in 1757 and its further very substantial developments thereafter in England led to the huge contribution of coal towards the industrial development. Britain thus arrived in the Industrial age.

Current Global Coal Reserve, Production and Reserve Life

The global coal economy involving coal mining, processing and consumption for power generation is of enormous size and India and China are playing very conspicuous roles in it. Unlike the petroleum oil, coal is also well distributed over large areas of the earth. The Table 11.3 shows the country wise distribution of coal reserves in 2006:

The second point we may note from Table 11.3 is that the coal reserves are also quite healthy. At the current rate of consumption, the world reserves may last 142 years, while the reserve life years for USA 224, India 114 and even for China with its excessive current coal production it is 41 years. Russia has an unbelievably long reserve life of 481 years. Hence as energy source, coal is a dependable energy resource. Hence if by using clean coal technology, we can control CO_2 emissions by Carbon Capture and Sequestration Technology (CCST) and indirectly by raising the energy conversion efficiency, the world economy will be saved from extreme energy dislocation by hunting for unknown energy sources, so fast and so soon.

The country wise coal production is shown in Table 11.4

Table 11.3:Proven Recoverable Coal Reserves end 2006 (million Tonnes)

Country	*Bituminous and Anthracite*	*Sub-Bituminous and Lignite*	*Total (Million Tonnes)*	*Share (%)*
U S A	111,339	135,305	246,6643	27.63
Russia	49,088	107,922	157,010	7.27
China	62,200	52,300	114,500	12.60
India	90,085	2360	92,445	10.17
Australia	38,600	39,900	78,500	8.63
South Africa	48,750	0	48,750	5.36
Ukraine	16,274	17,879	34,153	3.75
Kazakihstan	28,151	3,128	31,279	3.44
Poland	14,000	0	14,000	1.54
Brazil	0	10,113	10,113	1.11
Germany	183	6,556	6,739	0.74
Others				8.24
World	**478,771**	**430,293**	**909,064**	**100**

Table 11.4: Production of Coal by Country

Country	*2008 Million tons*	*Share %*	*Reserve Life Years*
China	2782.0	41.02	41
USA	1062.0	16.66	224
EU	587.7	8.67	51
India	521.7	7.69	114
Australia	401.5	5.92	190
Russia	326.5	4.81	481
South Africa	250.4	3.69	121
Indonesia	229.5	3.38	19
Germany	192.4	2.84	35
Poland	143.9	2.12	52
Others	283.6	4.18	
Total World	**6781.2**	**100.00**	**142**

Source: Nersesian 2008.

IB
Clean Coal Technology

Here are three basic ways to mitigate the environmental impacts of combusting coal:

1. Reducing or eliminating sulphur and nitrous oxides (SO_x and NO_x) and solid particulates;
2. To improve combustion technologies to increase efficiency, thereby to reduce coal input and hence the emission gets reduced;

3. To reduce or eliminate CO_2 emissions with Carbon Capture and Sequestration, which is an emerging technology.

Removal SO_x and NO_x

The Sulphur oxide removal is generally effected by spraying Lime solution on the exhaust gas, whereby nearly 90-95 per cent of the sulphur gets removed as slurry of Calcium Sulphate. It is a difficult proposition to isolate Nitrous Oxide (NOx). Nitrogen constitutes 80 per cent of air and at the high temperature of combustion it invariably gets oxidised. We will see that in the Fluidised Bed Combustion (FBC) method, because of lower temperature of combustion, the formation of NO_x is much controlled. In the Integrated Gasification Combined cycle (IGCC) operation, the oxygen of the air is separated and hence there is no NO_x pollution in the exhaust.

Choosing Combustion Technology to Increase Efficiency

There are today three efficient ways of generating electricity from coal combustion with high efficiency:

1. Pulverised coal combustion with supercritical boiler technology (SCPC);
2. Fluidised Bed Combustion (FBC);
3. Integrated Gasification Combined cycle (IGCC);

The Supercritical Pulverised coal (SCPC) boiler technologies have been used worldwide in recent times, even in developing countries like China, in large scale boilers of 800 Mwe capacities or above, offering efficiency of 45 per cent with hard coal; with further improvement in technology, the efficiency is likely to upgrade to 48 per cent. Even with lignite, supercritical pulverised combustion can attain 43 per cent efficiency. For the basic simplicity of technology with high efficiency, this is today the preferred choice for large scale plants for the coming decades.

Fluidised bed combustion (FBC) boilers (including bubbling, circulating and pressurised boilers) are suitable for smaller but high ash coal, presently operate at 40 per cent efficiency, which is likely to rise to 44 per cent. FBC technology has unique advantages in pollution control; it operates at much lower temperature than a conventional or a SCPC system, thus greatly reducing the amount of NO_x formed. It can also reduce the SO_x formation by injecting Lime in the combustion temperature.

The Integrated Gasification Combined Cycle (IGCC) is more complicated than the FBC. In the first stage, coal is gasified with steam and oxygen. For the process, oxygen is separated from nitrogen by a cryogenic air separation plant. The coal is milled and dried in prior to being treated with oxygen and steam. The synthetic gas produced is mixture of carbon monoxide and hydrogen, which are then cleaned of any solid and sulphur by the usual process. The combusted synthetic gas is then used to run gas turbine and the exhaust gas is used to generate steam which runs the steam turbine. The use of combined gas and steam cycles to generate electricity enable the high overall efficiency of 50 per cent with lesser generation of solid waste, sulphur, nitrous oxide and carbon dioxide. The carbon dioxide emission from an IGCC plant is pure and can even be sold and if not, it is to be captured, sequestrated and stored underground.

In Table 11.5 we show the average efficiencies and other Characteristics of different combustion systems in coal-based generation plants.

Table 11.5: Characteristics of Different Combustion Systems of Coal-based Power Plants

Generation Technology (Coal Combustion Technology)	*Energy Consumption KWh Fuel per KWh Electricity*	*Net Efficiency*	*CO_2 Emission Kg/KWh*	*CO_2 Redu-ction %*
Hard Pulverised Coal fired Power Plant (PC)	2.63	0.38	0.87	0
Above with Supercritical Boiler Technology (SCPC)	2.22	0.45-0.50		
Fluidized Bed Combustion with Pulverised coal PFBC	2.44	0.41-0.45	0.80	8
Integrated Gasification Combined Cycle (IGCC)	2.38	0.42-0.50	0.79	9

IC
Carbon Capture and Storage (CCS)

CCS is really a part of the Clean Coal or Gas Technology. The basic technology aims to sequester or separate the CO_2 from the emissions from a coal/gas power plant, to capture them, transport them to a distant or near site and store those safely on long term basis under ground. The CCS could be operated after the combustion, as in Conventional PC Boiler or before the combustion, as in FBC or IGCC.

Thus, there are in principle three basic processes of carbon separation Systems: (i) Post-Combustion; (ii) Pre-Combustion and (iii) Oxyfuel Process as briefly reported below.

Post-Combustion: In post-combustion CCS, only flue gas is processed for CO_2 sequestration. The disadvantage is the high volume flow of flue gases at very low concentration of carbon dioxide (3 to 15 per cent) and at low atmospheric pressure. The CCS set-up will necessarily be large with large compressor. The problem may be partially mitigated in advanced type of plants where combustion takes place under pressure. Such advanced plants are viable for larger coal, gas or oil fired power plants of capacity 5000 MWe and also natural gas combined cycle Plants (NGCC of more than M00 GWe) and also some pulverised coal plants of 300 MWe.

Pre-combustion: The pre-combustion CCS is applicable to gasification plants like IGCC, or NGCC. The advantage here is the operating pressure is high and CO_2 concentration could be as high as 40 per cent. In both, fuel is gasified using water vapour and air usually enriched in oxygen to produce Synthetic gas (mainly H_2 + CO). A so-called gas shift process is applied to the synthetic gas, reacting with water vapour and transforming most of CO into additional CO_2 and H_2. The CO_2 capture process may then be applied immediately downstream of the gas shift and upstream of the combustion turbine where the gas is still very concentrated (say 40% CO_2) and under pressure (say 15 bars).

Oxyfuel Process: Oxyfuel processes use oxygen instead of air for combustion thus producing a flue gas that is practically nitrogen free and highly concentrated in CO_2 (more than 80%), the other major component being water vapour. There is no need for a complex CO_2 plant capture process, since a simple condensation of water would suffice, obtained through cooling and compression. The recovery is naturally very high.

Carbon Dioxide Capture Technologies

Carbon dioxide capture has been applied in the past, not for the storage, but for sweetening of natural gas (which also often involves removal of H_2S and other Sulphur compounds).

In the important post –combustion process, the composition of the flue gas may have a strong influence on the choice of capture technology. The CO_2 content may be as low as 3% (a typical NGCC) to about 15% (coal-fired plants). Besides CO_2, N_2, O_2 and H_2O there may be impurities such as NO_x, SO_x, HCl, Mercury and other metal vapour, particles and various organic and mineral traces compounds etc. Accordingly, in the post- Combustion systems, other unit operations such as electrostatic precipitation, desulfurisation and washing may be used in addition to CO_2 recovery.

There are three broad processes for carbon capture and separation:

1. Sorption Process (includes sorption in liquid solvents and solid sorbent or adsorption);
2. Membrane Separation Process;
3. Cryogenic Distillation Process.

Absorption in Liquid Solvents

The process is based on continuous contact between the flue gas and a liquid solvent flowing in the counter direction to the gas flow. The solvent is regenerated by heat or depression in a stripper, where the pure CO_2 is released. The solvents may be chemical (usually an amine), where weak acid/ weak base reaction takes place; with physical solvents the capture of CO_2 is due to solubility of the media. Some typical absorbents are shown in Table 11.6.

Table 11.6: Some CO_2 Absorbents

Solvent Types	*Solvent Name*	*Commercial Name*	*Thermodynamic State*
Chemical	Mono-ethaloamine (in water)	MEA (Kerr-McGee)	Absorption: 40-50C
Chemical	Methyl di-ethylamine (in water)	MDEA (BASF)	Generation: 120-140C
Physical	Dimethyl Ethers of Polyethylene glycol	Selexel (Unilever)	Absorption: ambient temp. Regeneration: -10 C 0.2 bar

Adsorption: Adsorption on activated carbon and/or molecular sieves (zeolite) is a common technique for removing CO_2, together with sulphur compound and hydrocarbons present, whereas light components like Na, O_2, CO will be less retained. Adsorption takes place at room temperature and desorption can be achieved with superheated steam. The purity of carbon dioxide is high.

Membrane Separation: Membrane filtration allows permeation of unwanted gas and retention of the desired gas. A number of potential membranes are available as shown in Table 11.7. Inorganic and ceramic

Table 11.7: Some Membrane Characteristics

Membrane Material	*Type of Separation*	*Separation Factor*	*Temperature Range*
Alumina Ceramics	H_2 permeate CO_2 retentate	15	< 500
Silica Ceramics		15	400
Carbon Molecular Sieves		20	400
Zeolite		50	500
Metal		100	600
Perovskite Ceramics			> 800
Polyethylene-glycol	CO_2 permeate/ N_2 retentate	> 60	< 120

membrane can withstand high temperature and pressure and is also corrosion resistant. They can be made selective for CO_2/H_2 separation, with hydrogen permeating preferably.

In post-combustion, the key separation is CO_2 / N_2 and dense polymeric membrane must be considered. They can be made more selective at the cost of permeability.

The principles of carbon dioxide technologies – Absorption, Adsorption and Membrane separation are illustrated diagrammatically in Fig. 11.2A, 11.2B and 11.2C.

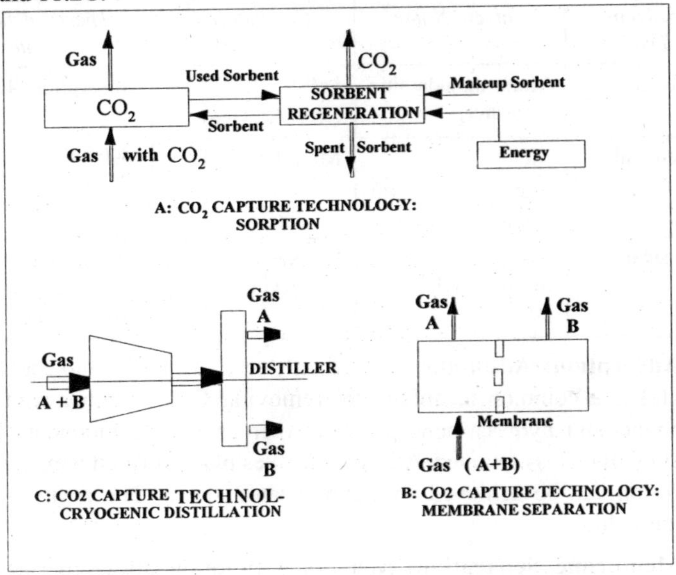

Fig. 11.2: Important Carbon Dioxide Sequestration Technologies

Geological Storage of CO_2

What is alternative to the CO_2 emission polluting the atmosphere gravely precipitating the climate change? The likely solution appears to store the captured carbon dioxide in a number of geological voids deep inside the earth. Such potential geological storages have been identified as the depleted oil and gas mines, empty coal mines and deep saline aquifers, etc. Some of these underground storages have already been tried in enhancing oil and gas recovery by pumping CO_2 in the depleted oil fields and coal mine fields. The character of carbon dioxide storage is somewhat different as the CO_2 has to be stored in higher pressure say about 2000 psi and the storage has to be on considerably long term basis. The IPCC has also tried to identify such geological storages globally as indicated in Table 11.8A.

Table 11.8A: Geological Storage Potential of CO_2

Storage Type	*Storage Technology*	*Global Potential (Gigatonne CO_2)*	*Cost USD/t CO_2*	*Technology Status*
Geological Storages	Depleted Gas and Oil Fields	675	0.5 - 8	Tried
	Enhanced Oil Recovery (EOR)	35		Tried
	Enhanced Gas Recovery (EGR)	80		Tried
	Coal-based Methane Recovery CB M	20		Tried
	Deep Saline Aquifer	1000		Speculative
Ocean Storages Mineral Carbonization		Not Applicable	unknown	Under research

Sources: IEA (2004): Prospects for CO_2 Capture / storage; IPCC (2005): Special Report: (Metz B *et al*)

Cost Consideration of CCS Technology

The UN Inter-Governmental Panel on Climate Change gives indicative price increases for incorporating CCS in New Power Plants in Table 11.8B (Tondier and Teng 2008).

Table 11.8B: CCS for New Power Plants: Cost

Type of Power Plants	*NGCC + CCS*	*PC + CCS*	*IGCC + CCS*
Capture Technology	MEA	MEA	Selexol (PEG)
Energy penalty %	10 – 25	25 – 50	15 – 25
Increase in Electricity costs %	35 – 80	60 – 100	20 – 80
Cost of CO_2 captured USD/t	35 – 60	30 -70	10 – 35

NGCC: Natural Gas Combined Cycle; PC: Pulverised coal; IGCC: Integrated Gasification Combined Cycle;

MEA: Meno-ethanol-amine; PEG: Polyethylene Glycol; *Source*: IPCC (2005)

There is as yet not much direct experience of combining capture, transport and storage of CO_2 in a fully integrated large scale power plant system. However, what IPCC provides in Table 9.6B is broad indicative costs, whose reliability is naturally unproven. The bases of IPCC projection data are believed to be drawn from the followings:

- Newly built large scale plants for electricity generation are considered with 85% recovery of CO_2.

- Reference plants without capture are of the same type.
- Storage is geological.
- Transport costs are of the order USD 1 – 8 per tonne of CO_2.
- Geological storage costs range from USD 1 to 10 per tonne.
- Energy penalty is indicated in the Table 9.6B.

II
Gas Power

Types of Gas Fuels

Natural Gas (this is the main source); Petroleum Gas and LPG; Coal Gas (Synthetic Gas); Coal Bed Methane Gas; Shale Gas and Methane Hydrate are the principal types. We will study the following aspects of the natural gas in the first instance:.

- Sources of Natural Gas
- Natural Gas: Processing, Transmission and Storage
- Trading of Natural Gases and Transmission Companies
- Natural Gas as Clean Energy Fuel
- World Reserve, Production and Consumption of Natural Gases
- Overview: Indian Gas Prospects

Natural Gas: Cleanest Fossil Fuel

Natural Gas is the cleanest fossil fuel. Its carbon footprint is about 40 per cent lower than that of coal and 20 per cent lower than that of oil as shown by the values of carbon emission per MJ of fuel energy (see Table 11.9A)

Table 11.9A: Carbon Footprint of Fossil Fuels

Fuel	*Coal*	*Oil*	*Gas*
Carbon Content gm/MJ	26.8	20.0	15.3

Source: IPCC, 1997

Hence by switching over to gas power, the world and India could reduce the carbon emissions very significantly in fulfillment of the incoming UNFCC mandate.

We also see from Table 11.1A that nearly 21 per cent of world power is already contributed by natural gas based thermal plants. Table11.9B gives the consumption of natural gas by some leading industrialised regions. Hence the use of natural gas in the generation of power shall increase significantly, along with clean coal power to meet the developmental needs, before viable renewable energy sources are adequately established.

Types of Gas Fuels

Presently natural gas, principally methane, is mainly produced from natural gas wells, which compared to oil, are relatively well distributed all over the world. The world reserve of natural gas, as it is, about twice the world oil reserves. But there are three other prospective sources of natural gases e.g. Coal Bed Methane, Shale gas and Methane Hydrate, which once developed could dramatically alter the world methane gas reserves. Thus, gas power could play a much bigger role in the new century.

The coal gas, a synthetic gas, is obtained by steam reforming of coal and is composed of carbon dioxide and hydrogen. The use of coal gas, for lighting and others, is older than the natural gas. But its heating value is however half of that natural gas, which limits its prospect.

Liquified Petroleum gas (LPG) is primarily propane and butane. LPG is produced as a byproduct of oil refining and is also stripped out of gas liquids from natural gas. The LPG has great application in the domestic cooking and heating of the developed world, where pipeline natural gas is not easily available see Table 11.9B.

Table 11.9B: World Consumption of LPG million tons (2000)

Countries	*Consumption*	*Imports*	*Exports*
USA	51	1	1
Europe	31	15	7
Japan	15	15	
Korea	7	5	
China	13	5	

Source: Nersesian 2009, p. 254

Natural Gas: Processing, Transmission and Storage Natural Gas is primarily composed of methane; when combusted it produces carbon dioxide and water vapour. But natural gas could come out of a well as an admixture of other hydrocarbons in smaller proportions. For instance, raw natural gas from the wells in Southwest United Sates contain 88 per cent methane, 5 per cent ethane, 2 per cent propane, 1 per cent butane and other heavier hydrocarbons in traces and some impurities. Heavier hydrocarbons of pentane, hexane and heptanes are in gaseous state under pressure in a natural gas reservoir but condense to a liquid when brought up. The condensates are separated out from the natural gas and diverted to oil refineries. Ethane is difficult to separate out because of its quite low temperature of liquefaction and allowed to remain in the natural gas. But propane and butane are more easily separated by fractionation or cooling of natural gas and are separated for conversion to LPG. The impurities such as hydrogen sulphide, carbon dioxide, nitrogen and water are required to be removed before marketing to consumers. Thus, the 'pipeline quality' natural gas is primarily methane and some ethane.

The processes of cleaning and stripping of the gas are carried out at the well head and finally the gas is compressed to the transmission system pressure and delivered to the transmission systems. The transmission pipelines typically are 24 to 36 inches in diameters and the operating pressures are 600 to 1200 psi (40 to 85 atmospheres). The compressor stations are located about every seventy miles and the speed of gas varies from 15 to 30 miles per hour. Monitoring facilities are strategically situated about every 5 to twenty miles to enable detecting pipe rupture, leakage and carry out routine maintenance.

Storage facilities are also built up to cater to the rise and fall in demand or even in prices in US. In US they are as high as 30,000 miles of pipelines.

Trading of Natural Gas and Transmission Companies The US has been a mature market for gas trading with very high consumption of gas and many miles of inter-connecting pipelines crisscrossing the country. Gas planning for a transmission company depends upon long and short term forecasting models. Long term forecasting determines investments to ensure that pipeline system can meet future demands. Short term forecasts ensure that the volume of gas can meet current demand.

In US, Natural gas can be procured from several options: directly from the gas producers or from the interstate transmission companies via their hubs. Hubs connect various interstate pipelines in a common system whereby natural gas consumers connected into one interstate pipeline can obtain supplies from the natural gas suppliers hooked into another interstate pipeline.

Hubs not only allow for common distribution but also common pricing. There are about a dozen major hubs in the United States, the most important being the 'Henry Hub' in Louisiana, a multitude of pipelines connect natural gas suppliers and interstate pipelines into what amounts to a huge common system. The Henry Hub is the most important distribution and pricing hub in the United States; its price is the base price for US. Other hubs are generally priced at the Henry hub price plus transmission costs with local market-related variations. The New York City is connected by three interstate pipelines allowing major gas purchasers to negotiate and bargain prices from those three pipelines.

Another pricing hub is in Boston where gas from two interstate pipelines from the US Southwest and two pipelines from the western and eastern provinces of Canada are interconnected, providing major buyers with opportunities to buy gas for the best price from suppliers in four natural gas producing regions.

Whereas natural gas consumers are doing their best to buy cheapest, the natural gas producers are doing their best to obtain the best and the highest price.

The demarcation point between the transmission and distribution is the city gate, where regulators reduce gas pressures, scrubber and filter any remaining traces of impurities and water vapour and finally *mercaptan* is added as an odorant to the gas to detect any gas leak.

Natural Gas as Clean Energy Fuel

The Natural Gas consumption market developed first in the United States and the global consumption pattern followed US market pattern as shown in Table 11.10 The only distinguishing feature between the two consumption markets is that the industrial consumption is higher in US at 35 per cent to 25 per cent globally, but the use of natural gas in electricity generation is higher at 35 per cent globally to 25 per cent in US.

The electricity generation is the fastest growing market segment of natural gas in both US and other world markets. The technologies of generation include conventional steam turbine, combustion turbine, and combined cycle plants. The steam turbines are similar to the conventional coal or oil burning units, but the boiler design details will vary; the thermal efficiency will be 25 – 35 per cent for the older plants and 35 – 40 for the newer plants. The combustion turbines are really modified versions of the aero jet engine running electric generators requiring low weight turbines for supplying peak load and require less investment but less efficient. The combined cycle plants direct the escaping exhaust gas from a combustion turbine through a steam generator to drive a second electric generator; their

Table 11.10: Consumption of Natural Gas: US & World

United States Various Sectoral Users	*Consumption*		*World Various Sectoral Users*	*Consu-mption*	
	tcf	*%*		*tcf*	*%*
1. Residential Customers (61 millions) (space and water heating, cooking, clothes drying, pool heating, gas fireplaces)	5	25	1. Residential Customers		25
2. Commercial Customers (restaurants, motels and hotels, retail establishments, Hospitals and healthcare facilities, offices, Government buildings)	3	15	2. Commercial Customers		15
3. Industrial Customers (21 thousand) (manufacturing industry, waste treatment, incineration, drying, humidification, process, fertilizer industry, chemical, petrochemical)	7	35	3. Industrial Customers		25
4. Electricity Generation (5,700 units)	5	25	4. Electricity Generation		35
5. Transportation	-		5. Transportation	-	-
US Total	20	100	World Total		100

Source: Developed from Nersesian 2009, pp. 247-248. Tcf – trillion cubic feet.

efficiency is higher at 50 per cent. Where there is possibility of using the warm water of the cooling system, the efficiency may go up to even 60%.

Overview on Natural Gases: World Reserve, Production and Consumption

The most important point to note from Table 11.11 is that the nominal reserve life of natural gas globally is presently only 60 years plus compared to over 150 years for coal. But the shorter life of natural gas is not really a matter of concern as reserves from the yet unexploited newer sources such as coal bed methane, methane from shale rock and methane hydrates are to be adequately assessed, which is likely to considerably increase the reserve level in the near future. Secondly, the present stock of natural gas, whose carbon footprint is 40 per cent less than anthracite coal and hence could very considerably reduce the GHG emission of power plants in the next three or four decades, till the world sees the emergence of the renewable energies and also nuclear fuels. Therefore natural gas energies are going to play a very significant role in combating carbon emission.

Table 11. 11: Natural Gas Reserve, Production, Consumption

Country and Region	*Reserve (2008) Trillion cu.m.*	*Production (2008) Billion cu.m.*	*Consumption (2008) Billion cu.m.*	*(Reserve/ Production) Years*
United States	6.93	574.4	657.7	12.1
Canada	1.75	173.4	98.2	10.1
Total North America	**9.18**	**801.8**	**822.0**	**11.4**
Venezuela	4.98	29.2	30.7	170.5
South and Central America	**7.32**	**157.1**	**141.0**	**46.2**
Russian Federation	43.30	601.7	416.0	72.0
Norway	2.20	99.2	4.3	22.2
Turkmenistan	8.10	66.1	19.0	122.5
Total Europe and Eurasia	**62.68**	**1086.3**	**1139.5**	**56.7**
Iran	29.61	116.3	119.3	254.6
Qatar	25.37	77.0	20.2	329.5
Total Middle East	**75.82**	**383.4**	**331.8**	**197.7**
Total Africa	**14.71**	**214.3**	**96.1**	**68.6**
Australia	3.08	38.3	25.5	80.4
China	2.46	90.3	81.3	27.3

(contd...)

(contd...)

Country and Region	*Reserve (2008) Trillion cu.m.*	*Production (2008) Billion cu.m.*	*Consumption (2008) Billion cu.m.*	*(Reserve/ Production) Years*
India	1.09	30.5	41.3	35.7
Indonesia	3.18	69.7	33.3	45.7
Japan	0	0	93.7	-
Malaysia	2.38	64.9	33.6	36.7
Total Asia Pacific	**16.00**	**417.9**	**481.4**	**38.2**
Total World	**185.28**	**3060.0**	**3010.0**	**60.5**

III
Overview on Indian Natural Gas Status:

From the above Table 11.11, we see that India has a national reserve natural gas stock of about a Trillion cubic metre of natural gas, which will last about 35 years at the current low level of consumption. From all indications of new gas finds, the Indian reserves may soon grow several times, as the consumption levels also go up substantially with the growth dynamism of the resurgent India.

Exploration and Production: While India may not be expected to be a significant contributor to the upstream oil sector, the outlook for the upstream natural gas sector is more positive, although the IEA forecasts that natural gas production in India will peak between 2020 and 2030. Presently most natural gas production in India comes from fields off the western coast, including the Mumbai High complex and the Tapti, Panna, and Mukti fields, while the major onshore fields are located in the northeast and in the arcas of Assam, Andhra Pradesh, and Gujarat. Of late, ONGC formed plans to invest $2.4 billion for E&P oil and gas projects in the Assam rcgion.

There have been some large natural gas finds in India over the last several years, predominantly offshore and particularly in the Krishna-Godavari basin in the Bay of Bengal. In December 2006, ONGC announced that it had found an estimated 21 to 22 Tcf of natural gas in place at the KG-DOWN-98/2 block off the coast of Andhra Pradesh in the Krishna Godavari basin. In addition, in August 2008, the company made four new finds in the GS-15-OA block in the same basin. ONGC also announced another find in 2006 in the Mahanadi basin off the coast of Orissa state, with an estimated 3 to 4 Tcf reserves in place. These discoveries fit into the recent trend of large upstream developments in the Bay of Bengal, especially in the Krishna Godavari basin. In addition, state-owned Gujarat State Petroleum Corporation (GSPC) holds an estimated 1.8 Tcf of natural gas reserves at the KG-OSN-

2001/3 block in the Krishna Godavari area, a substantial holding for the company.

Another key find is Reliance Industries' D-6 block in the Krishna-Godavari basin, which holds estimated reserves of 11.5 Tcf. Natural gas productions in this field began in late 2008. Reliance and its equity partner Niko Resources have reportedly invested $8.7 billion in the deepwater project, which is expected to supply 2.8 Bcf/d of natural gas at its peak. In order to transport the gas throughout the country, GAIL has recently completed a pipeline linking the Krishna-Godavari basin with existing gas transport infrastructure. The finds in the Krishna-Godavari basin are expected to double the country's current natural gas output in coming years. Companies are working to produce as much as possible domestically from current fields due to the widening supply and demand gap. ONGC has also worked to maximize its recovery rate at the Mumbai High Structure.

Natural Gas Imports LNG: Analysts expect that India's natural gas import demand will increase in the coming years. To help meet this growing demand, a number of import schemes including both LNG and pipeline projects have either been implemented or planned. India began importing liquefied natural gas (LNG) in 2004. In 2006, India imported 254 Bcf of LNG, making it the seventh largest importer of LNG in the world. India's LNG imports in 2006 came from Algeria, Egypt, Nigeria, Oman, Qatar, United Arab Emirates, Australia, and Malaysia. Qatar was by far the largest supplier in 2006, accounting for nearly 86 per cent of imports. India imports LNG through both long-term contracts and spot shipments.

Currently, India has two LNG import terminals, with several others that are planned or proposed. India started receiving LNG shipments in January 2004 with the start-up of the Dahej terminal in Gujarat state. Petronet LNG, a consortium of state-owned Indian companies and international investors, owns and operates the Dahej LNG facility with a capacity of 5 million tons per year (mta) (975 bcf/y). India's second terminal, Hazira LNG, started operations in April 2005, and is owned by a joint venture of Shell and Total. The facility has a capacity of 2.5 mta (488 Bcf/y), which may be expanded to 5 mta (975 Bcf/y) in the future.

In addition, Petronet LNG is currently finalising a deal with a Japanese consortium to build a 2.5 mta (488 Bcf/y) LNG import facility at Kochi. The facility was expected to be completed in March 2012 and will potentially be expanded to a capacity of 5 mta (975 Bcf/y). Petronet LNG plans to sign a long-term LNG supply deal with Australia's Gorgon LNG project for 2.5 mta (488 Bcf/y) in early 2009.

Another proposed LNG facility is the 5 mta (975 Bcf/y) LNG processing plant in Dabhol. In addition, several other companies are studying possible LNG import sites around India, such as GAIL's Ennore LNG terminal at Tamil Nadu, scheduled for commissioning in 2011.

IV

Alternative Sources of Methane: Coal Bed Methane, Shale Methane Gas and Methane Hydrate

Coal Bed Methane (CBM)

Methane (natural gas), while frequently developed with petroleum, also occurs in association with coal. Coalbed methane accounts for about 7.5 per cent of U.S. natural gas production. Recent U.S. estimates indicate more than 700 trillion cubic feet (TCF) of coalbed methane gas in place, with over 100 TCF economically recoverable—a five-year supply at present rates of consumption.

Coal bed methane is mining coal bed not for their coal but for their methane. Recovery of CBM works best for methane-rich coal beds that are fractured and submerged in water. Water surrounding the coal beds, located, say, 100 metres underground, absorbs and retains the methane as long as the water is under pressure. With low solubility in water, methane readily comes out of solution when the water pressure is dropped.

A well is drilled to the coal seam to allow the water to rise to the surface, where it is kept under pressure. From time to time, water is pumped out of the well to lower its pressure. The released methane is diverted to collection system for eventual distribution to a natural gas distribution system.

Coal Bed Methane in US

Coal bed methane accounts for about 7.5 per cent of U.S. natural gas production. Recent U.S. estimates indicate more than 700 trillion cubic feet (TCF) of coal bed methane gas in place, with over 100 TCF economically recoverable—a five-year supply at present rates of consumption. The principal region for CBM wells in US is the rocky mountains in Wyoming, Colorado and New Mexico.

USGS is developing models to accurately assess potential U.S. coalbed methanc resources. These models evaluate:

- Controls on the occurrence and recoverability of coal bed methane, including geologic, geochemical, engineering, technological and economic factors.
- Resource development issues and environmental implications.

The abundant occurrence of coal resources (30 trillion tonnes) worldwide is prone to generate at least 7.5 TCF of gas-in-place. USGS has on-going research and assessments of CBM provided through Technical Assistance Agreements in order to transfer information and technology.

Coal Bed Methane Programmes in India

Coal India

These are early days for CBM exploration in India. Coal India, the Indian

Public Sector giant and largest coal company in the world, is planning to invite CBM players to capture methane in the five existing underground mines in Jharia and Bokaro coal fields in Jharkhand State of India. The project will ensure mine safety, while expanding operation and use the captured gas for captive generation or other productive purposes. A tender in this regard will be floated shortly. The project proposed is similar to CBM operations but more complex. According to the plan, the coal major will invite operators to degasify the unexplored lower seams of five gassy mines and use the captured methane for productive end use on cost-plus basis.

The project will help CIL increase production from the mines whole ensuring safety aspects as well as avoiding the environmental consequences of releasing methane in the air. The basic process of the proposed project is similar to the CBM activity, where the operators drill wells in the virgin blocks, fracture the coal seams and collect methane for a 25-30 year period. Here in the proposed project, complications will arise on two counts; firstly the primary aim here is to facilitate further mining, methane gas will need to be evacuated within a period of 3-4 years, requiring higher concentration of production wells and hence higher investments. Secondly, the blocks are already mined; thus drilling technology and the depth and the types of wells will differ from that in a CBM asset. Thus, there is a risk of higher investment for a limited scope of commercialisation of gas. In the past, degassing was practiced by injecting fresh air and releasing diluted methane gas directly into the atmosphere. Now Methane is 22 times more harmful for global warming than carbon dioxide and such direct venting of methane into the atmosphere will be unacceptable internationally. CIL has now embarked on a pilot project at Muridih, Jharkhand to capture mine methane for generating electricity.

RIL Industries

Reliance Industries Limited (Reliance) is a private company in India with investments in upstream oil and gas production, petroleum refining, petrochemicals, and textile manufacture.

Reliance Industry, actually India's largest petro-energy industry in the private sector has also initiated a CBM project in Sohagpur, Madhya Pradesh State with the help of Consultant Advanced Resources International to assess the prospect of marketing coal bed methane.

Shale Methane Gas

Shale gas refers to natural gas that is trapped within Shale formations. Shales are fine-grained sedimentary rocks that can be rich sources of petroleum and natural gas.

Horizontal Drilling and Hydraulic Fracturing

Over the past decade, the combination of horizontal drilling and hydraulic fracturing has allowed access to large volumes of shale gas that were previously uneconomical to produce. The production of natural gas from shale formations has rejuvenated the natural gas industry in the United States.

The U.S. has Abundant Shale Gas Resources

Of the natural gas consumed in the United States in 2009, 87 per cent was produced domestically; thus, the supply of natural gas is not as dependent on foreign producers as is the supply of crude oil, and the delivery system is less subject to interruption. The availability of large quantities of shale gas will further allow the United States to consume a predominantly domestic supply of gas.

According to the EIA Annual Energy Outlook 2011, the United States possesses 2,552 trillion cubic feet (Tcf) of potential natural gas resources. Natural gas from shale resources, considered uneconomical just a few years ago, accounts for 827 Tcf of this resource estimate, more than double the estimate published last year.

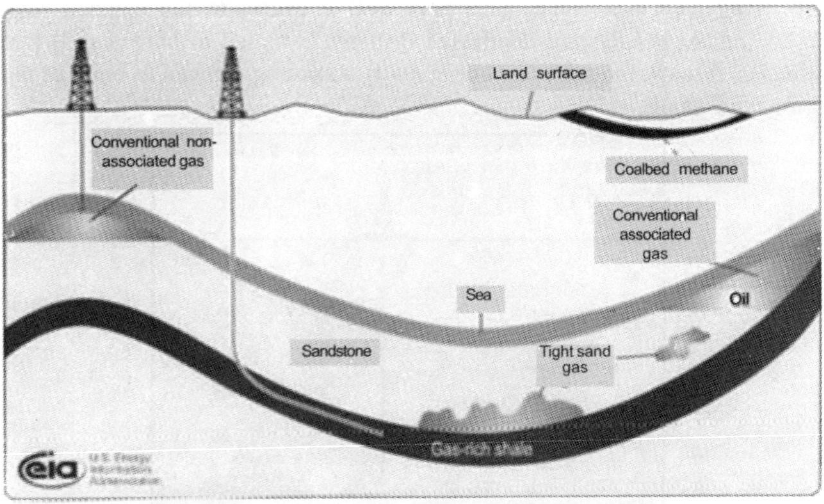

Fig. 11.3 Schematic Geology of Natural Gas Resources (Image from EIA)
Diagram showing the geometry of Conventional and Unconventional Natural gas.

Enough for 110 Years of Use

At the 2009 rate of U.S. consumption (about 22.8 Tcf per year), 2,552 Tcf of natural gas is enough to supply approximately 110 years of use. Shale gas resource and production estimates increased significantly between the 2010 and 2011 Outlook reports and are likely to increase further in the future.

In 2009 Shale gas constituted 14% of the total US natural gas supply. Production of shale gas is expected to continue to increase and constitute 45% of total US natural gas supply in 2035, as projected in EIA Annual Energy Outlook 2011.

Shale gas is found in shale "plays," which are shale formations containing significant accumulations of natural gas and which share similar geologic and geographic properties. A decade of production has come from the Barnett Shale play in Texas. Experience and information gained from developing the Barnett Shale have improved the efficiency of shale gas development around the country.

Other important plays are the Marcellus Shale and Utica Shale in the eastern United States; and, the Haynesville Shale and Fayetteville Shale in Louisiana and Arkansas. Surveyors and geologists identify suitable well locations in areas with potential for economical gas production by using both surface-level observation techniques and computer-generated maps of the subsurface.

Horizontal Drilling

Two major drilling techniques are used to produce shale gas. Horizontal drilling is used to provide greater access to the gas trapped deep in the producing formation. First, a vertical well is drilled to the targeted rock formation. At the desired depth, the drill bit is turned to bore a well that stretches through the reservoir horizontally, exposing the well to more of the producing shale.

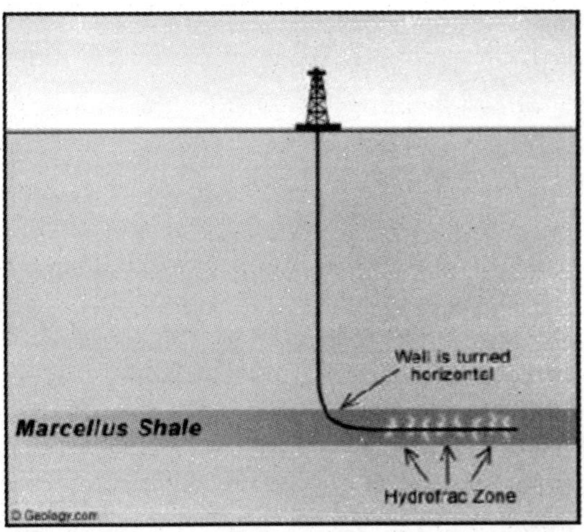

Fig. 11.4 Hydraulic Fracturing for Shale Gas

Hydraulic Fracturing

Hydraulic Fracturing (commonly called "fracking" or "hydrofracking") is a technique in which water, chemicals, and sand are pumped into the well to unlock the hydrocarbons trapped in shale formations by opening cracks (fractures) in the rock and allowing natural gas to flow from the shale into the well. When used in conjunction with horizontal drilling, hydraulic fracturing enables gas producers to extract shale gas at reasonable cost.

Without these techniques, natural gas does not flow to the well rapidly, and commercial quantities cannot be produced from shale.

Shale Gas vs. Conventional Gas

Conventional gas reservoirs are created when natural gas migrates toward the Earth's surface from an organic-rich source formation into highly permeable reservoir rock, where it is trapped by an overlying layer of impermeable rock. In contrast, shale gas resources form within the organic-rich shale source rock. The low permeability of the shale greatly inhibits the gas from migrating to more permeable reservoir rocks. Without horizontal drilling and hydraulic fracturing, shale gas production would not be economically feasible because the natural gas would not flow from the formation at high enough rates to justify the cost of drilling.

Methane Hydrate

What is Methane Hydrate and its Immense Energy Potential: Methane clathrate, also called methane hydrate, hydromethane, methane ice or "fire ice" is a solid clathrate compound (more specifically, a clathrate hydrate) in which a large amount of methane is trapped within a crystal structure of water, forming a solid similar to ice (see Fig. 11.5)

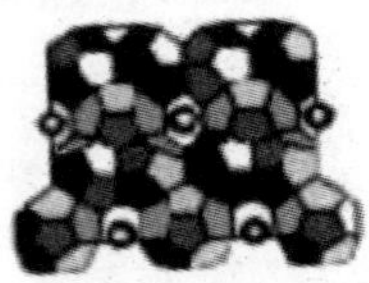

Fig. 11.5: Methane Hydrate.

Methane Hydrates are found in the shallow marine geosphere, and occur both in deep sedimentary structures, and as outcrops on the ocean floor. Methane hydrates are believed to form by migration of gas from depth along geological faults, followed by precipitation, or crystallisation, on contact of the rising gas stream with cold sea water. Methane clathrates are also present in deep Antarctic ice cores, and are believed to record a history of atmospheric methane concentrations, dating to 800,000 years ago. Possibly, the ice-core methane clathrate record is a primary source of data for global warming research, along with oxygen and carbon dioxide.

The immense future energy potential of Methane hydrate can be guessed from what Dr. William Dillon has to say:

> "Methane trapped in marine sediments as a hydrate represents such an immense carbon reservoir that it must be considered a dominant factor in estimating unconventional energy resources; the role of methane as a 'greenhouse' gas also must be carefully assessed." (Dr. William Dillon, U.S. Geological Survey).

Structure and Composition: The average methane clathrate hydrate composition is 1 mole of methane for every 5.75 moles of water, though this is dependent on how many methane molecules "fit" into the various cage structures of the water lattice. The observed density is around 0.9 g/cm^3. One litre of methane clathrate solid would therefore contain, on average, 168 litres of methane gas at STP.

Methane forms a structure of one hydrate with two dodecahedral (12 vertices, thus 12 water molecules) and six tetradecahedral (14 water molecules) water cages per unit cell. This compares with a hydration number of 20 for methane in aqueous solution.

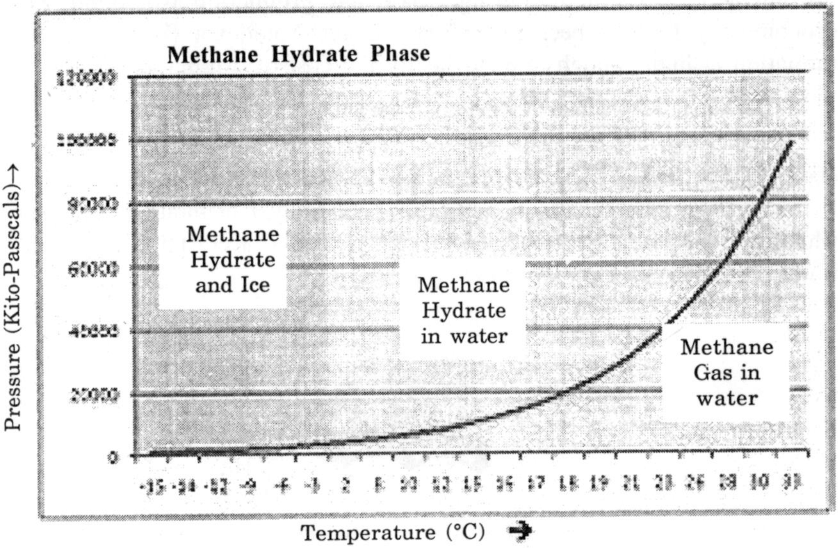

Fig. 11.6: Methane Hydrate Phase Diagram

In Methane hydrate phase diagram (see Fig. 11.6), the horizontal axis shows temperature from -15 to 33°C, the vertical axis shows pressure from 0 to 120,000 kilopascals (0 to 1,184 atmospheres). For example, at 4°C hydrate forms above a pressure of about 50 atmospheres.

Oceanic Deposits: There are two distinct types of oceanic deposits. The most common is that methane (more than 99% methane) contained in a 'structure I' clathrate and generally found at depth in the sediment. Here, the methane is in light isotope form, which indicates that it is derived from the microbial reduction of CO_2. Hydrates normally exclude the salt in the pore fluid from which it forms, and thus they have high electric resistivity just as ice and sediments containing hydrates have a higher resistivity compared to sediments without gas hydrates. In the less common second type found near the sediment surface some samples have a higher proportion of longer-chain hydrocarbons (less than 99% methane) contained in a 'structure II' clathrate. Carbon from this type of clathrate is isotopically heavier and is thought to have migrated upwards from deep sediments, where methane was formed by thermal decomposition of organic matter. Examples of this type of deposit have been found in the Gulf of Mexico and the Caspian Sea.

The size of the oceanic methane clathrate reservoir is poorly known, and estimates of its size decreased by roughly an order of magnitude per decade since it was first known that clathrates exist in the oceans since the 1960s and '70s. The highest estimates (e.g. 3×10^{18} m³) were based on the assumption that fully dense clathrates could litter the entire floor of the deep ocean.

Continental Deposits: Methane clathrates in continental rocks are trapped in beds of sandstone or siltstone at depths of less than 800 m. Sampling indicates they are formed from a mix of thermally and microbially derived gas from which the heavier hydrocarbons were later selectively removed. These occur in Alaska, Siberia as well as Northern Canada.

The sedimentary methane hydrate reservoir probably contains 2–10 times the currently known reserves of conventional natural gas. This represents a potentially important future source of hydrocarbon fuel. However, in the majority of sites, deposits are likely to be too dispersed for economic extraction. Other problems facing commercial exploitation are detection of viable reserves; and development of the technology for extracting methane gas from the hydrate deposits. To date, there has only been one field commercially produced where some of the gas is thought to have been from methane clathrates, Messoyakha Gas Field, supplying the nearby Russian city of Norilsk.

Hydrates formed during natural gas processing: Methane clathrates (hydrates) are also commonly formed during natural gas production operations, when liquid water is condensed in the presence of methane at high pressure. It is known that larger hydrocarbon molecules such as ethane and propane can also form hydrates, although as the molecule length increases (butanes, pentanes), they cannot fit into the water cage structure and tend to destabilise the formation of hydrates. Once formed, hydrates can block pipeline and processing equipment. They are generally then removed by reducing the pressure, heating them, or dissolving them by chemical means (methanol is commonly used.

Effects of Hydrate Phase Transition during deep water drilling

When drilling in oil- and gas-bearing formations submerged in deep water, the reservoir gas may flow into the well bore and form gas hydrates owing to the low temperatures and high pressures found during deep water drilling. The gas hydrates may then flow upward with drilling mud or other discharged fluids. As they rise, the pressure in the annulus decreases and the hydrates dissociate into gas and water. The rapid gas expansion ejects fluid from the well, reducing the pressure further, which leads to more hydrate dissociation and further fluid ejection. The resulting violent expulsion of fluid from the annulus is one potential cause or contributor to what is referred to as a "kick". (Kicks, which can cause blowouts, typically do not involve hydrates).

Measures which reduce the risk of hydrate formation include:

- High flow-rates, which limit the time for hydrate formation in a volume of fluid, thereby reducing the kick potential.
- Careful measuring of line flow to detect incipient hydrate plugging.
- Additional care in measuring when gas production rates are low and the possibility of hydrate formation is higher than at relatively high gas flow rates.
- Monitoring of well casing after it is "shut in" (isolated) may indicate hydrate formation. Following "shut in", the pressure rises as gas diffuses through the reservoir to the bore hole; the rate of pressure rise will exhibit a reduced rate of increase when hydrates are forming.
- Additions of energy (e.g., the energy released by setting cement used in well completion) can raise the temperature and convert hydrates to gas, producing a "kick".

Methane Clathrate and Climate Change: Methane is a powerful greenhouse gas. Despite its short atmospheric half life of 7 years, methane has a global warming potential. The sudden release of large amounts of natural gas from methane clathrate deposits has been hypothesised as a cause of past and possibly future climate changes. Events possibly linked in this way are the Permian-Triassic extinction event and the Paleocene-Eocene Thermal Maximum.

Natural Gas Hydrates (NGH) vs. Liquified Natural Gas (LNG) in Transportation

Since methane clathrates are stable at a higher temperature of -20° C compared to the liquified natural gas (LNG) at 162°C), there may be some interest in converting natural gas into clathrates rather than liquifying it when transporting it by seagoing vessels. A significant advantage would be that the production of natural gas hydrate (NGH) from natural gas at the terminal would require a smaller refrigeration plant and less energy than LNG would. Offsetting this, for 100 tonnes of methane transported, 750 tonnes of methane hydrate would have to be transported; since this would require a ship of 7.5 times greater displacement, or require more ships, it is unlikely to prove economic.

References

Max, Michael D. (2003). *Natural Gas Hydrate in Oceanic and Permafrost Environments*. Kluwer Academic Publishers.

Nersesian Roy L 2009, Energy for the 21st Century, Pentagon University Press, New Delhi, (M.E. Sharpe Inc 2007, New York).

Petersen, E.L. and Madsen, P.H. 2004, Wind Farms, *Encyclopedia of Energy*, Vol. 6, Ed. Cleveland, Elsevier.

Sen, A.K. 2010, Coal Beneficiation for Indian Thermal Power Plant, Symposium On Thermal Power Plants, Millennium Institute of Energy and Environment Management, Kolkata, Website www.mieem.org

Tester, Drake, Driscoll *et al.*, 2005, *Sustainable Energy – Choosing among Options*, MIT, Prentice Hall India, New Delhi.

Tondeur, D. and Teng, F. 2008, 'Carbon Capture and Storage for Greenhouse Effect Mitigation', *Future Energy*, Ed. Letcher T.M., Elsevier.

US Geological Survey, Gas hydrate: What is it?, accessed 27 February 2010 and Gas (Methane) Hydrates — A New Frontier, USGS fact sheet.

BP World Energy Data 2008, http://www.bp.com/assets/bp_internet/globalbp_uk_english/reports_and-publication/statistical_review_of_of_world_energyfull_report2011.pdf

IEA 2006-2010.

file:///E:/IEA%20Energy%20Data%20Electricty%20Coal%20 Naturalgas%2027July10/IEA%20Energy%20Statistics%20-%20Electricity%20for%20IEA%20Total.htm

IPCC 2007.

12

Hydropower: Reliable Non-fossil Power

Global Scenario

In 2007, hydropower constituted 16 per cent of the world's electricity generating capacity. The theoretical potential of worldwide hydropower is 6716 Twh and the actual power production was 2653 Twh, that is, constituting only about 40 per cent (see Table 12.1). About 44 % of the world's hydropower was generated in four countries in 2002, mostly large- and mid-scale plants.

Table 12.1: World Hydropower Potential and Installed

Continents	*Installed capacity GWe*	*Power Production TWh/y*	*Hydropower Potential TWh/y*	*Potential Hydropower Utilized %*
Asia	225	754	2673	28
North and Central America	157	702	1,016	69
South America	108	512	1040	33
Europe	173	567	756	75
Africa	20	76	1010	7
Australasia	13	42	221	19
World	**696**	**2653**	**6716**	**40**

Source: IEA 2007.

Asia accounted for 28 per cent of the world's hydro generation, with 754 TWh, followed by North and Central America with 26% (702 TWh) and Europe with 21per cent (567 TWh). The country wise hydropower capacity and generation are given in Table 12.2. Canada with 315 TWh is the largest producer of hydropower in the world followed by China with 309 TWh. Brazil with 282 GWh and the United States with 255 TWh comes after them. Even though Canadian hydro generation is growing, China will overtake Canada very soon, if it has not already done so, to become the largest hydro generator in the world. In Western Europe and the United States, the scope for additional hydropower is limited, as the most economic sites have already been developed and further expansion is hindered by environmental concerns. In North America, hydropower is the most widely used form of renewable energy.

Table 12.2: Hydropower Status by Country 2008

Country	*Installed Hydropower GWe*	*Hydropower Generation TWh*
USA	99	255
China	83	309
Canada	67	315
Brazil	64	282
India	36	144
Mexico	10	35
Russia	-	180
Norway	-	125

Source: WEC 2009.

Hydropower Technology: The basis of hydropower is first to create or identify a geodetic head of water (sere Fig. 12.1), which is used to run turbines to generate electricity.

There are also two Hydropower technologies to generate power:

•**Run-of-River:** Modern run-of-river plants usually unify the weir and the power house in a single structure, placed across the full width of the river in order to dam it and gain a higher head. Turbining and releasing of the incoming water takes place shortly after the intake, avoiding the environment problem of break in the continuous river flow downstream. In the other run-of-river system, water is diverted from the weir or built-up water intake to a distant power house through a canal or penstock.

•**Storage System Plant:** These storage plants make use of reservoirs to delink the time of water flow and electricity production. The storage schemes are mostly found in mountainous regions and reach head heights of several hundred metres. In contrast to run-of-river system, they use low flow with high head. Here the river in a valley is really impounded by the high rise dam. These types of high and heavy concrete dams are very capital intensive.

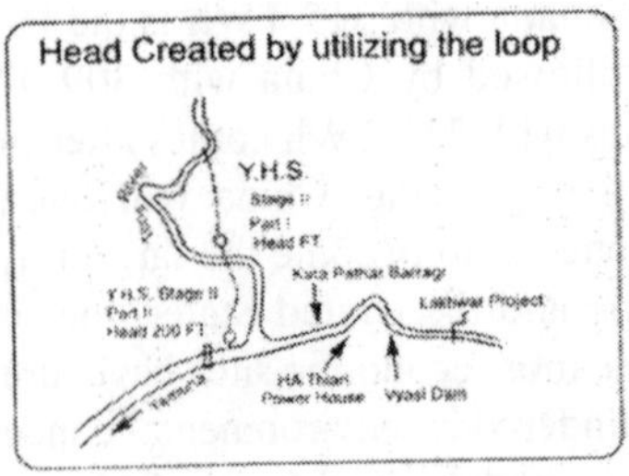

Fig. 12.1A: Yamuna Hydel Scheme

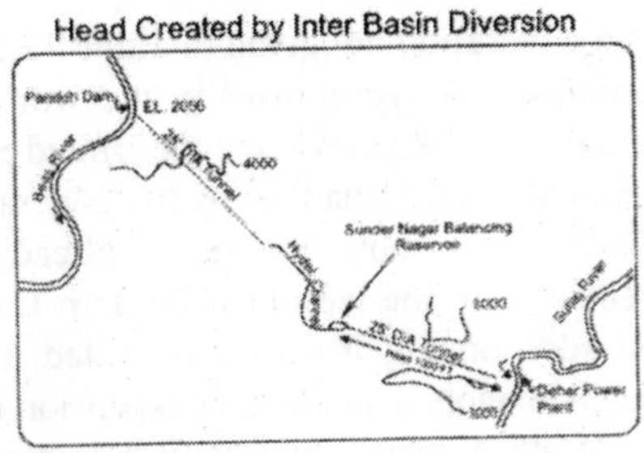

Fig. 12.1B: Beas-Sutlej Link Project

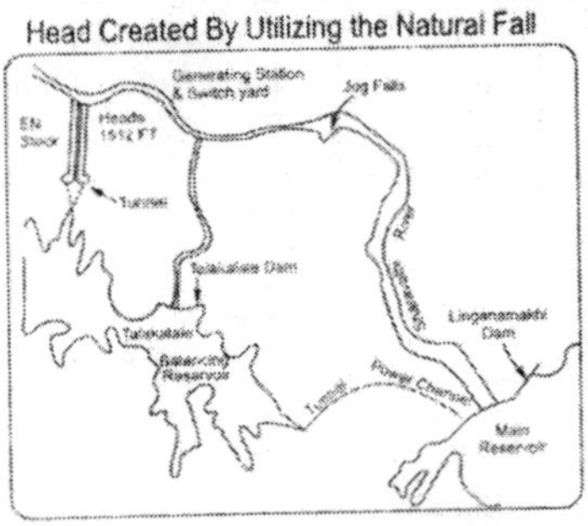

Fig. 12.1C: Sharavath Hydel Scheme

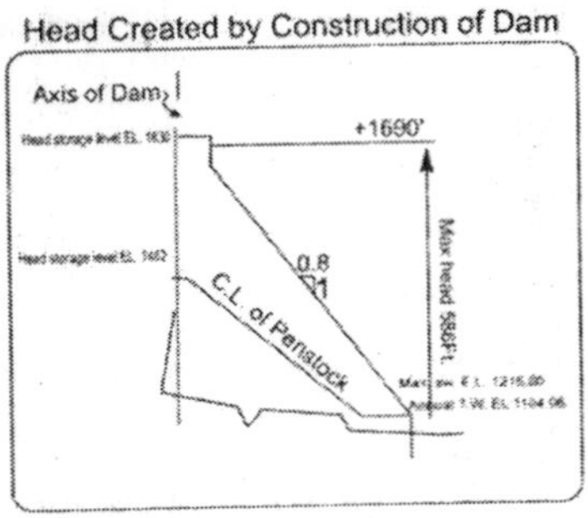

Fig. 12.1D: Bhakra Dam

Fig. 12.1: Different Types of Hydel Heads

Pumped Storage System: This is really not a direct system of power generation but really a storage of energy hydraulically. When the supply of electrical energy is in excess in the night or otherwise, the water is pumped up and stored at a sufficiently high head. Now as the energy demand is peaked at day time, the hydro-generator is activated by the release of stored water. In the pumped storage scheme, there are really an upper and a lower reservoir. Because the water is turbined and pumped through the same penstock and therefore over the same head, these plants use approximately 1.3 – 1.4 times more energy to pump than they generate.

Turbo-Generators: The created water-head is converted into mechanical energy by rotating water turbines, which in turn rotates electrical generator. There are three principal types of water turbines suiting high, medium and low water heads respectively (see Fig. 12.2).

- Pelton impulse wheel for high head water, receives high head jet impulse on its buckets which rotates the wheel at high speed;
- Francis Reaction Turbines for medium head is a mixed flow turbine, water entering the turbine radially and emerges axially and rotating the turbine.

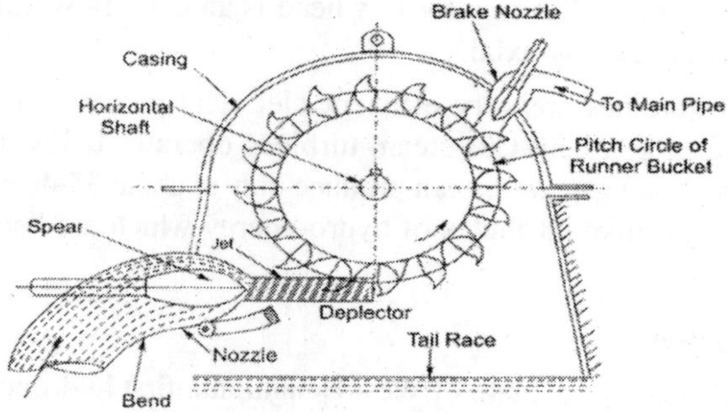

Fig. 12.2A: Pelton Turbine

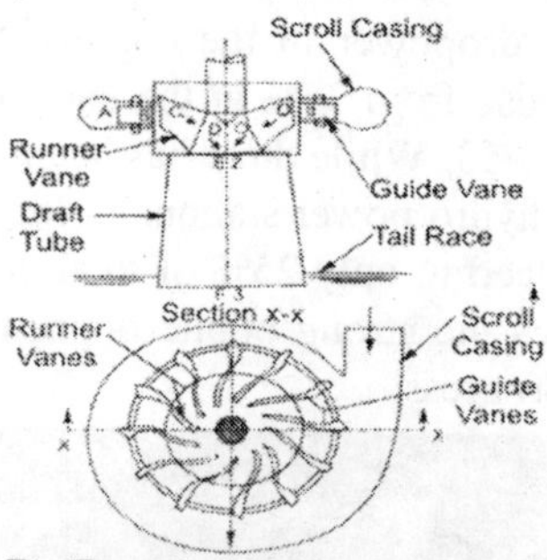

Fig. 12.2 B: Francis Turbine

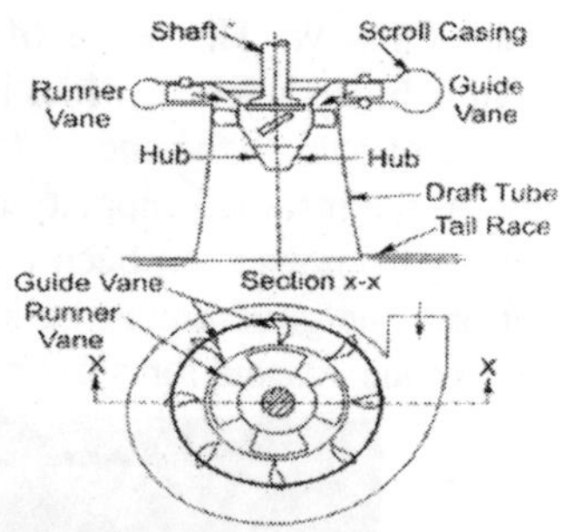

Fig. 12.2C: Kaplan Turbine

Fig. 12.2: Types of Modern Hydraulic Turbines

Table 12.3: Major Positive Attributes of Hydropower

Positive Characteristics	*Negative Characteristics*
1. Hydro-energy is renewable and is thus most welcome with dwindling stocks of fossil fuels.	1. Though the energy is renewable, supply may vary year to year due to vagaries of rainfall.
2. It is free of environmental pollution, not emitting GHG.	2. However, large projects lead to impoundment and inundation of large surface areas, which may lead to displacement of human habitation and fishing migration.
3. It is highly efficient, hydraulic conversion efficiency being on the order of 80% or above.	3. The capital cost is high.
4. It has low operation and maintenance cost.	4. Generally very long lead time in construction is required.
5. But it can have high initial cost, which can offset its multitude beneficent uses in Domestic and Agricultural irrigation water supplies.	
6. Long active life of the Hydro-plants.	

- Kaplan Propeller Turbine for low head is an axial flow turbine, water entering and exiting axially.

The hydro-turbines are very efficient electrical power generators with over 75% efficiency, where Gas-steam turbines operates below 60 per cent efficiency, Coal-steam generators can produce only at about 35-40% efficiency. There are other positive attributes of hydro-energy which are listed in Table 12.3.

Indian Hydropower

In 1887 at Darjeeling, State of West Bengal, the first hydropower station in India was commissioned. At the time of independence, out of total installed electrical capacity of 1,362 MW, hydro-power generation capacity stood at 508 MW. The share of hydropower in the country had a major thrust after Independence, when it rose from 37% at the end of 1947 to its peak share of 51% at the end of 1962/63. While there has been a continuous increase in the installed capacity of hydro power stations in India, today the share of hydro power has been reduced to only 25% of total electric power generation. The government believes the strong public opposition to dams in India is the reason for slower progress.

The Indian government considers hydropower as a renewable economic, non-polluting and environmentally benign source of energy. The exploitable hydro-electric potential in terms of installed capacity is estimated to be about 148,700 MW (See Table 12.4) out of which a capacity of 30,164 MW (20.3%) has been developed so far in 2007 and 13,616 MW (9.2 %) of

Table 12.4: India's Hydropower Potential

River Basin	*Potential at 60% LF Mwe*	*Probable Capacity Mwe*
Indus Basin	19,998	33,832
Brahmaputra Basin	34,920	66,065
Ganga Basin	10,715	20,710
Central India Basin	2,740	4,152
West Flowing River Systems	6,149	9,430
East flowing River Systems	9,532	14,511
Total	84,044	148,700

capacity is under construction. In addition, 6,782 MW in terms of installed capacity from small, mini and micro hydro schemes have been assessed. Also, 56 sites for pumped storage schemes with an aggregate installed capacity of 94,000 MW have been identified.

Small 100 KW hydro power project in Himachal Pradesh: ***MNES***

Small Hydro-Power: A Viable Option

Small and mini hydel projects have the potential to provide energy in remote and hilly areas where extension of an electrical transmission grid system is uneconomical. Realising this fact, the Indian government is encouraging development of small hydro power (SHP) projects in the country. Since 1994 the role of private sector for setting up of commercial SHP projects has been encouraged. So far 14 States in India have announced policies for setting up commercial SHP projects through private sector participation. Over 760 sites of about 2,000 MW capacities have already been offered / allotted. An estimated potential of about 15,000 MW of small hydropower (SHP) projects exists in India, of which 4,233 potential sites with an aggregate capacity of 10,071 MW for projects up to 25 MW capacities have been identified. In the last 10-12 years, the capacity of Small hydro projects up to 3MW has increased 4 fold from 63 MW to 240 MW. 420 small hydropower projects up to 25MW station capacity with an aggregate capacity of over 1,423 MW have been set up in the country and over 187 projects in this range with aggregate capacity of 521 MW are under construction.

References

Rao, K.L., 1979, *India's Water Wealth*, Orient Longman, New Delhi, India.

Tester, *et al.*, 2005, *Sustainable Energy*, MIT Press, PH, India.

Vaishwanar, R.S., 1997, *Fluid Power Engineering*, Jain Brothers, New Delhi.

13

Nuclear Energy and Aspects of Safety and Viability

I

Nuclear Power in the World Today

The first commercial nuclear power stations started operation in the 1950s. There are now some 436 commercial nuclear power reactors operating in 30 countries, with 3,72,000 MWe of total capacity. They provide about 21.6% (see also IEA2009) of the world's electricity as continuous, reliable base-load power (see Fig. 13.1); their efficiency is also increasing.

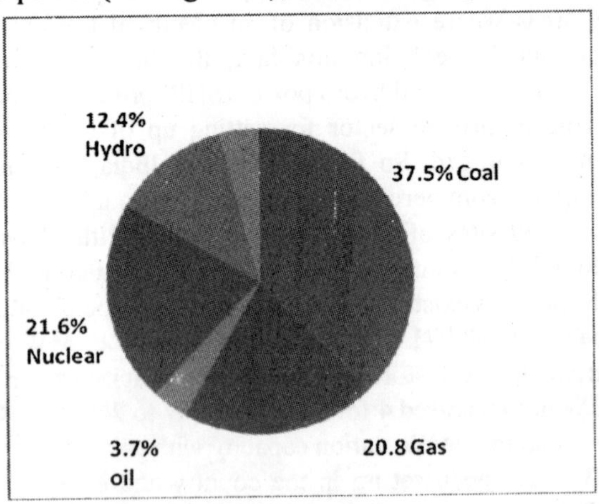

Fig. 13.1 Share of Nuclear Power (IEA 2009)

Today, the world produces as much electricity from nuclear energy as it did from all sources combined in 1960. Civil nuclear power can now boast over 13,000 reactor years of experience and supplies over 20 per cent of global electricity needs in 30 countries. Today one quarter of world's reactors have load factors of more than 90%, and nearly two thirds do better than 75%, compared with about a quarter of them in 1990. For 15 years Finnish plants topped the performance tables, but the USA reactors now dominate the top 25 positions, followed by South Korea. US nuclear power plant performance has shown a steady improvement over the past twenty years and the average load factor now stands at around 90%, up from 66% in 1990 and 56% in 1980. This places the USA as the performance leader with 12 of the top 25 reactors, the 25th achieving more than 97.5%. The USA accounts for nearly one third of the world's nuclear electricity. In 2007 and 2008 ten countries averaged better than 80% load factor, while French reactors averaged 76-77%, despite many being run in load-following mode, rather than purely for base-load power.

EIA World Outlook 2009: Period 2006-2030 Electricity generation from nuclear power worldwide is likely to increase from 2.7 trillion kilowatt-hours in 2009 to 3.0 trillion kilowatt-hours in 2015 and 3.8 trillion kilowatt-hours in 2030 in the EIA2009 reference case (see Table 13.1A), as concerns about rising fossil fuel prices, energy security, and greenhouse gas emissions support the development of new nuclear generating capacity. Higher capacity utilisation rates have been reported for many existing nuclear facilities and it is expected that most of the older plants now operating in OECD countries and in non-OECD Eurasia will be granted extensions to their operating lives.

Table 13.1A: Growth of Nuclear Energy Capacity GWe 2006-2030

Regions	2006	2015	2030
OECD Europe	132	127	121
OECD North America	115	121	132
OECD Asia	67	74	88
Non-OECD Europe/Eurasia	42	49	71
China	7	22	54
India	3	9	20
Rest of World	10	15	23

Source: EIA Energy Outlook 2007.

The Table 13.1B from the World Nuclear Association provides a comprehensive list of country wise nuclear power generation in 2009, number of power reactors operable in 2010 and number of power reactors under construction and further planned in 2010 and amount of uranium required at 68,546 tonnes to run the existing plants.

Table 13.1B: World Nuclear Power Reactors and Uranium Requirements

Country	*Nuclear electricity generation 2009*		*Reactors operable 1 July 2010*		*Reactors under construction 1 July 2010*		*Reactors planned July 2010*		*Reactors proposed July 2010*		*Uranium required 2010*
	Billion kWh	*Percentage*	*No.*	*MWe*	*No.*	*MWe*	*No.*	*MWe*	*No.*	*MWe*	*Tonnes U*
Argentina	7.6	7.0	2	935	1	692	2	767	1	740	123
Armenia	2.3	45	1	376	0	0	1	1060			55
Belgium	45	51.7	7	5943	0	0	0	0	0	0	1052
Brazil	12.2	3.0	2	1901	1	1270	0	0	4	4000	311
Bulgaria	14.2	35.9	2	1906	0	0	2	1900	0	0	272
Canada	85.3	14.8		12679	2	1500	4	4400	3	3800	1675
China	65.7	1.9	1	8587	24	26560	3	36910	1	120000	2875
Czech Republic	25.7	33.8	6	3686	0	0	2	2400	1	1200	678
Finland	22.6	32.9	4	2721	1	1600	0	0	2	3000	1149
France	391.7	75.2	5	63236	1	1630	1	1630	1	1630	10153
Germany	127.7	26.1	1	20339	0	0	0	0	0	0	3453
Hungary	14.3	43	4	1880	0	0	0	0	2	2200	295
India	14.8	2.2	19	4183	4	2572	20	16,740	40	49000	908
Japan	263.1	28.9	55	47348	2	2756	12	16532	1	1300	8003
South Korea	141.1	34.8	20	17716	6	6700	6	8190	0	0	3804
Mexico	10.1	4.8	2	1310	0	0	0	0	2	2000	253
Netherlands	4.0	3.7	1	485	0	0	0	0	1	1000	107
Pakistan	2.6	2.7	2	400	1	300	2	600	2	2000	68
Romania	10.8	20.6	2	1310	0	0	2	1310	1	655	175

(contd...)

(contd...)

Country	Nuclear electricity generation 2009		Reactors operable 1 July 2010		Reactors under construction 1 July 2010		Reactors planned July 2010		Reactors proposed July 2010		Uranium required 2010
	Billion kWh	% e	No.	MWe	No.	MWe	No.	MWe	No.	MWe	Tonnes U
Russia	152.8	17.8	32	23084	10	8960	14	16000	30	28000	4135
Slovakia	13.1	53.5	4	1760	2	840	0	0	1	1200	269
Slovenia	5.5	37.9	1	696	0	0	0	0	1	1000	145
South Africa	11.6	4.8	2	1842	0	0	3	3565	24	4000	321
Spain	50.6	17.5	8	7448	0	0	0	0	0	0	1458
Sweden	50.0	34.7	10	9399	0	0	0	0	0	0	1537
Switzerland	26.3	39.5	5	3252	0	0	0	0	3	4000	557
Ukraine	77.9	48.6	15	13168	0	0	2	1900	20	27000	2031
United Kingdom	62.9	17.9	19	11035	0	0	4	6600	6	8600	2235
USA	798.7	20.2	104	101263	1	1180	9	11800	22	31000	19538
WORLD**	2560	14	439	374.815	59	60.075	149	163,204	344	365,125	68,646

Sources: Reactor data: WNA to 1/7/10; IAEA- for nuclear electricity production & percentage of electricity (percentage) 3/5/10;
WNA: Global Nuclear Fuel Market (reference scenario) (World Nuclear Association Report)

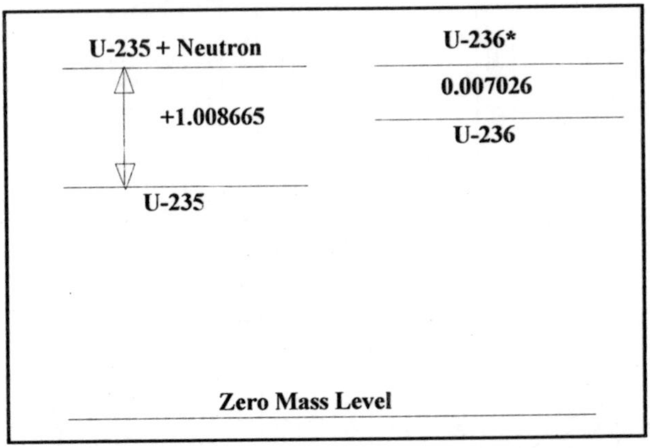

Fig. 13.2: Relation of Excitation Energy due to neutron absorption

Box 13.1: Basic Nuclear Fission Reaction

When a free neutron is absorbed by a uranium nucleus, there is created a U-236 nucleus in the first stage.

$^{235}U_{92} + {^1}n_0 = (^{236}U_{92})^*$

The U-236* that is created carries extra energy because of the summing up of the masses of two interacting particles is greater than that of a normal U-236 nucleus.

The mass of U-236* is given by 235.043923+ 1.008665 = 236.052588. But the mass of U-236 in the ground state has a mass of 236.0455622, lower by 0.007026 amu or 6.54 MeV. This amount of excess energy causes the U-236* to fission. Fig. 11.2 shows these energy relationships.

The above calculations did not include any kinetic energy brought to the reaction by the neutron, on the ground that fission can be induced by absorption in U-235 of very slow neutrons.

Eventually, the unstable U-236* undergoes fissions. The general reaction equation for the fission resulting from neutron absorption in U-235 is represented as follows:

$_{92}U\text{-}235 + {_0}n\text{-}1 \rightarrow [_ZF^A]_1 + [_ZF^A]_2 + N_u({_0}n\text{-}1) +$ Energy,

where F1 and F2 are fission fragments and N_u are number of neutrons. In a typical fission reaction, F1 and F2 could be Krypton and Barium as shown

$_{92}U\text{-}235 + {_0}n\text{-}1 \rightarrow {_{36}}Kr\text{-}90 + {_{56}}Ba\text{-}144 + 2\ {_0}n\text{-}1 + E$

The collection of isotopes, which are fission byproducts have an excess of neutrons or a deficiency of protons in comparison with naturally occurring elements. For example, the main isotope of barium is $_{56}$Ba-137 and a prominent element of mass 144 is 60Nd-144. Thus, there are extra neurons or four too few protons in barium isotope from fission and hence it is highly unstable.

It may be somewhat surprising that the introduction of only 6.54 MeV of excitation energy can produce a fission reaction yielding as much as 200 MeV (see Table B13.1)

Table B13.1: Energy from Fission of U-235

Processes	*Energy MeV*
Fission Fragments Kinetic energy	166
Neutrons	5
Prompt Gamma Rays	7
Fission Product Gamma Rays	7
Beta Particles	5
Neutrinos	10
Total Energy released	200

Source: R L Murray 2001.

II
Basic Nuclear Reactions and Reactors

Heavy atomic nuclei are not as stable as the light nuclei, because of the repulsive forces caused by the presence of the large number of protons in the nucleus. For this reason, it is possible to cause fission of a heavy nucleus of a fissile element uranium-235 by bombarding by a neutron in a nuclear Reactor.

Referring to Fig. 13.3, a U-235 neutron absorbs a neutron, converts temporally into an unstable U-236 and splits into high velocity fission products (notably, radioactive strontium, cesium-137, Iodine- 131, xenon etc.) and 2 or 3 fresh neutrons. The kinetic energy of high velocity fission products, which impacts on the surroundings and slows down, is converted into enormous heat energy. The fresh neutrons slowed down in a chosen moderator medium like water, heavy water, and graphite etc., impacts on other U-235 nuclei and create a chain reaction releasing an enormous energy of 200MeV per reaction equation E13.1 below:

235 U + n —— X+ Y +2.5 n + energy (200MeV) …… . (E13.1)

where X and Y represent fission products and n represents neutron.

The quantum of energy released actually amounts to 82 gigajoules per gm of U-235. In the Box13.1, more precise description of nuclear fission is presented for the technical readers.

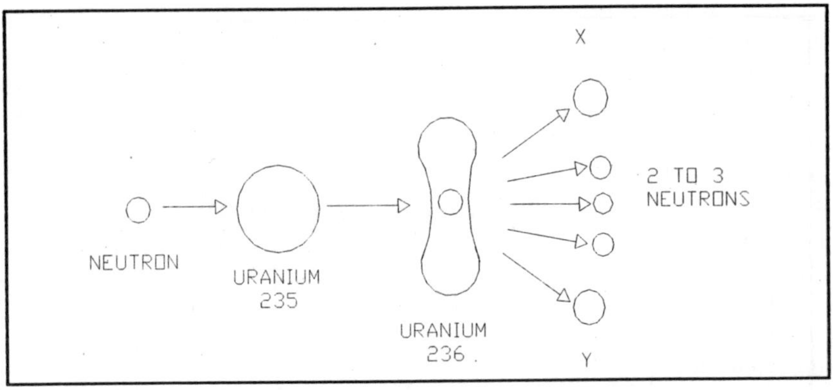

Fig. 13.3. Typical Nuclear Fission.

Reactor Types

Uranium-235 is the basic radioactive metal in the nuclear fuel, which is fissionable in a reactor for generating heat energy and finally electricity. The uranium ore, containing as little as 0.1 per cent uranium is milled producing uranium oxide in concentration of about 80 per cent. The uranium oxide product of a uranium mill is not directly usable as a fuel for a nuclear reactor. For the fissile uranium-235 constitutes only 0.7 per cent, the remainder is non-fissionable uranium-238 isotope. For the most common Light Water Reactor (LWR), the U-235 concentration needs to be raised to 3.5 per cent by a complex process known as enrichment. But for the Indian Pressurised Heavy Water Reactors (PWHR), no enrichment is necessary. The characteristics of these two major type reactors are shown in Table 13.2.

There are two broad classes of reactors – thermal and fast reactors. A fast reactor is one in which most of the neutrons are in the energy range of 0.1 – 1 MeV. In the fast reactor, neutrons remain at high energy in the absence of a moderating medium in the fission region; while in the thermal reactor the fission region is filled with a moderating medium and the bulk of the neutrons have energy less than 0.1 Mev.

Then depending upon the operating characteristics, the reactors could be classified into five major groups – Pressure Water Reactor PWR (also called Light water reactor), Boiling water reactor, Heavy Pressure Water reactor HPWR (also known as CANDU Canadian Deuterium Uranium Reactor), High temperature gas cooled reactor HTGR and Liquid Metal fast breeder reactor LMFBR.

PWR: It is a thermal reactor with light water at high pressure 150 kg/sq.cm and temperature 315°C serving both as moderator and coolant and the fuel is somewhat enriched (3%) uranium. It is also popularly known as Light Water Reactor LWR.

BWR: It differs from PWR in the sense that primary cooling water operates at low pressure and supplies steam directly to the Steam Turbine generating power.

PHWR/ CANDU: It uses heavy water under pressure to cool the fuel elements that are housed in many pressure tubes, while these tubes are housed at a larger Calendria, which also contains heavy water for moderation at lower pressure and temperature. The fuel element use natural uranium (0.7% U-235) that can be loaded and discharged during operation.

HTGR: It is characterised by use of highly enriched uranium with thorium and with graphite moderator and Helium gas cooling.

LMFBR: Being a fast breeder reactor it has no moderator and has liquid sodium coolant and plutonium fuel surrounded by natural uranium or depleted uranium.

Reactor Fuels

Nuclear Fuel belongs to one of the following three categories:

Fertile Isotopes – U-238, Th-232, which are convertible to fissionable isotopes;

Fissionable Isotopes – U-235, U-233, and Pu-239;

Natural Uranium - 0.7% U-235 and 99.3% U-238 as milled before enrichment.

Economically recoverable uranium in the world is estimated at 4 million tons. The reserves of thorium in the earth are far larger than that of uranium. Thus, if all the recoverable uranium and thorium are converted into fissionable isotopes, there would be an ample energy supply for the mankind for several millennia. In fact, each gram of U-235 releases 1 MW-day energy by reaction E13.1.

The natural uranium, which is generally mined, consists of 0.7% U-235 and the balance is U-238. For operation in the nuclear reactor, natural uranium is enriched to 3 -5% U-235.

Uranium that remains after the enrichment process is generally known as the depleted uranium (consisting mostly U-238), because U-235 concentration has been reduced to 0.3% or less. This depleted uranium can be converted to a fissionable isotope Pu-239 by a process called breeding. Thorium Th-232 can also be breeded to fissionable isotope U-233. The process involves absorption of a neutron by the nucleus of U-238 and Th-232 in a Breeder Reactor.

The reactor driver fuel consists of the elements or compounds of fissionable isotopes U-235, U-233 and Pu-239 along with fertile isotopes U-238 and Th-232.

The important nuclear processes that take place in the reactor are the following:

U-235 + n = X+Y + 2.5n + energy (200 Mev) E13.1;

Table13.2: Principal Power Reactor Types with Basic Specifications

Subsystems	*Pressurised Water Reactor PWR (Light Water Reactors)*	*Boiling Water Reactor BWR*	*Pressurised Heavy Water Reactor PHWR*	*High Temp. Gas Cooled Reactor HTGR*	*Liquid Metal Fast Breeder Reactor LMFBR*
Fuel Form	U_2O	U_2O	U_2O	UC, ThC_2	PuO_2,U_2O
Enrichment	3% U-235	2.5% U-235	0.7% U-235	93% U-235	
Moderator	Light Water	Light Water	Heavy Water	Graphite	None
Coolant	Light Water	Light Water	Heavy Water	Helium, CO_2	Liquid Sodium
Operating Temp	267 C	288	330 C	750 C	545 C
Cladding	Zircaloy	Zircaloy	Zircaloy	Graphite	Stainless Steel
Control	B_4C or Ag-In-Cd rods	B_4C Crosses	Moderator level	B_4C rods	B_4C or Tantalum rods
Vessel	Steel	Steel	Steel	Concrete	Steel

Source: Murray 2001; Tester et al 2005.

U-238 + n = Pu-239 .... E13.2;

Th-232 + n = U-233 ....E13.3;

where X and Y are called fission products, which are also other radioactive elements of the Periodic Table.

The general reactions in the nuclear reactor generally follows E 13.1, but as U-235 is fissioned, some U-238 also gets converted to Pu-239 and in fact some 40 per cent of the energy generated in the reactor is from the fission of U-238 in the 4-year life cycle of the fuel. Some Th-232, if present, can also get converted into U-233 by neutron absorption per E13.3. Another series of reaction by successive neutron capture similar to reactions E13.2 and E13.3 produces radioactive Actinide, Americium, Curium and Neptunium.

In LWR, only U-235 is burnt; most of the U-238 remains unutilised in the spent fuel. The spent fuel also contains some plutonium-239 (Pu-239), converted from U-238. In the Breeder Reactor, both Pu-239 and U-235 are used as charge fuel to convert U-238 into fissile U-235 to continue the normal fission process; so it is called a Breeder Reactor, as it breeds U-235 out of the non-fissile U-238.

Reactor Fuel Loading Arrangement

The Fig. 13.4 shows the blown- up arrangement of the loading of fresh fuels of a typical LWR. The fresh fuel installed in a typical PWR/LWR consists of cylindrical pellets of slightly enriched (3% U-235) uranium oxide (UO_2) of diameter 1cm and length 0.6 in (1.5 cm). A zircaloy tube of wall thickness 0.025 in (0.6 cm) is filled with pellets to an active length 12 in (365 cm) and sealed to form a fuel rod or pin. The metal tube serves to provide strength and rigidity for the column of pellets, to provide a cladding that retains radioactive fission products and to protect fuel from interaction with coolant.

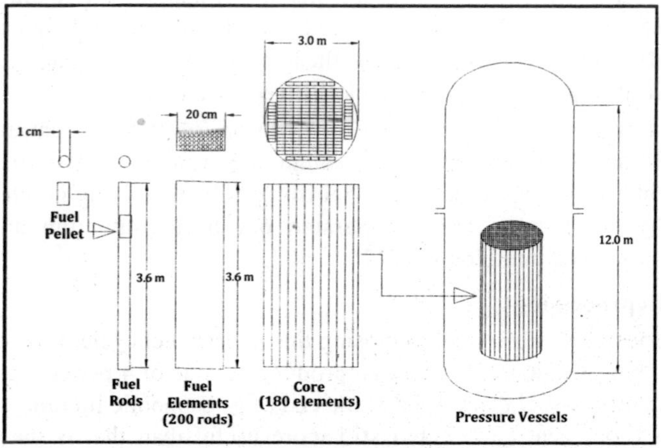

Fig. 13.4: Arrangement of nuclear fuels in a typical Reactor.

About 200 of fuel pins are grouped in a bundle called fuel elements of 8 cm on a side. Now about 180 fuel elements are assembled in an approximately cylindrical array to form a reactor core. The core is placed inside the pressure vessel of 16 ft (5 m) O.D., height 40 ft (12 m) and wall thickness 12 in (30 cm). Control rods, consisting of an alloy of cadmium, silver and indium provide the ability to change the amount of neutral absorption. The rods are inserted in some vacant fuel pins spaces and magnetically connected to a drive mechanism.

Fuel Processing and Reprocessing

The deployment of nuclear power generation plant will depend on having in place a fully functional and highly safe nuclear waste depository. To understand the problem of nuclear waste disposal, we have to follow up the nuclear fuel cycle as enumerated below and as shown schematically in Fig. 13.5:

(i) Uranium ore is mined, separated and milled to produce yellow cake U_3O_8 oxide.

(ii) The yellow cake is then chemically converted into a gas Uranium hexafluoride UF_6.

(iii) Enrichment through a process of gaseous diffusion or centrifuging: UF_6 is enriched to its form 235-UF6, which being lighter is separated out from 238-UF_6.

(iv) Once the permeate UF_6 is sufficiently enriched to 235-UF_6 (about 3%), the gas is oxidised to uranium dioxide pellets in the fuel fabrication plant

(v) Fuel pellets in the fuel rod are used in the nuclear plant for three years.

(vi) In USA, the spent fuel rods are stored in water storage to allow most intensely radioactive fission products to decay. Once the fuel rods are sufficiently decayed, they are then stored in a dry storage awaiting burial in a 300 m deep tunnel in a geologic medium.

(vii) Alternatively, the spent fuel which consists of 95.5 per cent uranium-238 and small amount of plutonium, are separated out from 4.5 per cent actinide and other industrial isotopes by dissolving in nitric acid. The uranium is finally separated out from plutonium and then reprocessed into Uranium-235.

Aim of Reprocessing

Reprocessing is essential to closing the nuclear fuel cycle. It takes more than 7 tonnes of uranium ore feed to produce 1 tonne of 4 per cent enriched uranium. Nuclear fuel discharged at the end of its economic lifetime contains less than 1 per cent U-235; but still more useful than that is the natural uranium it contains. The other naturally occurring isotope U-238 partly transmutes into Pu-239 and produces 40 per cent of the fission energy. It is the recovery of plutonium and to a lesser extent the uranium for use in the recycled fuel that is the focus of reprocessing.

Spent Nuclear Fuel

Average burn-up has increased with improvements in fuel technology, amounting to 40,000 MW-days/tonne being typical of spent fuel discharged. The mass of uranium contained in two fresh fuel assemblies for a PWR reactor is approximately 1 tonne. Each spent fuel assembly includes approximately 150 Kg of radioactive hardware that holds the fuel pellets in place. At 40,000 MW-d/tonne burn-up, the spent fuel assembly contains approximately 94.5 per cent unused U-238 uranium, 1 % plutonium, 0.2 % minor actinides (neptunium, americium and curium) and 4.3 per cent fission products by weight.

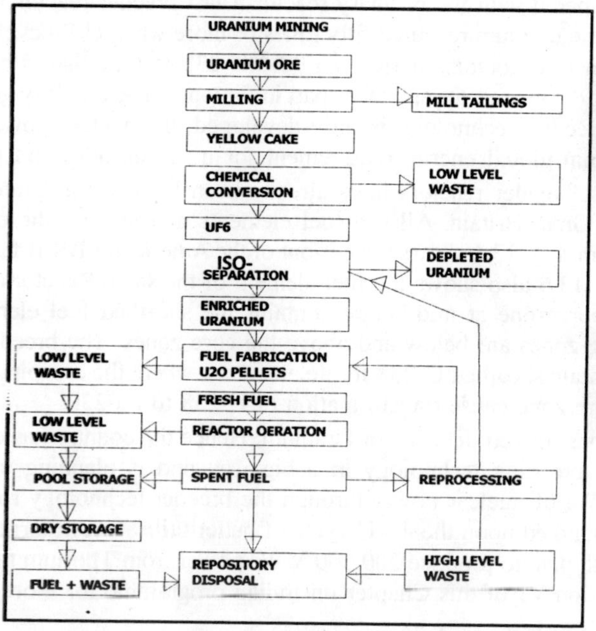

Fig. 13.5: Scheme of Nuclear Fuel Processing and Reprocessing

III
Breeder Reactor Technology

We know that in a nuclear fission reaction, for each neutron absorbed in a fertile nuclear fuel as U-235, more than two neutrons are released. In order to maintain the chain reaction, only one neutron is needed. Any extra neutron available can thus be used to produce to produce other fissile materials U-238 and Th-232 respectively. This is the basic principle of breeding technology.

In the Breeder Reactor, the moderator medium is dispensed with and the neutron from the neutron stream from U-235, in addition to causing fission of other U-235 nuclei, also impacts the fertile U-238 isotope of uranium and also Thorium isotope Th-232. Thus, two fissile isotopes Pu-239 and U-233 are breeded in addition to the initial fissile uranium-235.

In the first Breeder reactor, the reactor vessel is quite compact as the moderator is dispensed with. As indicated in the earlier paragraph, there are actually two breeding cycles:

Uranium-235 ⟶ Uranium-238 ⟶ Plutonium-239 (U-Pu cycle)
Uranium-235 ⟶ Thorium-232 ⟶ Uranium-233 (Th-U cycle)

However, in the Breeder reactor, the U-Pu cycle is mainly employed with Plutonium charge. The Thorium-Uranium cycle operated more favourably in the thermal energy range and employed in the advanced type of Thermal Reactor.

As compared with the ordinary reactor a fast breeder reactor is capable of utilising the uranium resource fifty times or more when U-Pu cycle is used. The uranium feed contains more than 90% U-238 isotope that is breeded to fissile Pu-239 in the reactor and takes part in fission and heat energy generation. Once the Breeder Technology is fully developed, the nuclear power will be able to sustain global energy requirement for many decades and centuries.

The fast breeder reactors have already been built in the United States, Russia and Great Britain. All use fuel elements as rods and the coolant as liquid sodium. Fig. 13.6 shows the cut-out of the American EBR II fast breeder reactor. Fig. 13.6 also shows the fuel element of the same Reactor alongside, where the core zone at mid-height contains the sheathed fuel elements and the breeding zones are below and above the core zones. The breeding zones likewise contain sheathed U-238 fertile elements, where the escaping neutrons from the core zone cause transformation of U-238 to Pu-239.

India has limited deposits of Uranium. Hence the country has developed U-Pu cycle breeding technology in a big way and is planning to develop 150,000 MWe of nuclear power through the breeder technology route. India has also embarked upon the Th-U cycle of better utilisation of large Thorium deposits with plan to produce 300,000 MWe power from Thorium fuel. Please see the Section VI of this Chapter on Indian programme for more details.

History of Fast Breeder Reactors

Liquid Metal Fast Breeder Reactors (LMFBR) have been operating successfully in some parts of the world. In the US, the Experimental Breeder Reactor I (EBR I) at the Idaho Falls was the first power reactor to generate electricity in 1951. The EBR II, the second of the series was used from 1963 to 1994 to test equipments and materials. The Fermi I reactor built near Detroit was the first intended to commercially generate power. It was started in 1963 but was damaged by the blockage of coolant flow passage and then operated for a brief period after repair.

The Fast Flux Test Facility (FFTF) of 400 Mwe capacity at Richmond, Washington, now shut down, did not generate electricity but provided valuable data on the performance of fuel, materials and coolant.

After a number of years of design work and construction the US Government cancelled the demonstration fast power reactor called Clinch River Breeder Reactor Project (CRBRP). There was great deal of debate before the CRBRP was abandoned. One reason of abandoning the project was the high cost of fuel, being only (1/5) the cost of producing the electricity. This political decision shifted the leadership of developing breeder technology from the US to other countries.

Thereafter, France led in the development of the breeder reactor for the production of commercial electric power in cooperation with other European countries. The reactor 'Superphoenix' was a full-scale pool-type breeder constructed with partial backing of Italy, West Germany, the Netherlands and Belgium. The reactor was shut down because of sodium leaks and unlikely to ever be restarted, despite of many billions of dollars invested in the facility.

With the discontinuance of the operation of Superphoenix, the leadership in the development of breeder reactor again changed. Japan put into operation the sodium-cooled loop-type breeder reactor 'Monju' 280 Mwe in 1993. It was Japan's plan to construct a number of Breeders starting around 2020. Unfortunately, Monju suffered a sodium leak and had to be shut down. That possibly signaled the end of Japan's venture with breeder reactor.

The current leader in the Breeder technology is Russia. The largest liquid metal fast breeder reactor in the world is BN-600 at Belonyasky plant in Russia, supplying electricity since 1981; it has operated more successfully than any other reactors in that country. The Table 13.3 gives the basic specification of the Russian BN-600.

Table 13.3: BN-600 Liquid Metal Fast Breeder Reactor

Electric Power	560 MWe
Sodium coolant temperature	377 – 550°C
Core Fuel Height	1.03 m
Core Diameter	2.05 m
Vessel Height and Diameter	12.6 m/ 12.86 m
Fuel (w/o U-235)	UO_2
Pin O D	6.9 mm
Cladding thickness	0.4 mm
Pin Pitch (triangular)	9.82 mm
Pins per assembly	127
Number of assemblies	369
Number of Boron Control rods	27
Average power density	413 KW/l
Operation cycle length	5 months

Source: World Nuclear Energy Handbook (INS)
http://www.insc.anl.gov

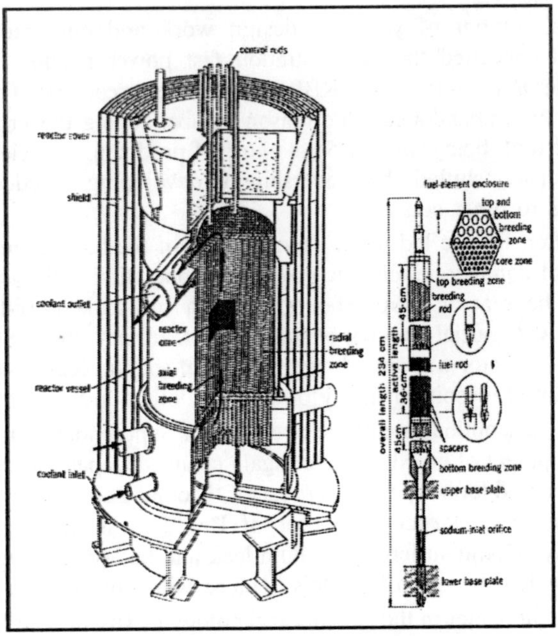

Fig. 13.6: An EBR II American Breeder Reactor.

III
Safety of Nuclear Reactor Systems

In any consideration of nuclear safety, we must review the operation of the nuclear power reactor, in which the nuclear fission of uranium or plutonium atoms is conducted so that the energy is released in a duly controlled process. To start with, a nuclear explosion in a nuclear power plant is impossible due to low enrichment. The worst possible accident is a severe loss-of-coolant accident (LOCA), leading to core-meltdown, for which there are many precautions. Firstly, the boron control rods will immediately act coming down and stopping the nuclear fission. Secondly, the zirconium cladding will stop the escape of neutron and radioactivity and also the reactor vessel wall, 25.4 mm thick steel with stainless cladding acts as the next line of containment. Thereafter, the very heavy concrete containment structure is the last barrier to prevent any radioactive leakage. We will now elaborate the emergency core cooling and containment system (See Fig. 13.7 Basic PWR Containment System).

In the event of the loss of coolant (LOCA) by way of damage to the primary cooling piping or the pump system or otherwise, the following emergency will activate automatically:

- Immediately, the reactor power would be reduced by the insertion of the control rods, but there will still be continuous supply of heat from the decaying fission products, that would raise the temperature of the core

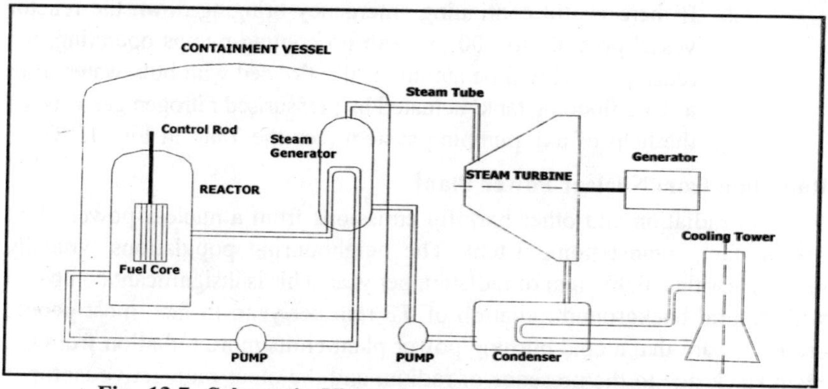

Fig. 13.7. Schematic View of a Pressurised Water Reactor

above the melting point of the fuel rods and cladding. In a severer situation, the fuel tubes may melt releasing considerable amount of fission products in the reactor vessel.

- To prevent the occurrence of the above assumed catastrophe, an 'emergency core cooling system' (ECCS) is provided in a water moderated reactors consisting of at least two sub-systems:

  1. When the reactor vessel pressure drops from 2250 psi to 1500 psi, an auxiliary system of pumps will carry water from the borated water tank to the reactor vessel which will both inhibit nuclear fission and also cool (see Fig. 13.8 Emergency Core Cooling System of a Typical PWR);

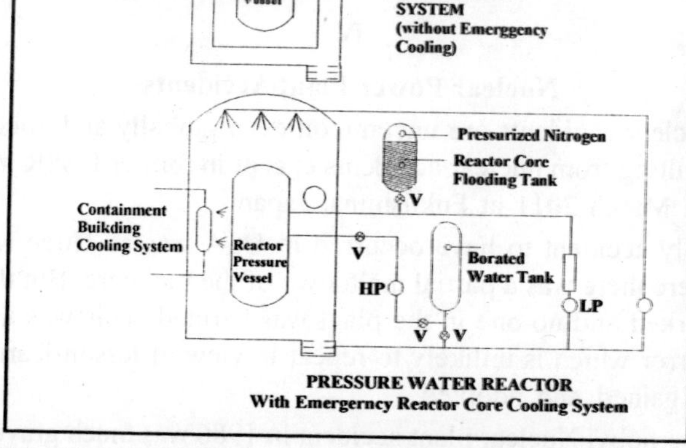

Fig. 13.8. Emergency Reactor Core Cooking System

2. If there is still continuing emergency bringing down the reactor vessel pressure to 600 psi with no cooling pumps operating, the reactor vessel will be automatically flooded with bulk water from a 'core flooding tank' actuated by pressurised nitrogen gas without the help of any pumping system (see the Tank in Fig. 13.8).

Radiation from Nuclear Power Plant

The radiation and other harmful emissions from a nuclear power plant are, in fact, almost non-existent. The neighbouring populations typically receive less than 0.001 rem of radiation per year. This is insignificant compared with natural background radiation of 0.3 rem per year. In fact, most people are not aware that a coal burning power plant emits more radiation from the plant stack due to the presence of radium and the other radioactive isotopes in coal.

Radioactive Nuclear Waste

The problems may be complex, but as on date, the problems of waste management and disposal have been largely solved in the technical sense. The major concern is, of course, with high level wastes, which include spent fuel rods from reactors which contain many radioactive isotopes.

Safety and Reprocessing of Spent Fuels

If one reprocesses the spent fuel, 99 per cent of the uranium and plutonium can be separated and can be reused, extending the useful lifetime of fuel 100 times. Equally important is the great reduction in the waste volume and the time for which the waste fuel must be sequestered.

Safe Sequestering of Nuclear Waste

One of the preferred options is deep land burial or geologic disposal. The waste is concentrated by dissolving in nitric acid and distilling and by converting into dry particles and by vitrifying with molten glass and pouring the same in stainless steel canister, duly sealed. Six vitrification plants are now in operation around the world but not in USA.

IV

Nuclear Power Plant Accidents

The nuclear accidents are uncommon even globally and nobody has so far died resulting from nuclear accidents except in former USSR in 1986 and presently in March 2011 at Fukushima, Japan.

The only accident to have occurred in USA was at Three Mile Island in 1979 where there was a partial meltdown of the fuel core. But the fail-safe systems worked and no one in the plant was harmed. This was a clear case of human error which is unlikely to repeat in view of lesson learnt and new knowledge gained and adopted.

The Chernobyl Nuclear plant accident in 1986 was much graver resulting in the reported direct loss of 31 lives arising out of a steam explosion and furious burning of the graphite moderator. The accident did not occur during the normal operation of the plant but during an unauthorised experiment the operators were trying to carry out. The experiment was to test if, in a station blackout, the "coasting-down" generators would generate enough electricity to run the cooling pumps until the standby diesel generators could take over.

The Fukushima nuclear power plant disaster is the most recent having been caused by the Tsunami waves that hit Japan coast on 11 March 2011. There were six plants, all old basic Boiling Water Reactors built by GE, USA. Tsunami damaged the power supply from both the main and backup generators. Battery backup did not last long. Because of lack of cooling water the core temperature and pressure built up in the reactor vessel which vented steam which included hydrogen which reacted with the oxygen in the air or cooling water causing explosion the reactor building blowing off the roof.

More details of the nuclear accidents at Three Mole Islands, Chernobyl and Fukushima are given in the Appendix.

V
Nuclear Power Economics

From the MIT study of 2005, based on a capacity factor of 85 per cent and 40 years of economic life, it was concluded that nuclear electricity at 6.7 US cents per kWh is not competitive with coal or gas based electrical power. However, if the construction costs could be brought down by 25 per cent and if the construction time could be brought down to 4 years from the present 5 years, the levelised nuclear power cost could come down to 4.2 cents per kWh, which will then be very much comparable to the coal and gas fired generation. The International Energy Agency study (2006) arrives at about similar price band of 4.9 US cents per kWh, if construction and operating risks are mitigated. In fact, nuclear power is projected to be cheaper if gas prices are held at $ 4.5 to 5.5 per million Btu. But the nuclear power will still be costlier than the coal-fired generation, unless the coal price goes above $ 70 per tonne or nuclear investment is less than $2000 per kW.

VI
INDIAN NUCLEAR PROGRAMME

It is gratifying that India has built up a very comprehensive capability in nuclear technology and research, well appreciated all over the world. India achieved the distinction of commissioning Asia's first research nuclear reactor Apsara in 1956 on her own but with fuel elements uranium-235 supplied by UK.

Dr. Homi Bhaba, the doyen of Indian nuclear science and the first Head of Indian Nuclear programme, recognised the inadequate reserve of uranium in India but with plentiful reserve of Thorium. Optimising this scenario, India adopted a three stage nuclear development policy.

(i) In the first stage, nuclear reactor using natural uranium with heavy water cooling and moderation is to be commissioned; this process will produce Plutonium-239 as feed fuel for the second stage.

(ii) In the second stage India will develop Fast Breeder Reactor, which will have feed of fissile PU-239 and fertile U-238 to convert U-238 to fissile PU-239; the specialty of the breeder technology is it will produce more fuel than the original feed of PU-239 it consumes; thus using the spent fuel stock of the stage (i), it will produce 30 to 40 times more energy in standard 500 MWe FBR, already under implementation.

(iii) In the third stage, plentifully available Thorium will be fed in the Advanced Heavy Water Reactor (AHWR), whose design is already completed and construction should be under way in 300 MWe Thorium Reactor.

The eventual optimum capacities of this programme were also being projected at the turn of the last century, as shown in Table 13.4.

Table 13.4: The Original Projected Capacities of Three-Stage Nuclear Programme

Reactor Type	*Nuclear Power Routes*	*Annual Power Capacities*
PHWR (Natural uranium Reactor)	Natural Uranium-235 Based	10,000 Mwe
FBR (Breeder U-238 Reactor)	Plutonium-239/U-238 Based	150,000 Mwe
AHWR (Thorium Reactor)	U-235/PU-239/Thorium Based	300,000 Mwe

Source: Ghosh Roy, M.K. 2010, *'Sustainable Development'*, Ane Books, New Delhi..

In the light of India signing the Civilian Nuclear Treaty with USA followed by Agreements with IAEA and other NSG countries, the situation has got transformed dramatically. Negotiations are in progress to set up in four new and one existing Nuclear Power Parks in Gujarat, Andhra, Tamilnadu, West Bengal and Orissa with capacities up to 10,000 Mwe each with imported Light Water Reactors (LWR), two Parks earmarked for US (GE and Westinghouse), two Russia (Atomstroyexport) and one for France (Areva).

New and enlarged nuclear power capacities are already being contemplated in 2007 and 2008 at the level of Dr. Anil Kakodkar, then Chairman AEC. Assuming the successful implementation of the imported above LWR installations of capacities of the order of 40,000 Mwe, he speculates –

"For example, if you set up 40,000 MWe light water Reactor capacity, using the spent fuel and using the fast breeder technology, we should be in a position to further augment electricity generating capacity without the need of additional fuel to the tune of 10 times i.e. 40,000 MWe LWR capacity can lead to 400,000 Mwe of fast breeder capacity." (*Nuclear India*, Vol. 42, No. 1-2, July-August, 2008).

In continuation, Dr. Kakodkar again expands;

> "... We are talking about reaching something like 200,000 to 250,000 MWe based on indigenous programme, the indigenous three stage programme with PHWR, fast breeder and thorium utilisation. We could have another 400,000 Mwe on the basis of the multiplier to the imported acquisition i.e. LWR through fast breeder reactors. Thus, we are talking of about 500,000 Mwe to 600,000 Mwe over the next 50 years or so." (*Nuclear India*, quoted ibid)

So, we conclude that theoretically the revised nuclear capacity could reach target of about 600,000 Mwe in 2052. In his lecture at the IAEA in 2007, Dr. Kakodkar estimates that this capacity will constitute around 25 per cent share of electricity generation in the country by the year 2050. (Nuclear India, Vol. 41, No. 03-04, Sept.-Oct. 2007).

Let us now review the probable scenario of nuclear power capacities in the short and medium terms i.e. in 2012, 2017, 2032 and 2052.

We know that 18 nos. PHWR and 2 nos. BWR are already operating with a total installed capacity of 4780 Mwe (see Table 13.5A). Construction activities are underway in full swing at six other reactors – three PHWR, two LWR and one 500 MWe PFBR, on completion of which by about 2010, there will be 23 reactors with total installed capacity of 7280 Mwe (refer to Dr. Kakodkar 2008 amended as on date).

The Government of India had also sanctioned the construction of 4 x 700 PHWR, whose design had been finalised as early as 2007/2008 and construction in hand and a further 4x700 MWe or 4x1000 MWe will be taken up from 2010. These eight reactors together add up another 5600/6000 MWe before 2017.

Of the imported LWRs in the five nuclear parks with maximum capacity of 40,000 MWe, it is expected that they will contribute at least 5000 MWe by 2017 in eight years and another 25,000 MWe by 2032 in further 15 years.

With the commissioning of the foreign LWR, it is expected newer generation Indian FBRs will come into being to utilize the spent fuel with twice generating capacity of the LWLs to 50,000 MWe and another 10,000 MWe from AHWR in the 15 years period ending in 2032.

The projected nuclear power capacities in the run up to 2032 and beyond are now be summarised in Table 13.5B:

Table 13.5A: Details of 20 No. Indian Nuclear Reactors March 2011

Nos.	*Atomic Power Station*	*Unit*	*Type*	*Capacity MWe*	*Date of Commissioning*
1	Tarapur (APS)	1	BWR	160	Oct 28, 69
2		2	BWR	160	Oct 28, 69
3		3	PHWR	540	Aug 18, 06
4		4	PHWR	540	Sept 12, 05
5	Rajasthan (RAPS)	1	PHWR	100	Dec 16, 73
6		2	PHWR	200	Apr 1, 81
7		3	PHWR	220	Jun 1, 81
8		4	PHWR	220	Dec 23, 00
9		5	PHWR	220	Feb 4, 10
10		6	PHWR	220	Mar 31, 10
11	Madras (MAPS)	1	PHWR	220	Jan 27, 84
12		2	PHWR	220	Mar 21, 86
13	Kaiga	1	PHWR	220	Nov 16, 00
14		2	PHWR	220	Mar 16, 00
15		3	PHWR	220	May 6, 07
16		4	PHWR	220	Jan 20, 11
17	Narora (NAPS)	1	PHWR	220	Jan 1, 91
18		2	PHWR	220	July 1, 92
19	Kakrapar	1	PHWR	220	May 6, 93
20		2	PHWR	220	Sept 1, 95
	TOTAL			4780	

Source: Nuclear Power Corporation of India.

Table 13.5B: Projected Indian Nuclear Power Capacities

Reactor Details	*Power Capacities MWe*	*Year of Target*
20 nos. PHWR and others per Table 5A	4780	
3 nos. PHWR / LWR	2500	
Total (2012)	7280	2012
8 nos. PHWR / LWR (Indian) 700 MWe	5600	
5 nos. PFBR / FBRs (Indian) 500 MWe	2500	
Add Power contributions from Foreign LWRs	5000	
Total (2017)	20,380	2017
Balance Contribution from Foreign LWRs	20,000	
Indian BFR / AHWR	23,000	
Total (2032)	63,000	2032
Total (2052) per Dr. Kakodkar projection	500,000	2052

India's Futuristic Programmes in Nuclear Power

In support of the above tentative targets of nuclear power generation, there are a number of very important areas of nuclear research and development programmes that have been undertaken.

- *AHWR Programmes:* The construction of the Advanced Heavy Water Reactor will enable India to proven technology of large-scale utilisation of thorium in a reactor and also demonstrate the efficiency of a number of pressure safety features. The design is finalised and construction will start any time.
- *High-Flux Research Reactor:* India has already a high-flux Research Reactor 100 MWe "Dhruv" which went critical in August 1985. Constructing another high flux reactor for testing materials and producing high specific radioactive isotopes is an important thrust goal.
- *Augmenting Reprocessing:* India has been reprocessing spent fuel and recovering plutonium, which is a fuel for the FBR. To meet the requirement of India's expanding FBR programme, it is essential to augment the scale of its operation. The overall growth of India's nuclear programme will depend on how fast India keeps building more FBRs. The augmentation of reprocessing is thus a thrust area.
- *Enrichment of Uranium:* India is working on uranium enrichment for the growing number of Indian Light Water Reactors. The target is now to improve the capacities and efficiencies of the isotopes separation processes and to deliver products at the internationally competitive prices.
- *Nuclear Waste Management:* This is an important thrust area of growth of nuclear power. To handle high-level waste, that is, radioactive material, BARC has already developed Waste Immobilisation Plant, vitrifying them into liquid glass and pouring and solidifying them into canister, to be finally placed in sealed stainless steel container in deep repositories.
- *Accelerator-driven systems (ADS):* It is required for two purposes. Firstly, ADS produce neutrons by spallation process, which means boiling. Neutrons in a reactor are produced by fission. When high energy protons (from an accelerator) are bombarded on heavy nuclei lead, the target nuclei get excited to level that some of the neutrons are liberated from the nucleus. This is known as spallation. Spallation neurons can augment the neutron supply, which can again be used for converting thorium to uranium-233. The second reason for development of ADS is that they are eminently suitable for transmuting the highly radioactive waste from the conventional nuclear reactors to short-lived radio-nuclides, which do not require long-term storage under surveillance.

However, ADS development will take time, because this involves a series of steps and each of them are technically challenging.

Development of compact High temperature Reactor:

The present generation of PHWR reactors uses heavy water as coolant and the maximum allowable temperature is 300° Celsius only. In FBR, it is 550° Celsius, hence coolant is liquid sodium. Now, India wants to develop a futuristic reactor where the core temperature can reach 950° Celsius, which is required to generate hydrogen from water. Hydrogen is the new fuel of the future hydrogen economy, when fossil fuel will be all exhausted. The objective of the future Compact High Temperature Reactor (CHTR) is to develop a reactor system that will produce both electricity and hydrogen. When fossil fuel will not be available, liquid hydrogen can be bottled in a car as automobile fuel either for fuel cell or even for the possible use in the internal combustion engine.

The production of hydrogen using sulphur-iodine process requires high temperature of 950°Celsius. The CHTR programme will require development of a variety of materials, because the reactor will become white-hot object at 950°C. The coolant has to be liquid lead or helium, preferably lead. The third option is to use molten salt which can extract heat at that high temperature. The entire structural material inside the reactor will be ceramic or carbon-based material. Very few materials can withstand CHTR temperature and retain their properties. India will need to develop special kind of fuel for CHTR.

APPENDIX
MAJOR NUCLEAR ACCIDENTS

THREE MILE ISLANDS

The nuclear mishaps at TMI near Harrisburg, Pennsylvania, US occurred on 28 March 1979. The plant was a typical PWR system. The reactor was working normally at nearly full power at 4 a. m; there was a malfunction in the steam generator's feed water system not returning condensate water to the steam generator. Because of this malfunction the turbine generator tripped and the control rods were driven into the core inhibiting fission reaction and slowing down thermal power. Three backup feed pumps should have provided the necessary feed water; but could not as the valve to the steam generator was left closed by oversight; nearly 8 minutes passed before the oversight was discovered and the valve opened. In the mean time the steam generator got dried.

Thus, the primary coolant temperature and pressure rose; the pressure increased to 2355 psi, causing a relief valve to the pressuriser to open up. The coolant then could escape to a vessel called the quench tank designed to condense and cool any releases from the reactor system. The pressure relief valve unfortunately remained stuck in the open condition, a fact unknown to the operators for two hours. Therefore a considerable amount of coolant escaped filling he quench tank and causing a rapture disk on the tank to blow out. The coolant water with some amount of radioactive contamination drained to the Containment building sump.

In the mean time, the reactor vessel pressure continued to fall. At 1600 psi, the emergency core cooling system (ECCS) actuated injecting makeup water into the reactor vessel. To the operator, the pressuriser appeared to be full of water and thus preventing it to function. The operators decided to shut off the ECCS and also the main reactor cooling pumps. The resulting severe lack of water caused the core to heat up and became uncovered. Although the main fission power had been shut off, there remained a large amount of residual heat emanating from the decaying fission products. The coolant flow in the core being inadequate, the fuel rods were damaged and melted, leading to the release of much radioactive noble gases xenon, krypton and iodine out of the reactor. The sump pumps automatically sent the radioactive water from the containment building into the tanks in the auxiliary building next door. The tanks over flowed, permitting the radioactive materials to escape through the filters into the atmosphere.

The reactor cooling system was finally turned on and the core temperature began to fall.

CHERNOBYL

The second most disastrous nuclear accident occurred on April 26, 1986 at the Chernobyl Reactor near Kiev in USSR Ukraine. The nuclear fuel core was severely damaged and melted, explosions occurred blowing away the building roof and large amount of radioactive cloud found way to the atmosphere.

The Chernobyl -4 reactor was of RBMK type, of which there were 18 of them in USSR. Its core was cylindrical, diameter 7 m and height 12 m, consisting of blocks of graphite acting as moderator and structure. The blocks were pierced with vertical holes, through which passed pressure tubes of 8.8 cm diameter. Clusters of 18 slightly enriched (2% U-235) uranium oxide fuel rods were placed inside the tubes and circulating ordinary water, while cooling the fuel elements, got to boiling enabling supplying steam to the turbo-generator. In all 1661 fuel elements formed square array with 25 cm spacing. Separate channels were provided for 222 control and shut-down rods. A refueling machine above the core allowed individual fuel elements assemblies to be changed during the reactor operation. A vapour suppression water pool is located beneath the reactor but was not connected to the core itself.

The sequence of events leading to the accident was revealed in August 1986 in a meeting of IAEA. An experiment involving supply of electricity to the reactor equipment in emergency situation was being performed. If the electricity supply system to the plant got disrupted, standby diesel generator were to take over the supply. To bridge the gap till the diesel generator started supply, an auxiliary power supply was needed. The test related to the use of electricity produced during the coasting down of the turbocharger. Emergency power was to be provided to the coolant pumps and feed water pumps of the steam generator.

The first step in the test was to reduce the power from 3200 MWt to the range 700-1000 MWt. In attempting to do so, the operators allowed the power to drop to 30 MWt. At this level there was too low a neutron flux to burn out the xenon-135 being produced. The build-up of the absorber made it very difficult to bring back the power level. In violation of all the rules, the operators pulled out most of the control rods, but could not still get the power back to higher than 200 MWt. At this power the rector system tended to get unstable.

At this point coolant pumps were run at the flow rate higher than required for the low power output and the coolant was brought near the boiling point. Later when the coolant flow was reduced, steam voids were created. The fatal flaw in the design of the RBMK reactor played its role at this point.

The graphite reactor had an inherent positive void coefficient, in contrast to the negative coefficient of the light water reactor. The reactivity produced by the steam voids caused the reactor to flash up around 30,000 MWt i.e. ten times the operating level. The power could not be reduced quickly as the control rods were too far away to have any effect. The excessive energy pulverised the fuel and caused the steam pressure to build up rapidly. The pressure increase ruptured the coolant tubes and the resultant explosion blew out a hole in the roof. The normal nitrogen-helium blanket of the core was lost and air and water contacted the hot graphite materials.

FUKUSHIMA, JAPAN

The nuclear disaster at the Fukushima Daiichi is the story of a series of equipment failures, nuclear meltdowns, and releases of radioactive materials at the Fukushima Nuclear Power Plant, following the Tôhoku earthquake and tsunami on 11 March 2011. The plant comprises six separate boiling water reactors designed and built by GE, and maintained by the Tokyo Electric Power Company (TEPCO). Fukushima disaster is the largest nuclear accident since the 1986 Chernobyl disaster, but it is more complex as multiple reactors and spent fuel pools are involved.

At the time of the quake, Reactor 4 had been de-fuelled while 5 and 6 were in cold shutdown for planned maintenance. The remaining reactors shut down automatically after the earthquake, with emergency generators starting up to run the control electronics and water pumps needed to cool reactors. The entire plant was flooded by the 15 m (49 ft) tsunami wave, including low-lying generators and electrical switchgear in reactor basements and external pumps for supplying cooling seawater. The connection to the electrical grid was broken as the Tsunami destroyed the power lines. All power for cooling was lost and reactors started to overheat, owing to natural decay of the fission products created before shutdown. The flooding and earthquake damage hindered external assistance.

In the hours and days that followed, reactors 1, 2 and 3 experienced full meltdown. It has been estimated that the upper 75 % of the core of unit one melted and slumped into the lower quarter of the core at 15:10 on 12 March; the core mass would have cooled again as it entered the water in the bottom part of the reactor tank before reheating during the time before sea water was added at 20:20. Hydrogen explosions destroyed the upper cladding of the buildings housing Reactors 1, 2, and 3, with the explosions at Reactors 1 and 3 damaging the secondary containment of Reactor 2. With the remnants of its reactor core fallen to the bottom of its damaged reactor vessel, Unit 1 continue to leak cooling water approaching three months after the initial events; similar conditions are hypothesised to exist at the other two melted-down reactors in the complex.

Fuel rods stored in pools in each reactor building began to overheat as water levels in the pools dropped. Fears of radioactivity releases led to a 20 km (12 mi) radius evacuation around the plant while workers suffered radiation exposure and were temporarily evacuated at various times. One generator at Unit 6 was restarted on 17 March allowing some cooling at Units 5 and 6 which were least damaged. Grid power was restored to parts of the plant on 20 March, but machinery for Reactors 1 through 4, damaged by floods, fires and explosions, remained inoperable. Flooding with radioactive water continued to prevent access to basement areas where repairs are needed. However, on 5 May, workers were able to enter reactor buildings for the first time since the accident.

Measurements taken by the Japanese science and education ministry in areas of northern Japan 30–50 km from the plant showed radioactive cesium levels high enough to cause concern. Food grown in the area was banned from sale. Based on worldwide measurements of iodine-131 and cesium-137, it was suggested that the releases of those isotopes from Fukushima are of the same order of magnitude as those from Chernobyl in 1986; Tokyo officials temporarily recommended that tap water should not be used to prepare food for infants. Plutonium contamination has been detected in the soil at two sites in the plant, although further analysis revealed that the detected density are within limits from fallout generated from previous atmospheric nuclear weapons tests. Two workers, hospitalised with non-life threatening radiation burns on 25 March, had been exposed to between 2000 and 6000 mSv of radiation at their ankles when standing in water in Unit 3. Due to frustration with Tokyo Electric Power Company (TEPCO) and the Japanese government "providing differing, confusing, and at times contradictory, information on critical health issues", a citizen's group called "Safecast" has been recording detailed radiation level data in Japan. The Japanese government "does not consider non-government readings to be authentic". The group uses off-the-shelf Geiger counter equipment. Members of the Air Monitoring station facility at the Department of Nuclear Engineering at the University of Berkeley, California have been doing extensive tests of environmental samples in Northern California.

References

Anderson, P.K. *et al.,* (2004), *Nuclear Waste*, New Mexico Institute of Mining, Encyclopedia of Energy, ed. C J Cleveland, Elsevier Press.

Banerjee, B.P. (2005), *Handbook of Energy and Environment in India*, Oxford University Press, YMCA Library Building, Jaisingh Road, New Delhi 110001.

Ghosh Roy, M.K. 2011, *Sustainable Development – Environment, Energy and Water Resources*, Anne Books, New Delhi.

Harper Collins India 1992, "*How Things Work–The Universal Encyclopedia of Machines*", Indus Publishers.

Jagger, J. (1994), The Natural World of Radiation, *Environmental Ethics* – Readings in Theory and Application, ed. L J Pojman, Jones and Barlet, Boston.

Jagger, J., (1994), Nuclear Power is Safe and We Need it, *Environmental Ethics*, ed. L Pojman, John and Barlet, Boston.

Kakodkar, Anil (2007), IAEA 51st General Conference, Vienna September 19, *Nuclear India*, Vol. 41, No. 03-04, September-October 2007.

Kakodkar, Anil (2008), Managing New Nuclear Paradigm, *Nuclear India*, Vol. 42, No.1-2, July-August 2008

Lenssen, N. (1994), Nuclear Waste: *The Problem Wont Go Away, Environmental Ethics*, ed. L Pojman, Jones and Barlet, Boston.

Magaud, P. and Marback, G. *et al.*, Nuclear Fusion Reactors, *Encyclopaedia of Energy*, ed. C.J. Cleveland, Elsvier.

Masters, G.M. 1991, *Introduction to Environmental Engineering and Science*, Prentice Hall, NJ, USA.

Mcfarlane, H.F. (2004), Nuclear Processing, Argonne Lab., Illinois, *Encyclopaedia of Energy*, ed. C J Cleveland, Vol. 4, Elsevier Press.

Ministry of Power, Government of India, Web Site www.powermin.nic.in

Reyce, Jr. J.N. and King, Jr. B. 2004, Nuclear Energy, *Encyclopaedia of Energy*, ed. Cleveland, Elsevier.

Subramanium, T.S. 2007, "Special Feature: BARC at 50", *Front Line Magazine*, Vol. 24, Number 8, April 21 - May 4, 2007, India.

Tester, G.M. Drake, E.M. and Driscoll, M.J. *et al.,* 2005, "Sustainable Energy – Choosing among Options", Massachuchusetts Institute of Technology, Prentice Hall India, New Delhi.

Vogt, D.K. 2004, Nuclear Fission Reactor, Lawrence Lab., California, *Encyclopaedia of Energy*, ed. Cleveland, Vol. 4, Elsvier.

14

Green Bio-Energy–Ethanol and Biodiesel

I

Introduction to Biomass Energy

The net energy available from Biomass when it is combusted ranges from 8 MJ/kg for green wood to 20 MJ/kg for dry plant matter compared to 27 MJ/kg for coal or 55 MJ/kg of methane gas. However, plant biomass can be converted to various forms of energy like biogas or producer gas, liquid fuel like bio-ethanol and bio-diesel which have higher energy densities and will be described in later sections.

Globally, Bioenergy constitutes about 11% of total primary energy consumed as per IEA annual data 2002 (see also Fig. 14.1). While Fig. 14.1 shows the percentage use of various energies, the Table 14.1 shows also the quantities of actual energy consumed along with the percentage values.

Table 14.1: Global Primary Energy Consumption

Energy forms	*Energy Quantities (EJ/Y)*	*% Energy Contribution*
Fossil Fuel	388	80
Nuclear	26	7
Hydro	28	2
Biomass	45 (+/- 10)	11

In developing countries, biomass is used more extensively in domestic consumption compared to the developed countries where the percentage usage is lower:

Industrial Countries < 10%;

Developing countries ~30%

Biomass Photosynthesis enables biomass to be created. The carbon in the atmospheric carbon dioxide (CO_2) is captured through the absorption of solar energy into organic carbohydrate. The chemical formulation is

$$CO_2 + H_2O + \text{sunlight} \longrightarrow (CH_2O) + O_2 \qquad \text{Eqn. 14.1}$$

The organic building block (CH_2O) is the source carbohydrate of all cellulose, sugar and starch, which constitute the plants, trees and all biomass. In world forest there is standing biomass carbon of 774 GT, while new biomass carbon is generated at the rate of approximately 33.3 GT of dry biomass per annum as a result of photosynthesis on a global scale. The biomass presently contributes 11% of the total annual energy consumption (see Fig. 14.1) and meets about 35% of the energy needs of the developing countries.

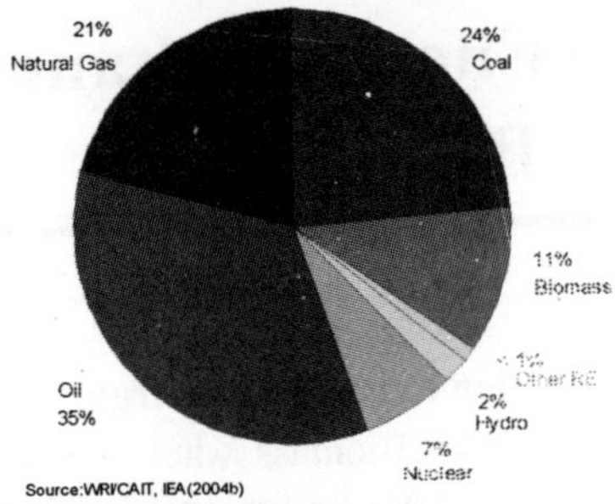

Fig. 14.1: World Primary Energy Sources.

Basic Composition of Biomass

From Eqn. 14.1, we observe that biomasses are basically carbohydrates. Specifically, the wood in biomass is made up of cellulose, hemi-cellulose and lignin in varying proportions (say 50%, 30% and 20% respectively). The major constituents - cellulose, a polyhexose and hemicelluloses, a polypentose are carbohydrates, whose origin again goes back to the basic carbohydrate building block in Eqn. 14.1. Lignin has a phenolic structure highly resistant to microorganisms.

In comparison fossil fuels are hydrocarbons. Small proportions of hydrocarbons in biomass are generally polyisoprenes (terpenes) – that is polymers of alkynes hydrocarbon, Isoprene C_5H_8. They are generally found in euphorbia such as rubber.

Biomasses being a carbohydrate, the vegetable oils are esters; whereas mineral oils are hydrocarbons. Vegetable oils can be directly used as a fuel by itself or can be transformed as bio-diesel by transesterification.

II
VARIOUS FORMS OF BIOENERGY

There are various ways of using biomass as fuel energy. The traditional sources of biomass energy is by combusting wood and agro-wastes at 64%, followed by municipal wastes (MSW) at 24%, agricultural wastes at 5% and land-fill gases at 5%.

In recent times, biomasses, constituted of cellulose, hemi-cellulose and lignin mainly, are getting converted into secondary energies like different liquid bio-chemicals and fuels (ethanol and biodiesel), gases (methane-based etc.) by three different conversion technologies:

- Biochemical processes;
- Thermo-chemical processes;
- Mechanical processes.

Biochemical Processes

1. *Anaerobic Digestion*: Anaerobic digestion is the biochemical conversion of biomass in which various species of bacteria participate in the decomposition of organic matter in the absence of oxygen to produce a biogas consisting approximately 50-75% methane and 25-45% carbon dioxide. The biogas produced typically has an energy content of 20-40% of the heating values of the biomass feedstock. Digester ranges from 1 m^3 for domestic units to 2000 m^3 for commercial plants.
2. *Hydrolysis and Fermentation*: Fermentation is used commercially on a large scale in various countries to produce ethanol from sugar and starch crops as well as lingo-cellulose biomass. Typically biomass is ground and cellulose and hemicelluloses are converted by enzymes to sugar; subsequently the sugars are converted by yeasts to ethanol. Sugar cane and beets are probably the best feed stocks. Wood and straw, basically of longer chain polysaccharides molecules require pre-treatments (acid, enzymatic or hypo-thermal hydrolysis) to depolymerise cellulose and hemi-cellulose to monomers, which will be subjected to yeast-fermentation later.

Thermo-chemical Processes

1. *Combustion*: It is a widely used process in which the biomass is burnt in excess air but in a controlled manner to produce heat at 800-1000° C. In efficient combustion process, flue gases are mostly composed of CO_2 and water vapour.
2. *Gasification*: The process is the partial oxidation of carbon-rich biomass feed stock into a combustible gas (Producer gas) at elevated temperatures typically in the range of 800-900°C in restricted atmosphere of air or oxygen. In using biomass, the resultant gas is a mixture of carbon monoxide, carbon dioxide, hydrogen, methane and water vapour, as well as small quantities of higher hydrocarbons. The producer gas has a lower calorific value of 4-6 MJ/m^3 compared with the calorific value of natural gas at about 39 MJ/m^3.
3. *Pyrolysis*: This is a thermo-chemical process that converts biomass into liquid (bio-oil or crude), solid (charcoal), non-condensable gases, acetic acid, acetone, methanol by heating the biomass at 400 – 800°C in the complete absence of oxygen. Pyrolysis bio-oil has a heating value of 17 MJ/kg or about 60% of the value of diesel, on volume basis.

Mechanical Conversion – Extraction Process to Bio-Diesel

Oil seed crops which contain high fractions of oil can be physically crushed and oil extracted as glyceride vegetable oil; this then is esterified to mono-alkyl ester using methanol in presence of a catalyst to bio-diesel. The energy content is around 40 MJ/kg comparable to petroleum-Diesel.

III

Bioenergy Contribution to World Energy System

Primary Energy: The biomass has the longest history of any energy source and still provides 44 EJ/Y amounting to 11 % of world's primary energy needs as shown in Table 14.1. Even in an advanced country and energy-intensive economy like US, it contributes 3.37 percentage points of primary energy (see Table 14.2).

Table 14.2: Total Primary Energy and Total Biomass Energy for Top Ten Energy Consuming Countries

Country	*Total Primary Energy (EJ/Y)*	*Total Biomass Energy (EJ/Y)*	*Bio-energy as % Total*
USA	99.85	3.373	3.37
China	47.25	9.244	19.56
Russia	26.18	0.326	1.24
Japan	22.78	0.242	1.06
India	20.84	8.596	41.25
Germany	14.74	0.366	2.48
France	11.16	0.496	4.44
Canada	10.90	0.487	4.47
U.K.	10.10	0.093	0.92
South Korea	8.41	0.092	1.09
Brazil	7.80	1.862	22.14
World	400.00	44.00	11.00

Source: Klass 2004; 1 EJ =277.8 TWH.

Table 14.3: Total Electricity & Biomass-based Electricity Generations For Top Ten Electricity Consuming Countries

Countries	*Total Electricity Energy TWH/Y*	*Biomass-based Electricity TWH/Y*	*Bio-Energy Electricity as percentage of Total*
USA	4003.5	68.805	1.718
Russia	1239.3	1.963	0.158
China	845.3	2.075	0.245
Japan	1081.9	16.518	1.526
India	527.3	-	0
Germany	567.1	10.121	1.785
France	535.8	3.290	0.614
Canada	605.1	7.379	0.012
UK	372.2	4.360	0.012
South Korea	292.4	0.396	0.135
Brazil	332.3	8.519	0.026
World			

1 TWH = 10^{12} x 3600 J = 3.6 x 10^{15} J = 0.0036 EJ ; 4003.5 TWH = 14.41 EJ.

Electrical energy: Even in electrical power generation, the biomass has a significant 1.7 per cent contribution in US (see Table 14.3).

Transportation fuel energy: In the developed countries, about two-third of Petro-energy is consumed in the Transportation sector comprising automobile, railways, shipping and aviation industries. This huge consumption of fossil fuels has given rise to unprecedented pollution of the global atmosphere. Fortunately, two liquid bio-fuels: bioethanol and biodiesel are getting technologically acceptable and commercially viable as the transportation energy.

IV
Viability of Bio-Ethanol and Bio-Diesel as alternative Transportation Energy

A. ETHANOL (C_2H_5OH)

The bio-fuel Bio-Ethanol offers the possibility of reducing the carbon dioxide reduction by about 30 per cent and therefore its application should also receive all encouragement. Still today fossil fuels are mostly used in the automotive sector, which account for more than 50 per cent of the global consumption of fossil fuels. The bio-fuels can be blended with the fossil fuel up to 20 per cent without requiring major modifications of the internal combustion engine; for higher proportion of bio-fuels can also be used with minor modifications of the heat engine.

Properties of Ethanol vis-à-vis Gasoline

Ethyl Alcohol or Ethanol C_2H_5OH is a colourless, clear liquid and looks like water. Bio-ethanol produced by fermentation from biomass is non-toxic and biodegradable. It has excellent fuel properties for spark ignition internal combustion engine and compares very favourably to the Petro-gasoline as seen from the Table 14.4.

Table 14.4: Fuel Properties of Ethanol

Properties	*Ethanol*	*Gasoline*
Octane number	99	88
Latent Heat of vaporization (KJ/L)	664	250
Calorific value (MJ/L)	21.2	30
Flash Point Celsius	78	27
Reid Vapour Pressure (Kilopascal)	16	71
Reid Vapour Pressure (Psi)	2.3	10.3

Source: Wyman 2004.

From the calorific value above, it will be observed that for the same engine energy output, more ethanol will be required than gasoline, but using higher compression and hence higher efficiency of ethanol engine would reduce the difference.

Considering the properties, economics and favourable energy balances achieved during production of ethanol, we may summarize the three principal concerns of using ethanol as an alternative transport fuel:

1. *Energy Balance Concern*: From Table14.6, it is seen that the input energy to get output of one unit of ethanol energy is only 25%, which may or may not be fully fossil energy. Secondly from the consideration of calorific values (Table 14.4), each litre of ethanol can replace the consumption of 0.63 litre of gasoline. Thus ethanol, from sugar cane in Brazil or corn (in US) is viable transport energy on energy balance consideration.
2. *GHG Emission Concern*: As it is, ethanol being an alcohol will emit about 30 – 40% less carbon dioxide than a fossil fuel, which is generally a heavy hydrocarbon. Though today, some amount of fossil fuel may have to be expended in growing sugar crops, Ethanol be fermented burning bagasse or bio-waste. Ethanol is therefore more carbon-economic than any fossil fuels. If we think more fundamentally, all biomass energy is 100% carbon-neutral, for whatever CO_2 is emitted on combustion is reabsorbed by the new crop season.
3. *Cost Advantage Concern*: Here is a riddle that as time goes, the petroleum prices will steeply increase because of scarcity, while the bio-fuel prices will steeply fall because of gradual change over to mass production and improving input technology. In Brazil, where even today ethanol is much batch-produced, the cost of bio-ethanol is not higher than gasoline (Table 14.7)

The production processes of Bio-ethanol are described in more details in Appendix A.

B. BIO-DIESEL (R-COO-CH3)

Properties of Bio-Diesel vis-à-vis Petro-Diesel

Like bio-ethanol, bio-diesel, an organic ester, also offers substantial reduction in the GHG emission.

Biodiesel is really an extract vegetable oil. Vegetable oil has properties strikingly similar to those of Petro-Diesel and hence may be considered as

Table 14.5. Comparative Physical Properties of Bio-diesel and Petro-diesel

Fuel Properties	*Jatropha Oil*	*Jatropha Methyl Ester*	*European Std. for Bio-diesel*	*Petro-Diesel*
Density at 15 C (kg /m^3)	920	880	860-900	820
Pour Point (C)	5	4	-	-23
Viscosity at 20C (m-Pa-s)	75.6	4.8	3.5 - 5	2.6 -4.1
Calorific Value (MJ/kg)	37	39	-	42
Flash Point (C)	340	191	>120	74
Acid value (mg KOH /g)	8	0.41	< 0.50	0.35
Carbon residue %	4.2	0.05	<0.3	0.17
Ash Content %	1.1	0.014	<0.02	0.17

a potential fuel for Internal combustion engine as an auto-fuel or other transportation system like Railways or Shipping.

In fact, Professor Diesel invented the Diesel Engine to be run with vegetable oil. The major problem associated with direct use of vegetable oil in the engine with its high viscosity (Table 14.5) due to three ways interlinking of fatty acid chain in triglyceride form of the raw oil. One possible way to break open the interlinking is by transesterification of the Triglyceride into mono-alkyl ester of single fatty acid chain. The standard reaction from vegetable oil (Triglyceride Oil) to Biodiesel (Mono-Alkyl-Ester) is depicted in Fig. 14.3 and may thus be represented by the chemical formula R –COO-CH_3, where R stands for a fatty acid chain radical. The detailed production from different feed stock is described in Appendix B.

Appendix A

BIO-ETHANOL FUELS

Basic Properties and Production Processes

Production methods of Bio-ethanol

The final process in the production of Bio-ethanol is the fermentation of sugar. But currently there could be three distinct source biomass feed stocks for the production of Bio-ethanol:

(i) Sugar cane crop,

(ii) Corn / beet starch crop,

(iii) Plant cellulose biomass.

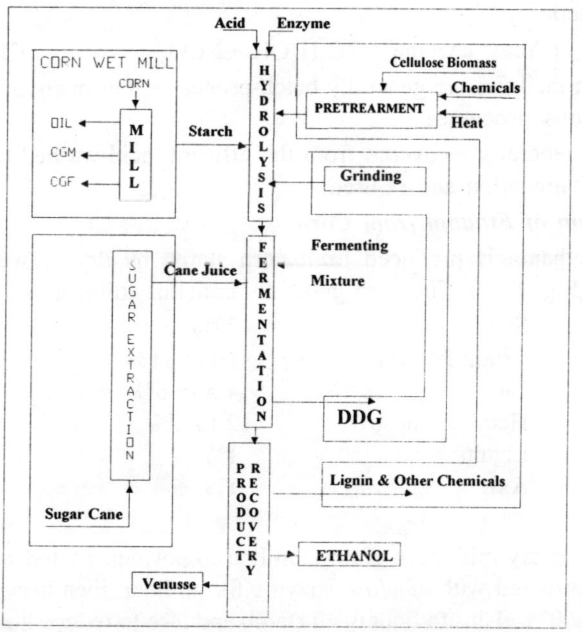

Fig. 14.2: Schematic Production Process of Bio-ethanol (Wyman 2004)

Brazil now is the largest producer of bio-ethanol and sugar cane is the main feedstock of producing ethanol in Brazil. In US, the next largest producer, corn starch is used. The cheapest source is the plant biomass conversion, for which the technology is presently under development. India cannot possibly use either sugar cane or corn because of high domestic demand as food. Hence, India should try the third process of using plant biomasses, which are abundantly available.

Fig. 14.2 shows the Schematic Process Flow diagram of producing bio-ethanol using all the above three feed stock sources.

(a) Ethanol Production from Brazilian Sugar Cane

Brazil produced 272 million tons of sugar cane in 2001-2002, second only to India. From the 272 million tons of sugar cane, they produced 13.9 billion litres of ethanol and also 13.5 million tons of sugar. Brazil produced about 5100 litres of bio-ethanol from a hectare of land in 1996. Brazilian sugar cane generally contains 12 to 17% sugar (90% sucrose, 10% glucose) and 68 to 72% moisture.

In the dedicated ethanol plant (see Fig. 14.2), cane juice is heated to about 105 to 110°C to reduce microbial contamination and is then fermented. In the combined ethanol- sugar production facilities, sucrose is removed by crystallisation and the remaining molasses containing 65% sugar is fermented.

In either process, sugar $C_6H_{12}O_6$ is readily fermented to C_2H_5OH ethanol in presence of common yeast (sacharomycos cerevisae) as catalyst by the following reaction:

$$C_6H_{12}O_6 + \text{Yeast enzyme} = 2C_2H_5OH + 2\ CO_2 \qquad \text{...E14.2}$$

Brazilian facilities are generally batch-processes, but many of them are also continuous processes.

Yeast is generally recovered from the effluent, acid-washed to remove bacterial contamination and re-used.

(b) Production of Ethanol from Corn

In US, ethanol is produced from corn starch by dry or wet milling process (see Fig. 14.2). The corn generally contains followings:

Starch	70%
Crude Protein	10 to 11%
Oil	4.5 to 6%
Hemi cellulose	2 to 3%
Lignin	1%
Ash	1%

Process

In a typical dry mill, corn grain is milled to powder, heated with water to 85° Celsius mixed with *amylase* enzyme for an hour, then heated further from 110 to 150° Celsius to liquefy all starch and also to reduce the bacterial contamination. Then the content is cooled back to 85° C steadily and more alpha-amylase is added. The mixture is cooled till complete conversion to dextrose sugar. This overall saccharification operation is:

$$(C_6H_{10}O_5)n + nH_2O + \text{Amylase} = n.C_6H_{12}O_6 \qquad \text{... E14.3}$$

The sugar is then fermented in presence of the catalyst Yeast, similarly to the cane sugar fermentation. The fermented broth, known as beer, contains 12 to 14% ethanol and is distilled to 95% purity. If necessary, the hydrous ethanol can be further dehydrated to anhydrous ethanol with molecular sieve.

'The solid remnants containing 27% protein are dried as Distillers' Dried Grain (DDG) or with soluble as (DDGS). A portion of the corn protein is sold as corn gluten meal (CGM), a 60% protein product.

(c) Production of Ethanol from Cellulose Biomass

Cellulose Material

Finally, Ethanol can also be made from the sugar contained in the structural part of the plant biomass.

Plants contain the following by deadweight percentage –

- 40 to 50% cellulose, which is 10,000 glucose molecules joined by beta 1-4 linkages as long line molecules, aligned parallel to one another joined by hydrogen bonding.
- 25 to 30 % hemi-cellulose, which is amorphous cellulose made up of following five sugars.

  Arabinose, Galactose
  Glucose, Mannose
  Xylose,

  and hexuorine acid, methoxyl acetyl, free carboxyl group.
- 15 to 20%, Lignin, which is complex phenylpropanic acid polymers. Lignin and Hemi cellulose are joined by covalent bonds and cellulose fibers into microbial bonds.

  Plants also contain lesser amounts of minerals, free sugar, protein, and other compounds that vary with species.

  Together, this composite of cellulose, hemicelluloses and lignin and other lesser ingredients are called lignocelluloses or just cellulose biomass.

  Cellulose biomass is available abundantly as

- Agriculture residues (e.g. corn stover, sugarcane bagasse),
- Forestry residues (e.g. forest thinnings)
- Industrial wastes (e.g. paper, sludge)
- Municipal bio-wastes
- Woods (e.g. trees and plants)

Basic Process

The method to convert cellulose to ethanol involves three basic processes.

(i) Sulphuric and other acids can convert both cellulose and hemicelluloses into sugars;

(ii) Alternatively, enzyme *cellulase* can hydrolyze cellulose into glucose. Similarly, a collection of enzyme loosely called *hemicellulase* can breakdown hemicelluloses into above five sugars. The balance lignin cannot be converted into sugar but can provide boiler fuel feed.

(iii) Sugars are then fermented into ethanol as given by equation 14.2

Detailed Processes

(i) Cellulose is reacted in 0.5 to 1% dilute Sulphuric acid at 160 to 190°C for approximately 10 minutes, releasing most hemi-cellulosic sugar by a hydrogen reaction similar to equation 14.2.

(ii) The content mixture is then quickly cooled down to 100°C to stop the reaction and also to remove acetic acid, furfural and other fermentation inhibitors which are released from the biomass.

(iii) The solid and the liquids are separated to remove further the remaining acetic acid and much of the sulphuric acid. Sulphuric acid is further neutralised by lime treatment and the solution pH increased to 10 by excess liming.

(iv) A small portion of the solid alone or also including the treated liquid is fed with a microbial fungus trichoderma reesai to convert the feed into the enzyme *cellulase.* The effluent is now combined with the bulk of the pretreated solid to hydrolyze cellulose into glucose. Enzyme Yeast is added to ferment the glucose to ethanol soon after the release into a simultaneous saccharification and fermentation called SSF approach.

(v) The fermentation beer broth containing 5% ethanol is then distilled to 95% ethanol and then could be further enriched to 100% anhydrous ethanol passing through molecular sieve.

Process Improvement

Further process research was carried out to achieve both economic and high productivity in following two routes:

(i) SSF approach was refined to achieve improved economies and productivity.

(ii) Secondly, ordinary yeast could not ferment all the different sugars of the biomass, which was needed for high productivity. In 1980, bio-scientists were successful to genetically engineer new bacteria to make ethanol from all five sugars.

Energy Balance

It will be interesting to know how much average energy input is required to produce a litre of the ethanol by any of the above three methods. From the detailed energy balance investigation, the observations have been derived and tabulated in Table 14.6.

Table 14.6: Average Energy Balance

Process Description	% Ethanol Fuel value
To grow biomass	10%
To transport to factory	5%
To factory processing	5%
To fuel distribution	5%
Total Energy input	25%

Cost of Production of Ethanol

The cost of production of ethanol from Cane Sugar in Brazil is outlined approximately in Table 14.7 and that from Corn starch in US in Table 14.8. Obviously Brazilian ethanol is cheaper than the US Ethanol.

Table 14.7 : Costing of Ethanol from Sugar in Brazil

Items of Costs	*Costs at USD per litre*
Labour	0.006 – 0.007
Maintenance	0.004 – 0.006
Chemicals	0.002
Energy	0.004-0.003
Others	0.004
Sugar Materials	0.127 – 0.134
Total Cash Costs	0.147 – 0.156
Capital Costs per annum litre	0.52
Interest on Capital & other financing	0.21 – 0.25
Total Indirect Cost per litre	0.53 – 0.77
Total Cost per Litre of Sugar Cane Ethanol	0.677 – 0.926

Table 14.8: Costing of Ethanol from Starch in US

Items of Costs	*Costs at USD per Litre*	
	Dry Process	*Wet Process*
Average Costs of all feedstock for dry and wet matters	0.23	0.23
Electricity and fuels	0.035	0.030
Operating labour, repairs and maintenances	0.029	0.033
Yeast, Enzymes, Chemicals and other ingredients	0.024	0.030
Management, Administration, insurance and layers	0.09	0.09
All other direct costs	0.013	0.019
Total Cash Costs	0.25	0.25
Average Capital Costs (annual litres)	0.50	0.60
Total Costs per litre of Corn Ethanol	0.75	0.85

Both the sugar and the starch are edible products and will attract higher economic price. The current effort is to convert cellulose in plants to ethanol by acid hydrolysis. Because of lower feedstock price of cellulose, ethanol from cellulose as and when commercially developed, will be most price competitive (see Table 14.9).

Table 14.9: Approximate cost of Cellulose Ethanol

Items of Costs	*Costs in USD per year*
Capital Costs	\$23.4 per year
Feed Stock Costs	\$19.3 per year
Other Variable Costs	\$5.7 per year
Fixed Operating Costs	\$7.5 per year
Total Costs incurred per annum	\$55.9 per year
Total Annual Production	200 million litres
Basic Costs per litre	\$0.28 per litre
Add transportation, storage, distribution	\$0.07 per litre
Total Cost per litre	\$ 0.35 per litre

Appendix B
BIO-DIESEL FUEL
Basic Properties and Production Processes

What is bio-diesel?

Vegetable oils are basically Triglycerides of Fatty acids as shown in Fig. 14.3, where R1 are Fatty acid hydrocarbon radicals. But vegetable oil has properties strikingly similar to those of Petro-diesel and hence it may be considered as a potential fuel for Internal Combustion engine as auto-fuel or other transportation system like Railways. In fact, Professor Diesel invented the Diesel Engine to be run with vegetable oil. The major problem associated with direct use of vegetable oil in the engine with its high viscosity due to three-way interlinking of fatty acid chain in triglyceride form of the raw oil. One possible way to break open the interlinking is by transesterification of the Triglyceride into mono-alkyl ester of single fatty acid chain by reaction shown in Fig. 14.3.

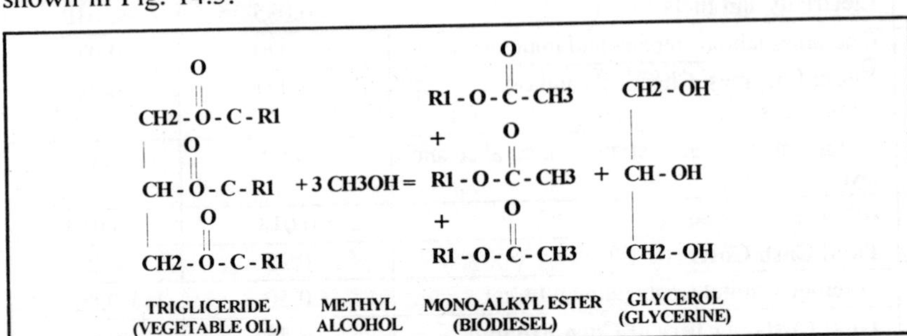

Fig. 14.3: Transesterification to Bio-Diesel

The process of manufacture of Bio-Diesel is schematically shown in Fig. 14.4.

Feed Stock

In USA, refined soybean oil and in the European Union refined Rapeseed and sunflower oil are used in large plants of capacity 3000 tonnes per day.

India, being an edible oil deficient country and net importer of such edible oil at high costs, it is uneconomical to use purified oil for bio-diesel production. Moreover, with a very large population to support, all irrigated and arable lands are engaged in food crop cultivation.

India therefore requires using waste arid and semi-arid lands to grow non-edible tree-borne oil (TBO) bio-diesel of the following botanical families:

— Jatropha Curcas (Ratanjot / Verenda),

— Pongamia Pinnata (Karanj),

— Azadiracta Indica (Neem),

— Madhuca Indica (Mohua).

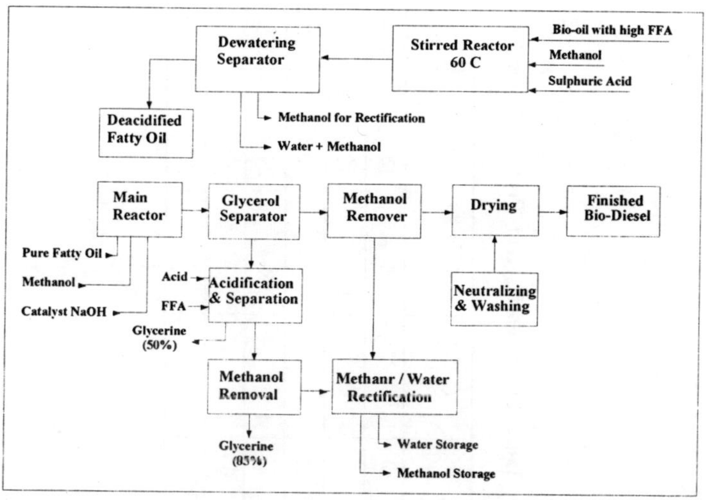

Fig. 14.4: Schematic Production Process of Bio-Diesel (Klass 2004)

There are also good possibilities of using rice bran and Sal fruits (Shorea robusta) as feedstock.

Of all the feed stock plants, Jatropha Curcas, most favoured in India, is a drought-resistant perennial growing in marginal soil. It grows quickly, starts producing fruits in 2 years, maturing in six years and living thereafter for 50 years. The annual output is 1 to 5 tons per hectare. The oil content is 37%, the residue cake makes good fertilizer.

India has about 60 million hectares of waste lands, of which, if 30 million hectares are utilised for energy cultivation with preferably Jatropha plants

Table 14.10: Fatty Acid Profile of Selected Vegetable Oil

Name of Feed Stock	*C8*	*C10*	*C12*	*C14*	*C16 Palmic*	*C16.1*	*C18 Stearic*	*C18.1 Oleic*	*C18.2*	*C18:3*
Jatropha	14-15				14-15	1.3-2.0	5.1=5.8	42-44	34-38	0.2-0.5
Palm				1.1	42.7		4.6	39.4	10.6	
Corn		4			10.3		2	25.5	59.3	1.1
Cotton			1	0.8	21.9	3		21.4	15.2	54.2
Peanut			1		9.4		2.7	48.7	31.1	
Sunflower					5.7		4.8	15.3	71.2	0.5

Source: da Rosa 2009.

producing 2 tons of oil per hectare per annum, amounting to a national output of 60 million tons of Bio-diesel in say, 10/15 years, being nearly equal to 50 per cent of total Indian consumption today.

Vegetable Fatty Acids

Among the common Fatty acids are Palmitic acid CH_3-$(CH_2)_{14}$-COOH, Stearic acid CH_3-$(CH_2)_{16}$-COOH, which are both saturated and Oleic acid CH_3-(CH2)$_7$-CH = CH_3-$(CH_2)_7$-COOH (unsaturated). Please also see Table 14.10 for some other fatty acid. Glyceride of a fatty acid is a vegetable oil.

Bio-Diesel Production

The process Flow diagram of Bio-diesel Production is shown schematically in Fig. 14.4 below. If the Triglyceride oil contains less than 5% free fatty acid (FFA), no pretreatment of the oil is necessary. In that case, the primary reactor is fed with the following proportions of inputs:

Triglyceride oil .... 60 kg.

Methanol ... ... 13.6 kg.

Catalyst ....0.6 kg

The mixture is heated to 70° C at 1.5 kg/sq. cm pressure for an hour to achieve transesterification of the glyceride to Bio-diesel (mono-methyl ester) and glycerol to about following proportion:

Bio-diesel (CH_3OOCR) 58.0 kg

Glycerol 6.5 kg

To obtain fuel grade bio-diesel to ASTM or Euro norm, proper quality control is necessary to remove / separate the following during the process shown diagrammatically in Fig. 14.4.

(i) Excess Methanol,

(ii) Catalyst (NAOH),

(iii) Process Water,

(iv) Soaps, if any,

(v) unconverted triglyceride, if any,

(vi) Unconverted trace of free fatty acid, (vii) glycerol.

Pre-Treatment

As mentioned earlier that if the triglyceride oil is doped with small amount of the respective fatty acid up to 3%, the same can be neutralised by adding enough alkali catalyst, excess alkali still active to catalyze the reaction.

But if the FFA is more than 5%, FFA must be converted to methyl ester catalysed by sulphuric acid with large excess of methanol (20:1) for just about an hour at 60° C. The byproduct water must be decanted with methanol.

Thereafter, the transesterification process must start with NAOH catalyst as indicated in the process scheme.

Physical Properties

The Physical properties of bio-diesel (transesterified from Jatropha Curcas oil) are now compared in Table 14.11 with the Jatropha oil itself and also with the European standard of Bio-diesel (ref. Raheman and Mondol)

Table 14.11: Comparative Physical Properties of Bio-diesel and Petro-diesel

Fuel Properties	*Jatropha Oil*	*Jatropha Methyl Ester*	*European Std. for Bio-diesel*	*Petro-Diesel*
Density at 15C (kg /m^3)	920	880	860--900	820
Pour Point (C)	5	4-	-	-23
Viscosity at 20C (m-Pa-s)	75.6	4.8	3.5 - 5	2.6 – 4.1
Calorific Value (MJ/kg)	37	39	-	42
Flash Point (C)	340	191	>120	74
Acid value (mg KOH /g)	8	0.41	< 0.50	0.35
Carbon residue %	4.2	0.05	<0.3	0.17
Ash Content %	1.1	0.014	<0.02	0.17

Source: Raheman *et al.*, 2006.

From the above Table 14.11, it would be seen that Bio-diesel has very similar properties to Petro-diesel. But keeping in view that bio-diesel has little higher viscosity than petro-diesel, somewhat higher pour point temperature and slightly lower calorific value, bio-diesel being widely used in diesel engine at 20% mixed with 80% petro-diesel (Mixture popularly known as B20 without any technical problem. So much so, a single cylinder diesel engine has been run on pure Jatropha bio-diesel (B100) and engine torque, power and specific fuel consumption has been found generally satisfactory.

We will now mention a few points of advantages and disadvantages of using bio-diesel in lieu of petro-diesel.

Advantages of Bio-Diesel

1. Bio-diesel has flash point of 191°C compared to 74°C of petro-diesel.
2. Bio-diesel has better lubricity.
3. Cetane number of bio-diesel is much higher.
4. Bio-diesel is the only alternative fuel that can be used in any conventional unmodified diesel engine.
5. Bio-diesel is non-toxic and biodegradable. It is a 100% domestic fuel.
6. Biodiesel produces significantly lower pollution carbon dioxide, Co, and hydrocarbons. It has no sulphur dioxide emission.

Disadvantages of Bio-diesel

- Bio-diesel may give rise to winter starting and operational problem because of high pouring point and somewhat higher viscosity.

- Bio-diesel is highly solvent. Hence to avoid to filter deposits being dissolved and carried to the engine. The filter deposit is to be changed more often.
- Quality of bio-diesel can easily be tampered.
- Neat bio-diesel demands compatible elastomer (Hoses, brackets, etc.)

Economics of Bio-diesel

In the USA and European Union, the bio-diesel is produced from edible oil and hence the production cost of bio-diesel may still be higher than the petro-diesel, which is processed economically in large refineries working continuously.

However, in India, bio-diesel is proposed to be produced from non-edible oil (e.g., Jatropha, Karanja, etc.) grown on fallow arid wasteland. Hence the seed price will be quite low and the final bio-diesel product will definitely be cheaper and conspicuously competitive to petro-diesel.

Further, cultivation of 30 million hectares of waste land producing about 2 tons of bio-diesel per hectare i.e. 60 million tons will immensely enrich the poverty-ridden Indian agriculture, creating nearly US 20 billion dollars per annum agro-based bio-diesel industries. Bio-diesel thus promises to bring about both economic prosperity and equity in 10 years time.

Finally, the availability of petro-diesel is concentrated in the politically unstable Middle East, however long it may last. But bio-diesel being produced in all parts of the world will ensure energy security and energy independence.

Costing of Bio-diesels

The cost of feed stocks for bio-diesel tends to be high compared to that of the cost of petro-diesel. The end result is that the cost of bio-diesel is generally higher than petro-diesel fuel.

The feed stock costs vary from crop to crop e.g. soya vs. rapeseed and season to season. For example, Soya oil was priced from 14 to 26 cents per pound in 1995 and peanut oil from 27 to 49 cents per pound during the same period. However, use of non-edible oil like Jatropha in India will bring down the price considerably.

Economics of scale is another factor affecting price as bio-diesel is not yet available on pipeline network.

Finally the market price of the major by product Glycerol, mostly used in the pharmaceutical industry, will affect the production cost of the main product bio-diesel.

Appendix C
BIO-REFINERIES

Development Scheme

We are all acquainted with Petroleum Refineries which produce various categories of energy fuels, e.g., Aviation, Petrol /gasoline, light diesel, heavy diesel, furnace fuels etc. besides a host of chemicals. Fig. 14.5 shows a schematic of a Bio-Refinery which also covers all common bio-processes including those of Fig. 14.2 and Fig. 14.4.

<table>
<tr><th>Refining Processes</th><th>Products / Byproducts</th></tr>
<tr><td>Fermentation of Sugars: Sugar cane crop</td><td rowspan="3">Ethanol
Lactic Acid
Acetone
Butanol</td></tr>
<tr><td>Fermentation of Sugars:
- Dry Corn Mill
- Wet Corn Mill</td></tr>
<tr><td>Fermentation of Sugars –
Acidolysis of Cellulose, Hemi-celluloses and Lignin</td></tr>
<tr><td>Extraction of Bio-oil:
-Triglyceride of Fatty Acid, Transesterification (Monoalkylester)</td><td>Bio-Diesel</td></tr>
<tr><td>Wood Pulp Mill</td><td>Paper</td></tr>
<tr><td>Chemical Processes:
- Hydrolysis,
- Hydro-generation
- Oxidation</td><td>Furfural
Levuline Acid,
Polyol</td></tr>
<tr><td>Gasification:
- Steam Reforming
- Pyrolysis,
- Combustion</td><td>Synthetic Gas,
Tar Chemical,
Methanol</td></tr>
</table>

Fig. 14.5: Bio-Refinery Scheme.

Source: Elliot 2004.

References

Da Rosa, A.V., 2009, *Fundamentals of Renewable Energy Processes* (Chapter 13), Academic Process.

Elliot, D.C. 2004, Biomass Chemicals, *Encyclopedia of Energy*, Ed. C.J. Cleveland, Elsevier Press.

Ghosh Roy, M.K. 2010, Sustainable Development – *Environment, Energy and Water Resources*, Ane Books, New Delhi.

Klass, D.L. 2004, Biomass for Renewable Energy and Fuels, *Encyclopedia of Energy*, Ed. C.J. Cleveland, Elsevier Press.

McElroy, Michael B., 2010, *Energy - Perspectives, Problems and Prospects*, Oxford.

Schumacher.L.G. 2004, Bio-Diesel Fuel, *Encyclopedia of Energy*, Ed. C.J. Cleveland, Elsevier Press.

Waghray, A.P., Appropriate Technology for Economic Production of Bio-Diesel from Vegetable Oil Seeds, Dhopeswar Engg. Pvt. Ltd. (dhopeswar@sanchernet.in), Hyderabad, India.

Wyman, C.E. 2004, Ethanol Fuel, Dartmouth College, *Encyclopedia of Energy*, Ed. C.J. Cleveland, Elsevier Press.

15

In Search of Viable Green Energy – Solar Energy

Overview of Solar Energy

Considering that the non-renewable fossil fuels coals, oil, and gas, which have been sustaining the global energy-intensive industrial-urban-infrastructure so far, are on the wane, the alternative energy system like Solar must start contributing substantially in about 50 years' time. The huge energy reservoir of the sun releases about (3.8×10^{20}) MWe energy at 5500 °C, of which the earth receives only a tiny fraction (1.6×10^{11}) MWe. But the total electricity generated globally is approximately (1.5×10^{13}) KWh, again a very small fraction of the solar energy received by the earth. It is said that assuming full utilisation of solar radiation, a radiation-collecting Solar Power Station measuring 30,000 km^2 in area could theoretically meet the global energy demand. (Please note that Taiwan measures about 30,000 km^2).

Now there are several technical limitations to be overcome in the fuller utilisation of solar energy, which is enumerated below:

- Solar radiation is of quite low average intensity @1000 watt per sq. m. approximately.
- Sun shines intermittently in a daily cycle, which again changes by seasonal variation. Further the sun's position by azimuth and elevation in the sky and the angle of incidence on the earth vary continuously.
- Solar radiation cannot be stored directly and must be converted immediately into heat, current or chemical reactions to store and use.

(A) Thermal Concentrating Solar Power (CSP)

However, modern engineering has been able to surmount some of these problems that have led to increasing utilisation of solar energy into electrical power. In current CSP plants, several types of solar concentrators like parabolic trough, paraboloid dish or self-positioning mirror (Heliostat) are used to collect solar heat at high temperature (see Table 15.1 and Fig. 15.1).

Table 15.1:Performance Characteristics of Various CSP Technologies

Solar Concentrator Types	*Typical Operating Tempe-ratures °C*	*Concen-Tration Ratios*	*Tracking Axes*	*Maximum Carnot Conversion Efficiencies %*
Flat Plate Collector	30 – 100	1	-	21
Parabolic Trough Collector	260 - 400	8 - 80	One axis	56
Heliostat Field + Central Receiver	500 -800	600 - 1000	Two axes	73
Paraboloid Dish Concentrator	500 - 1200	800 - 8000	Two axes	80

Source: Tester *et al* (2005).

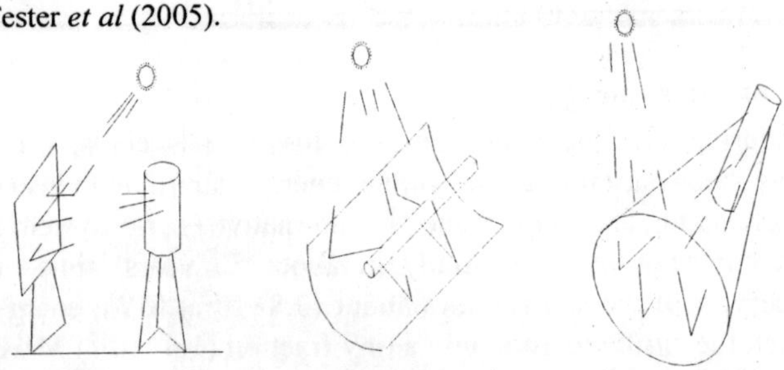

Heliostat **Parabolic Concentrator** **Paraboloid Concentrator**

Fig. 15.1: Different Types of Solar energy Concentrators.

The first commercial CSP plants, with a total electricity capacity of 354 MWe, the largest being of 80 MWe rating have been built in the Mojave desert in California; the plants have already generated a total of 15 TWh power and accumulated experience of 20 years of commercial experience (see Table 15.2).

Table 15.2: Nine Parabolic Trough Solar Electricity Generating Systems (SEGS), California

Plant Configuration	*SEGS I - II*	*SEGS III - VII*	*SEGS VIII - IX*
Site	*Daggett*	*Kramer Junction*	*Harper Lake*
Capacity MWe	14 + 30	5 x 30	2 x 80
Commissioning Years	1985-86	1987-89	1990-91
Annual Solar Electric Efficiency	9.5 -10.5	11.0 – 12.5	13.8
Maximum Working Temp C	305 -307	370 - 390	390
Investment $ / KWh	3800 - 4500	3200 - 3800	2890
Electricity cost $ / KWh	0.18 – 0.27	0.12 – 0.18	0.11 – 0.14
Annual Output GWh / y	30 + 80	5 x 92	2 x 250

Source: Muller-Steinhagen H & F Trieb.

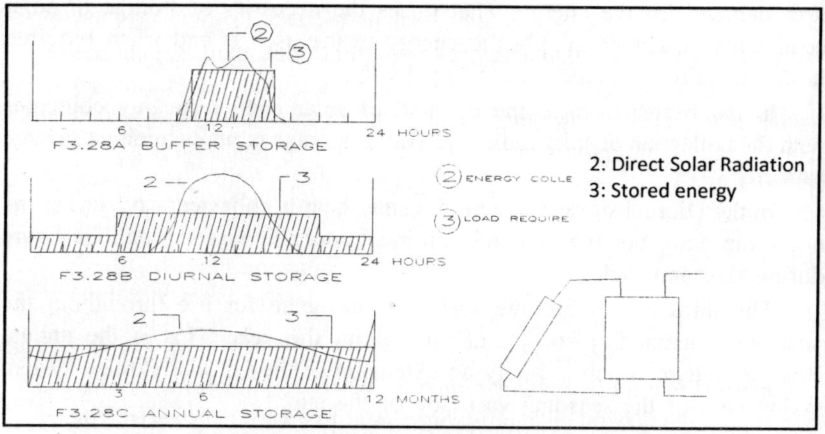

Fig. 15.2 Different Types of Heat Storage Systems.

As already indicated, the heat storage is very important in a CSP power generation as the duration of sunlight is limited in a day. The provision of heat storage systems (see Fig. 15.2) are required to be adopted to improve the economics of the plant. The molten salt can provide heat storage at as high as 566°C and is the preferred storage medium. Fig. 15.5 shows the system line diagram of the 50 Mwe CSP Andasol I in Spain.

Economics of Solar Thermal Power: We see from Table15.2 that the SEGS VIII-IX plants produce electricity at almost the commercially viable rate of 11 to14 US cents. It is thus appropriate to infer that after the world builds up a total of say about a 50 GWe of CSP plants and in the process helps develop the mass production technologies, the unit electricity costs will further come down.

Solar Energy Collection Systems

There are three important elements in the heat collection system, namely the heat collector, the concentrator and the receiver. If the temperature desired is less than 100°C, the heat concentrator may be eliminated. The concentrator further increases the temperature at which energy can be utilised.

There are three principal types of concentrators shown in Fig. 15.1.

The Heliostat employs a series of mirrors that rotates continuously to direct solar radiation into an absorbing surface of the receiver.

The parabolic concentrator directs all the incident solar rays to its focal point axis at which there runs water or other liquid pipe.

In the three dimensional paraboloid concentrator, all solar rays are likewise directed on to pipe which runs through its focal axis.

Solar Energy Storage Systems

Because of variable and uncertain nature of solar radiation, there is almost always a mismatch between the availability of solar energy and the user demand for the energy. That raises the necessity of storage of solar heat, which can then supply the energy to the user as and when required in the quantities demanded.(See Fig. 15.2)

In the buffer storage, the demand of solar energy roughly coincides with the collection of solar radiation. Buffer storage requires minimal storage capacity.

In the Diurnal storage system the solar heat is collected, say, during the noontime daily but the demand continues more or less through 24 hours during day and night.

The third energy storage system is designed for use throughout the year with demands also spread throughout the year. This is the annual storage system, which is really an extension of the Diurnal storage system taking care of the seasonal variation of the sun.

Solar Thermal Power Generation Systems

We have already mentioned about the solar power station of capacity 80 MWe. In fact, the application of solar thermal energy for the generation of electric power is the most important economic recovery and use of solar energy.

Depending upon the type of solar thermal energy collection system, the Power station may be grouped into three categories:

(i) Low temperature Turbo-Generator Power Station using Flat-Plate Heat Collectors (see Fig. 15.4 A):

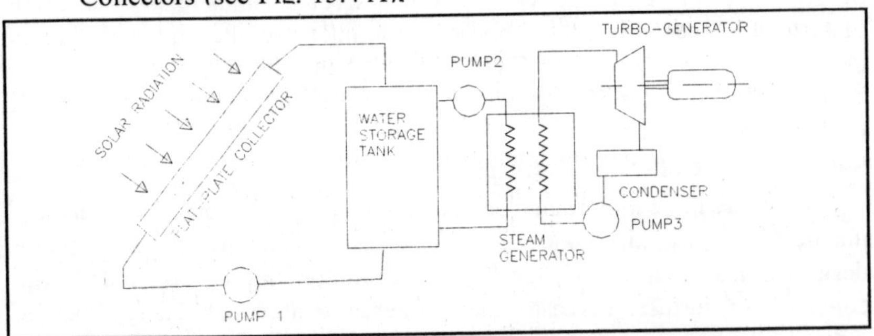

Fig. 15.4A: Low Temperature SEGS Plant (water-heated).

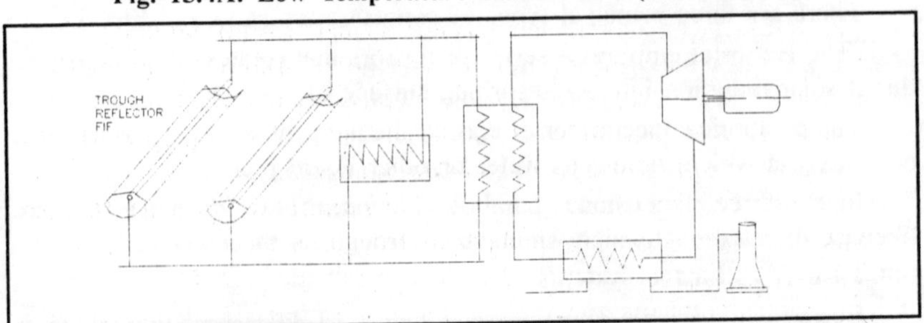

Fig. 15.4B: Medium Temperature SEGS Plant (oil-heated).

(ii) Medium Temperature Turbo-Generator Power Station using Parabolic Heat Collectors (see Fig. 15.4B);

(iii) High Temperature Turbo-Generator Power Station using either Paraboloid Heat Concentrator-cum-Collectors or the Central Heat Receiver energised by Heliostat Heat-Concentrator (see Fig. 15.4C)

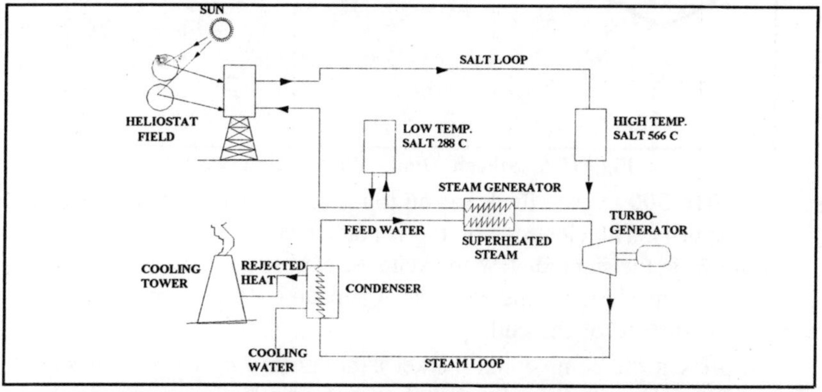

Fig. 15.4C: High Temperature Storage System (Saline Water-heated).

The above three illustrations show Solar Electrical Generating Stations without any heat storage system (SEGS). In Fig. 15.5 we show a 50 MWe SEGS at Andasol in Spain which can store heat by saline solution for seven hours.

Parabolic Solar Panel

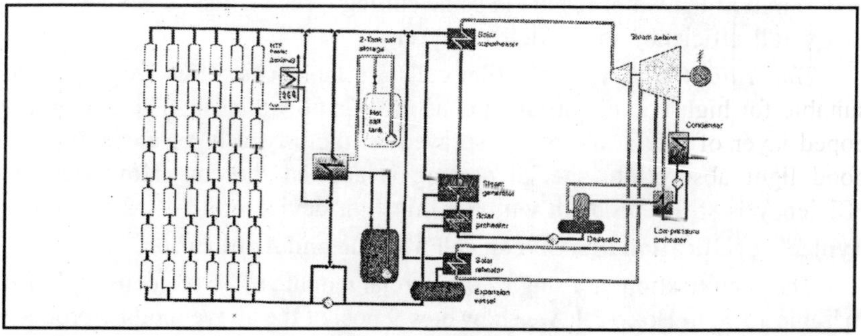

Fig. 15.5: Schematic Layout of a 50 Mwe Solar Thermal Plant (Andasol I).

B
SOLAR PHOTOVOLTAIC ENERGY

Photovoltaic Technology: The photovoltaic process is the direct conversion of solar radiation to electricity; it is a solid state device process and is thus highly robust and reliable. All photovoltaic devices, basically a silicon crystal diode, convert incident photons to free charge carriers through the process of absorption. Referring to Fig. 15.6 P-typed single crystal

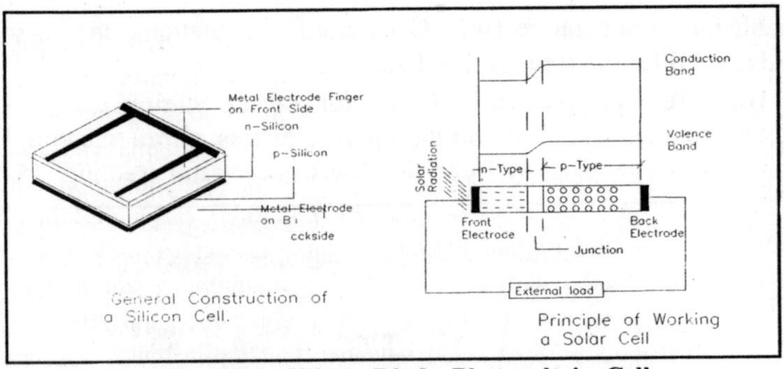

Fig. 15.6. Silicon Diode Photovoltaic Cell.

silicon wafer, 300 micron thick has on top a very shallow negatively doped N-type impurity. The electrode on top is made of twin-fingered Ti-Ag metal solder allowing the sun radiation to excite the exposed silicon crystal surface. The other electrode is of the same Ti-Ag metal solder sheet covering the entire back surface of the cell.

At present the commercial market is dominated by PV modules made of crystalline silicon (c-Si) cells, though technologically superior and cheaper thin film cells of amorphous silicon (a-Si) or cadmium Telluride (CdTe) or even Copper Indium Selenide (CuIS) already form 10 per cent market share.

Crystalline Silicon Cells: The first stage in the manufacturing process for both the mono- and multi-crystalline cells is to cut the bulk silicon into wafers; the wafer slicing is accomplished high speed wire line cutting in silicon carbide abrasive slurry. The wire thickness being 100 microns compared to 250 mm of the wafer results in some cutting losses. Today c-Si device has an overall efficiency of about 16 per cent.

Thin Film Cells: The thin film cells are more cost effective and more suitable for high volume production as there is no wafering. The negatively doped layer of a few microns is sprayed on the p-type layer and to ensure good light absorption, special coating is applied thereon. However, the efficiency is still a problem with the thin film device of 10 – 12 per cent.

Typical specification of a Silicon cell Module and Application

The specification of a single commercial module of 36 no. cells is shown in Table 15.3. In Box 15.1, we show how 9 nos. of the above module produce about 425.8 watt energising a pump delivering water at 10 tonne per day to head of 50 metres. This demonstrates the viability of using solar cell in a remote agricultural land.

Then there are esoteric applications, where portability of photovoltaic power wins over its natural high cost. It is believed that solar cells were extensively used in the Mariner and Skylab Satellite Missions of USA. The cost in the early Mariner mission was several thousand US dollars per peak watt, while in the later Skylab Mission, the cost camc down to USD 100.00 per watt. For terrestrial application like the above pumping system where the demand for quality and reliability are not as stringent, the cost may today range around USD 1 to 2 per watt.

Table 15.3: Typical Specification of a Silicon cell Module (Sukhatme)

Module Size	100 × 50 cm
Cell size	10.4 × 10.4 cm
Number of cells	36
Maximum output	50 W
Maximum Voltage	17.1
Maximum Current	2.92 A
Open Circuit Voltage	21.6 V
Short Circuit Current	3.22 A
Conversion Efficiency	12.8 %

Box 15.1

Example X 3.3

Example of Use of PV Energy for Rural Water Pumping

Problem: Design a Photovoltaic cell to raise underground water at 10 tonne per day with a head of 50 metres. Pump works 8 hours per day. Use PV of Table 15.3.

Assume Pump efficiency $E_p = 0.4$; Water density d = 995 kg/ m^3;

Pump capacity Q = [10,000 / (995×8×3600)] = 0.349 × 10^3 cu.m. per second.

Power demand of the Pump P_{net} = [(Q.d.g.H) / Ep] W = [0.349 × 10^{-3}] × 995 × 9.81x × 50/0.4 = 425.8 watt.

To use the PV module of Table 15.3, assume,

Inverter efficiency E_i =0.8;

Solar cell efficiency E_{sc} = 12.8 % = 0.128 (given);

Solar radiation = 1000 w/m^2;

A.C. power output of a single module of PV cells = 1000 × (0.104 × 0.104) × 36 × 0.128 = 49.84 watt.

Number of such modules required = 425.8 / 49.84 = 8.55 units. Therefore an array of 9 modules, each module containing 36 cells will be required to supply approximately 450 watts.

Future Outlook of Photovoltaic Energy: The photovoltaic cell energy remain substantially expensive, may be in the range of Rs. 20 to 25 per KWh in India (or even a dollar a unit in US). Basically the manufacturing cost is very high with low conversion efficiency. With the thin-film cells with special coating technology, the efficiency could come up to 25 per cent with substantial reduction in capital cost in not too distant future. The high point of the PV energy is portability; it is eminently suitable for isolated consumer group. For example, it is more economical to install a stand-alone PV system instead of a transmission line to a village in India having a load of say, 10 or 20 KW and which is situated some 20 or 40 km away from a main high tension line. The solar agricultural pumps are very useful in boosting Indian agricultural productivity. In many special applications like in remote satellite, there is no substitute to solar PV cells; similar special applications are in the isolated high transmission towers, communication towers and in hand-held calculators. The final score that PV energy scores the highest is its pure carbon-free energy supply; it requires no water, which is one of the most precious world resources which have become scarce today.

Appendix

Creating Abundance of Solar Energy in US*

(*based on *"A Solar Grand Plan"* by Ken Zweibel, James Mason and Vasilis Fthenakis)

The energy in solar radiation is so plentiful that the total energy consumed globally for one full year is available from sunlight striking earth for just 40 minutes.

The US is also lucky to be endowed with abundant sunlight over 250,000 sq. miles of Southwest alone, which could produce more than 4,500 quadrillion Btu of solar radiation in a year. Converting just 2.5 % of that radiation into electricity would equal the nation's total energy consumption in 2006 at 98 quadrillion Btu. Some of the major developmental initiatives may be enumerated as under:

Developmental Milestones

- Of course, a direct-current (DC) transmission backbone line would require to be built up to distribute that energy across the country.
- The basic technology is already available as indicated in the earlier sections. Technological commercialisation and further maturing up of technology will have to be financed under liberal federal grants and subsidies and matching industrial capacity has to be created.
- Under the Zweibel plan, by 2050, 69 % of U.S. electricity and 35 % of its total energy (which includes all transportation energy) would have been supplied by solar energy. This energy would be supplied to the consumer at today's rate of the conventional power through fossil energy sources. Further, if wind, biomass and geothermal energy sources would be developed and used simultaneously, 100 % of nation's electricity and 90 % of all its energy needs would be met from renewable energy sources by the same 2050.
- To achieve the above targets (see Table A15.1), Zweibel estimates the federal government will have to invest more than $ 400 billion (Rs 16

Table A15.1 : Zweibel Solar Plan for U.S. 2050

Technology	Parameters	2007	2050	Policy Initiatives
Photovoltaics (PV)	Land Area	10 sq. miles	30,000 sq. miles	To develop public land banks
	Module Efficiency	10 %	14 %	Developing Technology
	Installed cost	$4 (Rs. 60)/W	$1.2(Rs. 48)/W	High module efficiency; volume production
	Electricity Price	$0.16 (Rs. 6.4)/KWh	$0.05 (Rs.2)/KWh	Lower installation
	Total Capacity	0.5 GW	2940 GW	Future National solar energy plan
Compressed air energy storage (for photovoltaic)	Volume	0	535 billion cft	Developing sites with natural gas industry
	Installed cost	$5.8 (Rs. 230)/W	$3.9(Rs. 150/W)	Lower installation cost
	Electricity Price	$0.20 (R8/KWh)	$0.09(Rs. 4)/KWh	Lower installation cost
	Total capacity	$0.1 GW	558 GW	National Solar Energy Plan
Concentrated Solar Power (CS)	Land Area	10 sq. miles	16,000 sq. miles	To develop land bank
	Solar-to-electric efficiency	13 %	17 %	Improvement of technology
	Installed cost	$5.3(Rs. 212)/W	$3.7(Rs. 150)/W	Economies of scale; single tank storage
	Electricity Price	$0.18(Rs. 7.2)/KWh	$0.09 (Rs. 4) KWh	Lower installation cost
	Total capacity	0.5 GW	558 GW	National energy policy
DC Transmission	Length	500 miles	100,000 – 500,000 miles	New high voltage DC grid from Southwest to rest of U.S.

lakh crores) annually over the next 40 years to complete the 2050 solar developmental plan.

Major Technology Development landscapes

1. Photovoltaics Cells: The easiest to manufacture and cheapest today are the thin-film cell of cadmium telluride. To provide electricity at 6 cents per KWh by 2020, cadmium telluride modules would have to convert electricity with 14 per cent efficiency and systems would have to be installed at $ 1.2 (Rs 48) per watt of energy. Current modules have 10 % efficiency and an installed system cost of about $ 4 (Rs 160) per watt. In Zweibel plan, by 2050 photovoltaic technology would provide almost 3000 GWe power.

2. PV Energy Storage by compressed air in underground Caverns: The well-known limitation of solar power is that it generates very little electricity when sky is cloudy and none at all at night. Excess power must therefore be produced during sunny hours and stored for use during not so sunny or non-sunny hours. The power batteries are very costly and other storage systems are inefficient. Compressed air energy storage has emerged as a successful alternative. Electricity from photovoltaic plants compresses air and pumps it into vacant underground caverns, abandoned mines, aquifers and depleted natural gas wells. Compressed air energy storage plants have been operating reliably in Huntorf, Germany, since 1987 and in McIntosh, Alaska since 1991. The turbines burn only 40 per cent natural gas or 30 per cent with superior heat recovery technology to be developed than they would without compressed air. Studies by the Electric Power Institute in Palo Alto, California indicate that the cost of energy storage today is about half that of lead acid batteries and would add 3 or 4 cents per KWh to photovoltaic generation, bringing the total 2020 cost to eight or nine cents (Rs. 4) per kWh.

3. CS Energy Storage by Hot Salt: In the Zweibel model, one fifth of the solar energy comes from the concentrated solar thermal power generation system. For energy storage, the pipes run into a large, insulated tank filled with molten salt, which retains heat efficiently. Heat is extracted at night, creating steam. The molten salt does slowly cool however, so the energy stored must be tapped within a day.

Nine concentrated solar power plants with a total capacity of 354 megawatts have been generating power successfully for years in the US. A new 64-MW Plant in Nevada came online in March 2007. None of these plants have any heat storage. The first commercial installation to incorporate heat storage – 50-MW plant with seven hours of storage is being constructed in Spain. In the Zweibel plan, 16 hours of storage would be needed so that electricity can be generated 24 hours a day.

4. Direct Current Transmissions: Unlike the existing fossil fuel power stations distributed all over the US landmass, the solar plants will of necessity be situated geographically in Southwest. The existing system of alternating current (AC) power lines is not robust enough to carry power to consumer centres all over the United States. Hence a new high voltage direct current

(HVDC) power lines would have to be built. This view is also corroborated by the Oak Ridge National Laboratory. There are already several proposals to erect such robust DC transmission for conveying wind power from the Southwest to other parts.

5. Drastic Cut in US Fossil Fuel Consumption: The most valuable benefit of the grand plan will be closure of 600 mammoth coal-based and gas based power plants spewing enormous amcunt of carbon dioxide to the atmosphere. This will create a cleaner atmosphere and lead to a conspicuous slowdown of the process of global warming. The US would also be able to reduce politically sensitive Middle East oil import. These beneficiary outcomes of the grand plan are reflected in Table A15.2.

Table A15.2: Drastic Cut in US Fossil Fuel Consumption

Energy Fuels	*2007*	*2050 (Existing Energy Path)*	*2050 (under the Solar Grand Plan)*
Oil (billion barrels)	6.9	10.9	2.7
Natural Gas (trillion cubic ft)	22.2	35.4	11.4
Coal (billion tons)	1.2	1.9	0.5
US CO_2 Emission (billion tons)	6.1	9.4	2.3

Stage by Stage Implementation

The implementation of the grand development plan of solar energy is recommended to be implemented in three phases:

Stage I: 2011-2020

- In this phase the aim is to create a mass production base for solar power equipments, which will help to make all equipments available at competitive prices vis-à-vis conventional power equipments, e.g. coal and other fossil fuel accessories. This programme will have to be generously supported y US Federal subsidies @ USD 10.0 billion annually.
- About 84 GW of photovoltaic (PV) and concentrated solar (CS) power plants would be built up in sub-phases:
  (i) For the first 5 years till 2015, new solar energy capacity shall be built up at the rate of 1.5 GW of photovoltaic and 1.5 GW of CS every year totaling about 15 GW;
  (ii) For the next 5 years till 2020, new solar capacities at the rate of 10 GW (5 GW PV and 5 GW CS) will be built totaling additional 50 GW.

The above production schedule will bring down the electricity price to six cents (Rs. 3) per kWh.

- In parallel, required high voltage DC transmission lines would have been built up with State financial support.

Stage II: 2020- 2050

This phase of development should basically be self-sustaining, for the mass technology base has been built up in the first phase under generous subsidies. But some amount of Governmental support might still be needed where market weakness or inabilities are observed or technology shortcoming may still have to be made good. The various assumptions in the development scenarios may be mentioned on the following lines:

- The US national energy demand will increase by 1 per cent a year. But even with this increases, the 2006 national primary consumption at 100 quadrillion Btu would decrease to 93 quadrillion Btu by 2050, because of very significant energy saved from oil exploration, extraction and processing.
- In this scenario, by 2050 solar power plants will contribute 69 per cent of US electricity and 35 per cent of total US energy. This quantity includes enough to supply all electricity consumed by 344 million plug-in hybrid vehicles, which will replace all gasoline vehicles.
- To meet the 2050 projection of power, 46,000 sq. miles of land will be needed for photovoltaic and concentrated solar installations, which cover just 19 per cent of Southwest land, basically fallen land.

Post-Stage II: 2051-2100

In such long term of time, the estimates are basically intelligent projections but the exercise is well worth it. The renewable energy Scenario in 2100 may be projected on the following order:

Photovoltaic:	2.9 TW (connected directly to the grid)
Photovoltaic:	7.5 TW (conserved as compressed-air storage)
Photovoltaic:	1.3 TW (distributed installations)
Wind Farms:	1 TW
Geothermal:	0.2 TW
Biomass:	0.25 TW

The solar farms would require land of 195,000 acres, which is much less than the land available in the Southwest.

References

Ghosh Roy, M.K. 2011, *Sustainable Development*, Ane Books, New Delhi.

Luzi, A. and Grove, K.L. 2004, *Solar Thermal Power Generation, Encyclopedia of Energy*, Ed. Cleveland, Elsevier.

Mottershed, A. 1988, *Electronic Devices*, Prentice Hall India, New Delhi.

Nersessian Roy, L. 2009, *Energy for 21st Century*, Pentagon University Press, New Delhi (M.E. Sharpe Inc., 2007, New York).

Partin, J. 2004, History of Solar Energy, *Encyclopedia of Energy*, Ed. C. J. Cleveland, Elsevier Press.

Sukhatme, S.P. 1996, Solar Energy 2nd ed., Tata-Mcgraw Hill, New Delhi.

Tester, J.W., Drake, E.M., Driscoll, M.J. *et al.*, 2005, *Sustainable Energy- Choosing among Options*, MIT Press/PHI.

Zweibel, K., Mason, J. and Fthenakis. V. (2008), *Scientific American India*, January 2008, www.sciam.co.in

16

Wind Energy : Most Widely Used Green Energy Today

I

Wind Power Development Perspectives

Historical Development

The history of the wind energy dates back 2500 years ago to China when wind was used to pump water. Wind mills of various types continued to pump water and grind grain in Persia until 900 CE. Wind mills entered Europe in the eleventh century via the Crusade and was used extensively in Holland and England. Holland had 8,000 wind mills in 1650, England had 10,000 wind mills in 1800s and Germany had no less than 18,000 mills in the late 1800s (Nersesian 2009).

Modern Developments: The American dream of abundant nuclear power was very much shaken with the accident at the Three Mile Island plant in April 1979. This trauma was followed by the Chernobyl catastrophe in Russia in 1986 and led to the stoppage of all in nuclear power development projects in the United States. In the midst came the Middle East oil crisis in1973. Thus, the power development vacuum saw the fast growth in the wind turbine power initiatives in California. In Germany, wind power development followed a Government Law in 1991 requiring all power utilities to purchase at 90 per cent of their average retail rates renewable energy sources such as wind turbines. This led to very fast development of wind turbine energy all over Germany. In India too wind energy has seen fast growth along the long 7400 km coast line with strong winds.

Latest Growth in Wind Power: No doubt, the greatest incentive to growth of wind-fed electrical power in the 21st century has been to reduce carbon emission rates by all countries induced under the Kyoto Protocol 1997 of UN Framework Convention on Climate Change 1992. In some of the leading countries of the world, the wind electrical energy has grown about 9 times in nine years (see Table 16.1). Of course, this growth in wind power

Table 16.1: Growth in Global Wind Power Capacity MWe

Country	*1999*	*2008*
Germany	4072	23,903
United States	2502	25,170
Spain	1722	16,754
China	300	12,210
India	1077	8,896

has been made possible by the very fast technological development of wind turbo-generators so much so, the rotor diameters have increased from 15 m in 1985 to about 160 m in 2005, while the wing span of the Airbus 380 is only 80 m (Staudt 2008).

Modern Wind Power Technology

Modern Turbo-Generator: Fig. 16.1 shows a modern Propeller type wind turbine and the Table 16.2 shows the typical specification and values of all its design features. The power rating of the turbo-generator in the table is rated 0.5 MWe, but the bigger modern wind turbines are already delivering 5 MWe each.

Table 16.2: Principal Specifications of a Modern Propeller Type Turbo-Generator

Turbine Items	*Specifications*	*Turbine items*	*Specif-ications*
Turbine Type	Horizontal axis	Generator Fixed speed	4-pole Induction 660V/60
Rated Power Rotor Diameter	500 kwe 40 m	Rate of rotation Gear Box	
Swept Area	1257 sq. m.	Yaw Control	
Hub Centre above ground	50 m	Over-load Protection	
Wind Speed-related spec:		Mechanical Brake	
Cut-in	4 m/s	Tower	
Rated	13 m/s	Site wind: Weibull Factor	
Cut-out	25 m/s	Average wind speed	
Survival	60 m/s	Specific power available	
Rate of rotation	30 rpm constant	Overall performance:	
Tip Speed 62.8m/s		Total power coefficient	
Tip speed/rated wind speed	5.2	Power Availability	
Installed Cost	1000 $/ kwe rated	Specific Energy delivered	
Hence specific cost	398 $ / m2	Total Energy per annum	
Rotor Fibre glass	3-bladed variable pitch		

Source: Adjusted from Tester *et al.*, (2005).

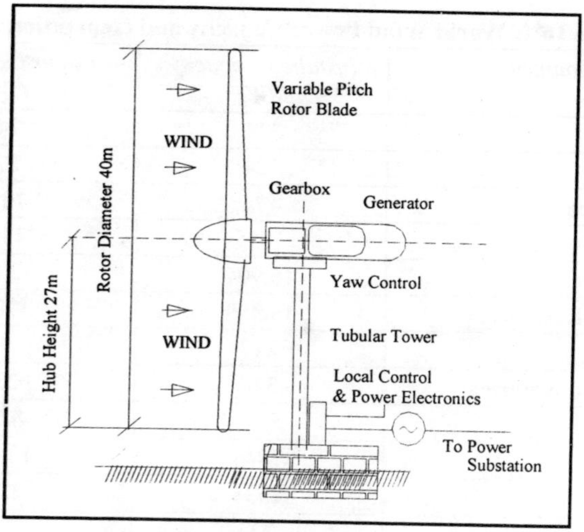

Fig. 16.1: A Modern Wind Turbine.

Modern Wind Farm: The modern Wind turbines are not arranged stand-alone but many of them are clustered together to deliver sizable quantity of energy to the grid and to facilitate operation and maintenance. In the Table 16.3 we present the principal particulars of two Danish wind farms, one includes 80 turbines and the second 72 turbines and each of them is rated to total 160 Mwe.

Table 16.3: Construction Data of Two Large Offshore Wind Farms

Wind Firm Particulars	*Horn Rev Farm*	*Nysted Wind Farm*
Installed Capacity Mwe	160	160
Number of Turbines	80	72
Wind Turbine Type	Vestas 2 MWe	Bonus 2.2 MWe
Expected Annual Production GWh	600	595
Hub Height m	70	70
Water Depth m	6.5 - 13.5	6 – 9.5
Wind farm Area km2	20	24
Distance to Shore km	14 - 20	10
Distance between Rows m	560	850
Distance betw. Turbines in a Row m	560	480
Internal Grid Voltage KV	34	33
Transmission to shore Voltage KV	150	132

Source: Petersen *et al.*, (2004).

Table 16.4: World Wind Power Capacity and Generation 2007

Country	*Installed Capacity MWe*	*Power Generation GWh*
Egypt	310	741
Morocco	124	279
Total Africa	**474**	**1093**
Canada	1845	4300
United States	16,904	32,143
Total North America	**18,959**	**37090**
Brazil	247	97
Total South America	**340**	**239**
China	5905	8200
India	7844	13,600
Japan	1538	2207
Taiwan	224	444
Total Asia	**15,929**	**25,184**
Austria	982	1875
Denmark	3,124	7,171
France	2,200	4,050
Germany	22,247	39,500
Netherlands	1,745	3,400
Spain	15,145	27,028
United Kingdom	2,477	5,274
Total Europe	56,802	104,904
Total Middle East	**76**	**144**
Australia	824	2526
Total Oceania	**1,170**	**3,419**
Total World	**93,750**	**172,000**

Source: WEC Survey Report 2009

II

Indian Wind Power Development

Although a relative newcomer to the wind industry compared with Denmark or the US, a combination of domestic policy support for wind power and the rise of Suzlon (a leading global wind turbine manufacturer) have led India to become the country with the fifth largest installed wind power capacity in the world. As of 31st December, 2010 the installed capacity of wind power in India was 13,065.37 MW, mainly spread across several States as shown in Table 16.5. It is estimated that 6,000 MW of additional wind power capacity will be installed in India by 2012. Wind power accounts for 6% of India's total installed power capacity, and it generates 1.6% of the country's power.

Table 16.5 : Shares of Wind Power of Indian States

States	*Capacity MW*	*States*	*Capacity MW*	*States*	*Capacity MW*
Tamil Nadu	**4906.74**	Rajasthan	**1088.37**	Orissa	**2.00**
Maharashtra	**2077.70**	Madhya Pradesh	**229.39**	West Bengal	**1.10**
Gujarat	**1863.64**	Andhra Pradesh	**136.65**	Other States	**1189.78**
Karnataka	**1472.25**	Kerala	**27.75**	Indian Total	**13,065.4**

Source: Wikipedia 2010.

Suzlon, foremost among other Indian owned companies like Shriram and Enercon, emerged on the global scene in the past decade and by 2006 had captured almost 7 per cent of market share in the global wind turbine sales. Suzlon is currently the leading manufacturer of wind turbines for the Indian market, holding some 52 per cent of the market share in India. Suzlon success has made India the developing country leader in advanced wind turbine technology.

Developments in Indian States

There is growing wind energy installations in the number of the States across India.

Tamil Nadu (4906.74 MW)

Fig. 16.2A. A wind farm in Muppandal, Tamil Nadu.

Tamil Nadu is the state with the most wind generating capacity: 4906.74 MW at the end of March 2010. Not far from Aralvaimozhi, the Muppandal wind farm, the largest in the subcontinent, is located near the once impoverished village of Muppandal, supplying the villagers with electricity for work. The village had been selected as the showcase for India's $2 billion clean energy program which provides foreign companies with tax breaks for establishing fields of wind turbines in the area.

In February 2009, Shriram EPC bagged INR 700 million contracts for setting up of 60 units of 250 KW (totaling 15 MW) wind turbines in Tirunelveli district by Cape Energy. Enercon is also playing a major role in development of wind energy in India. In Tamil Nadu, Coimbatore and Tiruppur Districts have number of wind mills from 2002 onwards.

Fig. 16.2B Suzlon Wind Turbines.

Maharashtra (2077.70 MW)

Maharashtra is second only to Tamil Nadu in terms of generating capacity. Suzlon has been heavily involved. Suzlon operates what was once Asia's largest wind farm, the Vankusawade Wind Park (201 MW), near the Koyna reservoir in Satara district of Maharashtra.

Gujarat (1863.64 MW)

Samana and Sadodar in Jamnagar district is set to host energy companies like China Light Power (CLP) and Tata Power have pledged to invest up to Rs. 8.15 billion ($189.5 million) in different projects in the area. CLP, through its India subsidiary CLP India, is investing close to 5 billion for installing 126 wind turbines in Samana that will generate 100.8 MW power. Tata Power has installed wind turbines in the same area for generating 50 MW power at a cost of 3.15 billion. Both projects are expected to become operational by early next year, according to government sources. The Gujarat government, which is banking heavily on wind power, has identified Samana as an ideal location for installation of 450 turbines that can generate a total of 360 MW. To encourage investment in wind energy development in the state, the government has introduced a host of incentives including a higher wind energy tariff. Samana has a high tension transmission grid and electricity generated by wind turbines can be fed into it. For this purpose, a substation at Sadodar has been installed.

ONGC Ltd has commissioned its first wind power project. The 51 MW project is located at Motisindholi in Kutch district of Gujarat. ONGC had placed the EPC order on Suzlon Energy in January 2008, for setting up the wind farm comprising 34 turbines of 1.5 MW each. Work on the project had begun in February 2008, and it is learnt that the first three turbines had begun production within 43 days of starting construction work. Power from this 308 crore captive wind farm will be wheeled to the Gujarat state grid for onward use by ONGC at its Ankleshwar, Ahmedabad, Mehsana and Vadodara centres. ONGC has targeted to develop a captive wind power capacity of around 200 MW in the next two years.

Karnataka (1472.75 MW)

There are many small wind farms in Karnataka, making it one of the states in India which has a high number of wind mill farms. Chitradurga and

Gadag are some of the districts where there are a large number of Windmills. Chitradurga alone has over 20,000 wind turbines.

The 13.2 MW Arasinagundi (ARA) and 16.5 MW Anaburu (ANA) wind farms are ACCIONA'S first in India. Located in the Davangere district (Karnataka State), they have a total installed capacity of 29.7 MW and comprise a total 18 Vestas 1.65 MW wind turbines supplied by Vestas Wind Technology India Pvt. Ltd.

The ARA wind farm was commissioned in June 2008 and the ANA wind farm, in September 2008. Each facility has signed a 20-year Power Purchase Agreement (PPA) with Bangalore Electricity Supply Company (BESCOM) for off-take of 100% of the output. ARA and ANA are Acciona's first wind farms eligible for CER credits under the Clean Development Mechanism (CDM).

Rajasthan (1088.37 MW)

Gurgaon-headquartered Gujarat Fluorochemicals Ltd (GFL) is in an advanced stage of commissioning a large wind farm in Jodhpur district of Rajasthan. A senior official told Project monitor that out of the total 31.5 MW capacity, 12 MW had been completed so far. The remaining capacity would come on line shortly, he added. For the INOX Group Company, this would be the largest wind farm. In 2006-07, GFL commissioned a 23.1-MW wind power project at Gudhe village near Panchgani in Satara district of Maharashtra. Both the wind farms will be grid-connected and will earn carbon credits for the company. The cement major ACC Ltd has proposed to set up a new wind power project in Rajasthan with a capacity of around 11 MW. Rajasthan is emerging as an important destination for new wind farms, although it is currently not amongst the top five states in terms of installed capacity. As of 2007 end, this northern state had a total of 496 mw, accounting for a 6.3 per cent share in India's total capacity.

Madhya Pradesh (229.39 MW)

In consideration of unique concept, Govt. of Madhya Pradesh has sanctioned another 15 MW project to MPWL at Nagda Hills near Dewas. All the 25 WEGs have been commissioned on 31.03.2008 and under successful operation.

Kerala (27.75 MW)

The first wind farm of the state was set up at Kanjikode in Palakkad district. It has a generating capacity of 23.00 MW. A new wind farm project was launched with private participation at Ramakkalmedu in Idukki district. The project, which was inaugurated by chief minister V. S. Achuthanandan in April 2008, aims at generating 10.5 MW of electricity The Agency for Non-Conventional Energy and Rural Technology (ANERT), an autonomous body under the Department of Power, Government of Kerala, is setting up wind farms on private land in various parts of the state to generate a total of 600 mw of power. The agency has identified 16 sites for setting up wind farms through private developers. ANERT is engaged in the field of development and promotion of renewable sources of energy in Kerala. It is also the nodal agency for implementing renewable energy programmes of the Union ministry of non-conventional energy sources.

West Bengal (1.10MW)

The total installation in West Bengal is just 1.10 MW as there was addition of only 0.5 MW in 2006-2007 and none between 2007–2008 and 2008–2009.

Bengal - Mega wind energy project of capacity 50 MW is soon to be launched by Suzlon Energy Ltd in West Bengal. Suzlon Energy is looking for sites at coastal Midnapore and South 24-Parganas districts. Suzlon aims to generate the power solely for commercial purpose and sell it to local power distribution outfits like the West Bengal State Electricity Board (WBSEB). Suzlon will invest around 250 crore initially, without taking recourse to the funding available from the Indian Renewable Energy Development Agency (IREDA).There are five wind-power units in West Bengal, at Frazerganj, generating a total of around 1 MW. At Sagar Island, there is a composite wind-diesel plant generating 1 MW. In West Bengal, power companies are being encouraged to buy power generated by units based on renewable energy. The generating units are being offered special rates.

Prospects

Initial cost for wind turbines is greater than that of conventional fossil fuel generators per MW installed. Noise is produced by the rotor blades. This is not normally an issue in the locations chosen for most wind farms and research by Salford University shows that noise complaints for wind farms in the UK are almost non-existent.

The Ministry of New and Renewable Energy (MNRE) has fixed a target of 10,500 MW between 2007–12, but an additional generation capacity of only about 6,000 MW might be available for commercial use by 2012.

Some of the larger wind turbines in operation in India are listed in Table 16.6.

Table 16.6: India's Largest Wind power production facilities (10MW and above)

Power Plant	*Producer*	*Location*	*State*	*Total Capacity (MWe)*
Vankusawade Wind Park	Suzlon Energy Ltd.	Satara Dist.	Maharashtra	259
Cape Comorin	Aban Loyd Chiles Offshore Ltd.	Kanyakumari	Tamil Nadu	33
Kayathar Subhash	Subhash Ltd.	Kayathar	Tamil Nadu	30
Ramakkalmedu	Subhash Ltd.	Ramakkalmedu	Kerala	25
Muppandal Wind	Muppandal Wind Farm	Muppandal	Tamil Nadu	22
Gudimangalam	Gudimangalam Wind Farm	Gudimangalam	Tamil Nadu	21
Puthlur RCI	Wescare (India) Ltd.	Puthlur	Andhra Pradesh	20
Lamda Danida	Danida India Ltd.	Lamda	Gujarat	15

(contd...)

Power Plant	*Producer*	*Location*	*State*	*Total Capacity (MWe)*
Chennai Mohan	Mohan Breweries and Distilleries Ltd.	Chennai	Tamil Nadu	15
Jamgudrani MP	MP Wind farms Ltd.	Dewas	Madhya Pradesh	14
Jogmatti BSES	BSES Ltd.	Chitradurga Dist	Karnataka	14
Perungudi Newam	Newam Power Company Ltd.	Perungudi	Tamil Nadu	12
Kethanur Wind Farm	Kethanur Wind Farm	Kethanur	Tamil Nadu	11
Hyderabad APSRTC	Andhra Pradesh State Road Transport Corp.	Hyderabad	Andhra Pradesh	10
Muppandal Madras	Madras Cements Ltd.	Muppandal	Tamil Nadu	10
Poolavadi Chettinad	Chettinad Cement Corp. Ltd.	Poolavadi	Tamil Nadu	10
Shalivahana Wind	Shalivahana Green Energy. Ltd.	Tirupur	Tamil Nadu	20.4

Source: www.salivahanagreengroup.com

III
Review of Global Wind Power

- Global Capacity

  Global wind power Installed capacity end 2010 — 196,630 MW

  Capacity added during 2010 — 37,642 MW

  Growth rate in 2010 — 23.6%

  Wind Power generating capacity — 430 TWh/annum
- Germany keeps its number one position in Europe with 27,215 MW followed by Spain at 20, 676 MW.
- Wind power constitutes the highest share of electricity production in Europe: Denmark 21%; Portugal 18%; Spain 16%.
- WWEA (World Wind Energy Association) considers a global capacity of 600,000 MW as feasible by year 2015 and more than 1.5 TW by year 2020.
- Table 16.7 shows the top ten countries in installed capacities (MW) in 2010:

Table 16.7: Top Ten Wind Power Generating Countries

Countries	*Installed Capacities MW*	*Countries*	*Installed capacities MW*
China	44,773	Italy	5,797
USA	40,180	France	5660
Germany	27,215	UK	5,203
Spain	20,676	Canada	4,008
India	13,065	Denmark	3,465

Wind Electric Power Generation

All wind turbines installed globally by end of 2010 contribute potentially 430 TWh units of power to the world electric supply, which represents 2.5% of the total global electric supply.

The total wind electricity power is actually higher than the total electricity needs of the United Kingdom, an industrialised country with more than 60 million inhabitants and the sixth largest economy of the world.

In some countries and regions of the world wind has become one of the largest electricity sources. Again in terms of wind share, Denmark is the world leader at 21% followed by Portugal (18%), Spain (16%), and Germany (9%).

In China, wind contributes 1.2% to the overall electricity supply, while in USA the wind share had touched about 2%.

Offshore Wind Energy

The top five countries possessing offshore wind power capacities are given in Table 16.8.

Table 16.8: Offshore Wind Power

	Sweden	*Belgium*	*Nether-lands*	*Denmark*	*United Kingdom*
Capacity up to 2009	164	30	247	663.6	688
Additions in 2009	0	165	2	`190.4	1341
Total in 2010	164	1959	249	854.0	1341

Offshore wind capacities continue to grow in 2010. Like in the previous year, wind farms installed in the sea could be found in twelve countries, ten of them in Europe and the other two countries are China and Japan in Asia.

Total installed offshore capacities amounted to 3117.6 MW, out of which as much as 1161.7 MW were installed in 2010. This represents a growth rate of 59%, far above the average growth rate of the wind sector as a whole.

China installed its second major offshore farm near Shanghai with a capacity of 100 MW.

Japan added the Kamasu near shore wind farm (14 MW), which survived the earthquake and the tsunami which ravaged Japan on 11 March 2011.

Continental Distribution of Wind Power (% 2010)

Asia became the new continental leader accounting 54.6% of the newly installed wind turbines, displacing Europe.

Table 16.9: Continental Distribution of Wind Power

Asia	*Europe*	*North America*	*Asia-Pacific*	*Africa*	*Latin America*
54.6	27	16.7	0.04	0.4	1.2

Future Challenges and Prospects Worldwide

There are at least six drivers promoting the spread of wind power in the medium and long term:

- Climate Change is now known to be caused by the accumulation of carbon dioxide and other Green House Gases (GHG). Wind power is one energy source which is hundred per cent free from GHG pollution;
- Fossil fuel is an exhaustible resource and its availability is restricted to few countries; its supply and pricing are subject to undue volatility and geopolitical pressures and influences; Whereas wind power is a renewable natural resource and theoretically inexhaustible;
- Wind power has no adverse effect or influences on ecology like recent oil spill in the Gulf of Mexico;
- Unlike the nuclear power whose serious radioactive hazards arising out of accidents at Three Mile Island in US and Chernobyl in Ukraine or natural disaster as in Fukushima, wind power is a benign power source;
- Thus, wind power possesses high sustainability economically, socially and ecologically incomparable with any other power resources;
- Further improvements in wind energy and related technologies including storage will further improve its reliability and price competitiveness.

The World Energy Report 2010

This Report forecasts the potentiality of wind energy to reach the following capacities:

2015 – 600,000 Mwe

2020 – 1,500,000 Mwe

Substantial growth can especially be expected in China, India, Europe and North America.

The above growth graphics of wind power is displayed in Fig. 16.3.

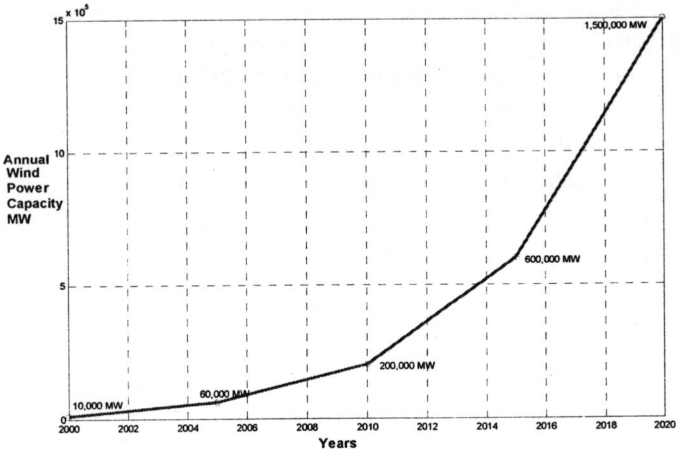

Fig. 16.3 Growth of Wind Power in Coming Decades.

APPENDIX

Sustainable Energy
Wind, Water and Solar Technologies

(based on Mark Z. Jacobson and Mark A. Delucchi, *Scientific American India*, November 2009, www.sciam.co.in)

The utilizable energy of Wind and Solar radiation on accessible land far outstrip the world demand of energy of all forms. A study in 2009 at the Stanford University concluded that to avoid the problems of global warming and other pollution hazards, energies from wind, solar, geothermal, tidal and hydroelectric – all of which are driven by wind, water and sunlight (henceforth referred to as WWS) are conspicuously better than energies derived from nuclear power, coal and gas with carbon capture and biofuels ethanol and biodiesel.

In the Jacobson-Delucchi plan, WWS will supply electric power for both heating and transportation—industries that will have to revamp if the world has to avoid climate change. They have assumed that most fossil fuel heating, including ovens and stoves, shall be replaced by electric energy and that most fossil fuel transportation can be replaced by battery and fuel-cell vehicles. Hydrogen, produced by using WWS electricity to split water (electrolysis), would power fuel cells and be burned in aircraft, ships and industries.

WWS Energy Supply Capacity

Today maximum power consumed worldwide at any given moment is about 12.5 TW per US EIA. The Agency projects that in 2030, the world will require 16.9 TW of power as global population and living standards rise, with about 2.8 TW in the US. This projection is based on the identical mix of energies, heavily dependent on fossil fuel as today. If however this planet were entirely powered by WWS, with no fossil fuel or biomass energy used, an interesting savings of energy will accrue. The global power demand will then gross to only 11.5 TW and the US demand would only be 1.8 TW. That decline in power consumption occurs for electric prime mover is more efficient in the use of energy. For example, only 17 to 25 per cent of energy in gasoline is used up to move a vehicle, whereas 75 to 86 per cent of electricity delivered to an electric vehicle goes into motion.

Detailed studies indicate the gross wind energy worldwide is 1,700 TW and solar alone offers 6,500 TW. But accessible and economically developable wind power may amount to 40 to 85 TW and solar power may amount to 580 TW, which combined or otherwise far exceeds the human requirement in the foreseeable future. Only thing that is known now is the present utilisation is mere 0.02 TW of wind and 0.008 TW of solar.

Among many feasible solutions, Jacobson-Delucchi has chosen the plan of utilising WWS energy sources as shown in Table A16.1. It may be said how to build and install as many as 3.8 million wind turbines? The answer is simple that if the world industries can build as many as 73 million cars and light trucks, the same industrial complex can definitely build 3.8 million turbines over a period of say 10 years or so. In the same fashion it is thought feasible to build 89,000 solar power stations

Appendix Table A16.1: Jacobson-Delucchi Plan of WWS Energy Supply for Future World

Renewable Energy Type	*Energy Contribution*		*Means of Energy Recovery*
	TW	*%*	
Wind	5.8	51	To employ 3.8 million large wind turbines of 5 MW each
Solar	4.6	40	To employ 89,000 PV and CS solar power plants, 300 MW each
Water	1.1	9	To deploy 900 hydroelectric stations world wide ,of which 70% already exist
Total	11.5	100	

Land Space Requirement

The land requirement for 3.8 million turbines would be less than 50 sq. km. (area less than Manhattan). When the spacing between the individual turbines is considered, the gross area may constitute about 1 per cent of the earth's surface, the empty space around the turbine could be used for agriculture or ranching or as open space or ocean.

For the solar projects, 30 per cent of the photovoltaic output will possibly come from rooftop panels on homes and commercial buildings. For the rest PV and SC solar plants the estimated land requirements may be about 0.33 per cent of the earth land. Further, this infrastructure will also take time to build up. But the current crop of coal and gas power plants did also take a very long time to build and if fossil fuel alone will have to cater for 16.9 MW power capacity in 2030, it will require more land space than the total space for 11.5 MW solar power, considering the mining space.

Material Constraints

There are some critical materials which may affect the rapid implementation of the WWS grand plan. Of course, the main and the bulk materials like steel and concrete will be available in enough quantities. The scarce will probably be rare-earth metals such as neodymium used in turbine gear boxes, and other photovoltaic amorphous or crystalline silicon, cadmium telluride or copper indium selenide and silver that cells require. However, the material problems could be resolved through research on substitutes or alloy products which will require minimal amount of scarce and costly metals.

Reliability of WWS Supplies through Judicious Mix

A novel power infrastructure must provide energy on demand at least as reliably as the existing infrastructure. WWS systems suffer less downtime than traditional fossil power sources. The average US coal plants is offline 12.5 per cent of the year for scheduled and unscheduled maintenance. Modern wind turbines have downtimes of less than 2 per cent on land and less than 5 per cent at sea. Photovoltaic systems also have less than 2 per cent downtime.

The main WWS challenge is that the wind does not always blow and the sun does not always shine in a given location. These intermittency problems can be mitigated by a smart and judicious balance of sources, relying on wind at night when it is plentiful, using solar by day and turning to a reliable source such as hydroelectric that can be turned on and off quickly to smooth out supply or to meet peak demand.

References

Chambers, A. 2000, "Wind Power Spins into Contentions", *Power Engineering* 104(2.)

Elliot, D., Swartz, M. and Scott, G. 2004, Wind Resource Basics, *Encyclopedia. of Energy*, Ed. Cleveland, Elsevier.

Kar, M. 2007, "Friendly Mining through Technology Integration" Seminar on Environmental Aspects in Mining Operations in India, Millennium Institute of Energy and Environment, Kolkata, India, March 2007.

Mani, A. and Mooley D.A. 1983, *Wind Energy Data for India*, Allied Publishers, India.

Nersessian Roy, L. 2009, *Energy for 21st Century*, Pentagon Press, New Delhi

Pasquate, M., Righter, R. and Gipe, P. 2004, History of Wind Energy, *Encyclopedia of Energy*, Ed. Cleveland. Vol. 6, Elsevier.

Petersen, E.L. and Madsen, P.H. 2004, Wind Farms, *Encyclopedia of Energy*, Vol. 6, Ed. Cleveland, Elsevier.

Sukhatme, S.P. 1996, *Solar Energy-Principles of Thermal Collection and Storage*, 2nd Ed., Tata–McGraw Hill.

Staudt, Lawrence 2008, 'Wind Energy' in *Future Energy*, Ed. T.M. Letcher, Elsevier

Tester, J.W. *et al.*, 2005, *Sustainable Energy*, MIT Press/PH India.

"World Wind Energy Report 2010", World Wind Energy Association, March 2011.

17

Hydrogen and Fuel Cells in Search of Viable Green Energy

A HYDROGEN

History of Hydrogen as an Energy Fuel

The recognition of Hydrogen as energy fuel has a celebrated history. In Jules Verne's fictional masterpiece "The Mysterious Island", published in 1874, one of the Character states – "Yes, my friends, I believe that water will one day be employed as a fuel, that hydrogen and oxygen, which constitute it, either singly or together, will furnish an inexhaustible source of heat and light, of an intensity, of which coal is not capable of." In 1874, the journey seemed so near, yet today it is still far, at least yet some distance to go before the tryst reaches its finale. A point should be remembered that hydrogen is available aplenty in the seawater and all organic matters. But hydrogen does not occur as free gas anywhere. Thus for hydrogen to meet its tryst as a fossil fuel, must be able to be produced with expense of energy per kg considerably less than its own specific energy. Secondly, the price of hydrogen fuel should be lower than that of the presently dominant fossil fuel. Hydrogen was first used as a fuel in combination with carbon monoxide (CO), distributed by pipeline and sometimes stored in underground caverns. This mixture of H_2 and CO, known popularly as 'Town gas' were produced by steam reforming of coal and hence also known as coal gas; even today coal gas is distributed in Europe.

I Hydrogen as a Fuel

Energy Density of Hydrogen

On the Lower Heating Value basis, 1 kg of Hydrogen has energy content of 120 MJ compared to only 45 MJ of 1 kg of gasoline. That means that gravimetrically, hydrogen is 2.7 times more energy-rich than gasoline, kg for kg. But at room temperature, hydrogen is the lightest gas known weighing only 0.1 gm per litre, while a litre of liquid gasoline weighs 670 gm; that means that for the same amount of energy, the storage tank of hydrogen has to be (6700 /2.7) = 2450 times larger. This is technically not acceptable. Hence the next proposition is to use liquid hydrogen, which for the same energy content, LH_2 occupies approximately only 4 times more space than gasoline, but weighs one-third of gasoline.

Table 17.1 : Energy Density of Various Types of Fuels

Type of Fuel	*Volumetric Energy Density (KWh/litre)*	*Gravimetric Energy Density (KWh/kg)*
Liquid H_2 (20 K, 1 atm)	**2.3**	**33.3**
H_2 (1500 atm)	2.0	33
H_2 (700 atm)	1.3	33
CH_4 (350 atm)	3.2	14
CH_4 (1000 at)	5.0	14
Liquid CH_4 (112 K, 1 atm)	6.0	14
C_8H_{18} (Gasoline)	**8.7**	**12.3**
CH_3CH_2OH (Ethanol)	6.0	7.4
CH_3OH (Methanol)	4.4	5.5
NiMH Battery ($LaNi_5H_6$)	1.45	0.28
Lead Acid Battery (PbO_2-H_2SO4)	0.84	0.25
Liquid Air (79K, 1 atm)	-	0.22
Compressed Air (700 atm)	0.1	0.16
Compressed Air (350 atm)	0.05	0.14

1 Gallon US =3.8 litres

On a lower heating value (LHV) basis, 3.75 gallons (1 kg) of liquid hydrogen (LH_2) has energy content of 120 MJ (33.3 KWh), same as 1 gallon of gasoline (2.6 kg), making H_2 by far the lightest and hence the least energy dense volumetrically as indicated in Table 17.1; however, gravimetrically LH_2 has highest energy density at 120 MJ/kg. (1 gallon US=3.8 litres)

Liquid Hydrogen (LH_2)

The use of liquid hydrogen has today become routine for launching spacecraft, though liquefying hydrogen is a very energy-intensive process even after decades of industrial experience. The process gets complicated because of the peculiar properties of hydrogen which gets warmer when expanded at ordinary temperature (negative Joule-Thompson effect). Hence the hydrogen is first pre-cooled by a closed-loop LN_2 refrigerator, cyclically compressed to moderate pressure (less than 100 atmospheres) and expanded before liquefying by throttling through a Joule-Thompson expansion valve.

Despite energy intensity of cryogenics, LH_2 is the favoured method of large scale hydrogen storage (e.g. about 270,000 kg LH_2 at Cape Kennedy, energy equivalent to 32 billion Btu). Liquid hydrogen is especially suitable for aircraft, high gravimetric energy density of hydrogen (120 MJ/kg for LH_2 vs. 45 MJ/kg for jet fuel) counterbalances its roughly four times volume.

Typical automotive LH_2 tanks occupy a volume 1.2 to 1.5 times greater and weighs 5 to 10 times more than the LH_2 they contain. One fundamental difficulty is the extreme thermal expansion coefficient of LH_2 at temperature and pressure below the critical point (38 K, 13 atmospheres) where LH_2 and H_2 vapour coexist. LH_2 tanks are usually double-walled alumunium or steel, sandwiching approximately 2 cm (70 layers) of multiple layer insulation in vacuo.

LH_2 is the favoured fuel of fighter aircraft, because of higher gravimetric energy density of hydrogen, despite its higher volume to the order of 4.

Compressed Hydrogen (GH_2)

Alternatively, hydrogen has to be compressed to high pressure to reduce the unusually large space required for its storage.

Thus, the principal alternative to LH_2 is to compress gaseous hydrogen (GH_2). Use of gaseous fuels onboard heavier type vehicle is quite common. In the United States, there are today more than 250 thousand compressed natural gas (CNG) vehicle cylinders. Under ambient condition (1 atm, 300 K), hydrogen has a density of only 0.082 kg/m3. The compression of H_2 up to 250 atm, a standard pressure for CNG vehicle, is thought to be insufficient for GH_2 vehicle. GH_2 at 250 atm is four times bulkier than LH_2 and 15 times bulkier than the energy equivalent volume of gasoline.

Hydrogen as Chemical Hydride

Hydrogen can be packed more densely than the density of liquid LH_2 when it exists as a hydride of a complex material. Recent developments in nano-scale engineering have yielded a host of new high surface-area materials, some with more than 5,000 sq. m. of surface area per gram. In particular, a cheap material, activated carbon, can store up to about 5 times its weight of hydrogen.

Most metal hydrides can contain 90 to 110 kg H_2 per cubic metre that is 60 per cent denser than LH_2. However, hydride storage system can achieve only 50 % volumetric efficiency, because the hydrides are in the powder form and expand when they chemically absorb H_2.

The field of Hydrogen storage materials was expanded from elements to compounds with the discovery of reversible hydride $ZrNiH_3$ by Libowitz in 1958. During 1960s, two prototypes practical hydride LaNiH6 and $TiFeH_2$ were developed to absorb H2 interstitially with low bond energy (about 30 KJ/mol H_2) and to release H_2 between 273 and 373 K and to absorb H_2 quickly at low pressure (1-10 atm).

Storage of Hydrogen Fuel for Transportation

In spite of the problems of being the lightest gas and also being the least energy-intense volumetrically, the hydrogen fortunately offers possibility of different storage configurations applicable to different modes of transportation.

For example, the preferred mode of storage in aircraft, especially for intercontinental and perhaps future hypersonic air travel is in liquid Hydrogen (LH_2) form. In aviation, the advantage of LH_2 lies at its highest energy intensity per kg weight, saving precious weight despite some penalty in volumetric loss.

Economics of scale and infrequent refueling stations recommends LH2 in shipping. For Railways, LH_2 could be preferred fuel.

Hybrid-electric vehicles have been commercially introduced despite some disadvantages on higher capital cost, limited battery capacity and hence range limitation. Now Hybrid-FC vehicles (Fig. 17.1) fed with hydrogen fuel are on the anvil of wider commercial launch. Either technology enables the hybrid-electric or Fuel-celled automobiles to achieve fuel economies equivalent to 60 to 100 miles per gallon (25-40 km per litre). This leap in fuel economy permits H_2 vehicles using as little as 5 kg H_2 on board storage to achieve a 300 to 500 mile driving range.

Decentralised H_2 production and particularly slow overnight fueling recharge would also suit metal hydride storage, easing refueling times and reducing rush for electrolysis. The lighter but higher temperature hydrides (e.g. $NaAlH_4$) are not yet sufficiently developed but the heavier ambient temperature hydrides ($TiFeH_2$) will be increasingly feasible as H_2 vehicle fuel economy increases.

The ultimate shape of the refueling infrastructure and hydrogen energy economy will determine whether LH_2 or GH_2 will be the dominant fuel of the future large commercial truck or the metal hydride storage.

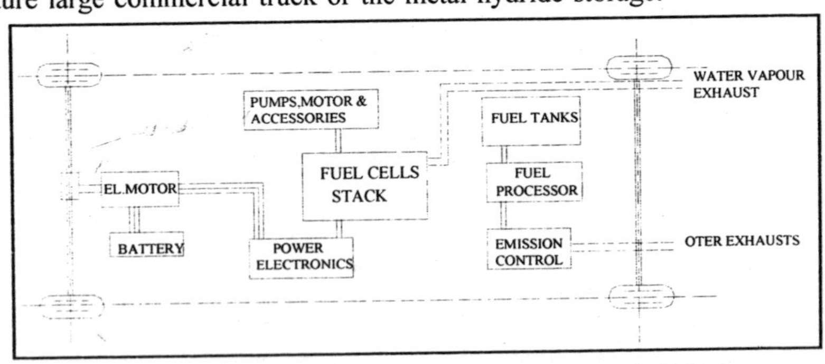

Fig. 17.1 System Outline Diagram of Hybrid FC Vehicle.

Application of Hydrogen as an Energy Fuel: How can hydrogen be harnessed to generate electrical power or any form of energy in general. We may consider here three possible uses of hydrogen among others:

(i) Hydrogen might be used as the combustion fuel for the internal combustion engine just like CNG, bio-fuels and of course petro-fuel. It is believed that IIT (Delhi) is possibly working in this field with some success, albeit low energy density of hydrogen gas. Large-scale power generation by this method is more practical.

(ii) Hydrogen is the ideal gas for Fuel Cells, where hydrogen gets oxidised with the oxygen of the air in presence of a catalyst at some elevated temperature, generating electrical power. The highlight of this electro-chemical process is that energy conversion efficiency is above 70 per cent, about twice that of the thermodynamic cycle.

(iii) Another very highly potential use of hydrogen fuel is possibly in the metallurgical industry like Iron and Steel etc. Huge amount of coal and coke charge is used to reduce the metal oxides to the parent metal and to provide heat energy. This role might admirably be discharged by hydrogen gas without polluting the atmosphere.

II
Production of Hydrogen

Hydrogen is plentiful all over the world in water and almost in all organic matter. But it is not available in free state as oxygen or nitrogen. Thus, hydrogen has to be separated or produced and the most common source is water.

Now we want hydrogen as an energy fuel. Hence the energy, that has to be expended in producing hydrogen, shall not exceed 20 to 25 per cent of the specific energy value of hydrogen. We briefly review some possible production schemes:

(i) Steam Reforming of Coal/ Methane: This is a traditional method of production of Town gas or coal gas. This involves use of fossil fuels or other GHG emitting fuel sources.

(ii) Thermal Dissociation of water: At high temperature water dissociates directly to hydrogen and oxygen. The Indian Atomic Energy Establishment is presently designing a High Temperature Reactor (HTR), operating at 900 to 1100°C, whose prime function is believed to be production of hydrogen with cogeneration of electric power. The process also promises energy economy.

(iii) Three-step Thermo-Chemical Cycle for Water Decomposition: At sufficiently high temperature (> 2500°K), the thermodynamic work for steam decomposition will shrink to zero; steam will decompose spontaneously, theoretically permitting production of hydrogen and oxygen using only thermal energy. But at these high temperatures, the molecules of hydrogen and oxygen themselves become unstable turning into atomic H and O, by reactions that generate more entropy than does steam decomposition.. This presents fundamental difficulty of producing Hydrogen in a single step.

In a three-step reaction (see Fig. 17.2), hydrogen sulphate at about 800°C is decomposed into H_2O, S_2O and O_2 and oxygen is removed. In a second lower temperature reaction (hydrolysis) at 150°C, hydrogen sulphate is regenerated and in the process recaptures Iodine from a third step. The third step at 400°C and 20 atmosphere pressure, hydrogen iodide is decomposed to release hydrogen. The conversion efficiency is low.

(iv) Electrolysis of sea-water: This is an energy-intensive process but perhaps most simple and common. But how this excess quantity of energy will be made available? Not by combusting fossil fuel, for the whole purpose of generating hydrogen will then be defeated. If the process can be energised by solar, wind or nuclear energy, then there is some logic, for the process is really clean and simple.

As mentioned the process is very simple. In the electrolysis cell, the anode terminal will be connected to the positive electricity supply line, while the cathode terminal will be connected to the negative supply. The slightly acidified ionised water will be split into anion oxygen atoms collecting at the anode and cation hydrogen will collect at the cathode. The gases are collected and dried.

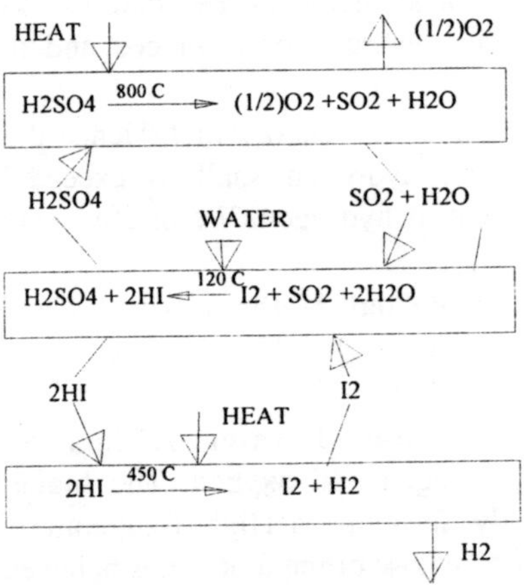

Fig. 17.2 Three-step Thermo-chemical Dissociation for Hydrogen Production.

B
FUEL CELLS

Fuel Cells - Principle of Operation and Types of Fuel Cells

A Fuel Cell is similar to a secondary battery; but a battery has a limited life unless rechargeable; unlike a battery, a Fuel cell can generate direct current as long as the fuel supply to the cell continues.

Thus, a Fuel Cell is an open electrochemical conversion, generating electricity at high efficiency than the electromechanical engine-generator device. During operation, a fuel (e.g. hydrogen) and an oxidant are fed separately at two electrodes anode and cathode respectively of a FC continuously. An electrolyte offers a path for ions resulting generally from the electrochemical reactions to migrate from the fuel (anode compartment) to the oxidant (cathode compartment), while forcing the electrons produced in the process to move through the external circuit and so doing produce electricity. A fuel cell using hydrogen and oxygen as reactants produce electricity, water and some heat. In Fig. 17.3A, we show a Proton exchange PEMC (Proton Exchange Membrane Cell) and in Fig. 17.3B, we show a Cation Exchange Fuel Cell (SOFC).

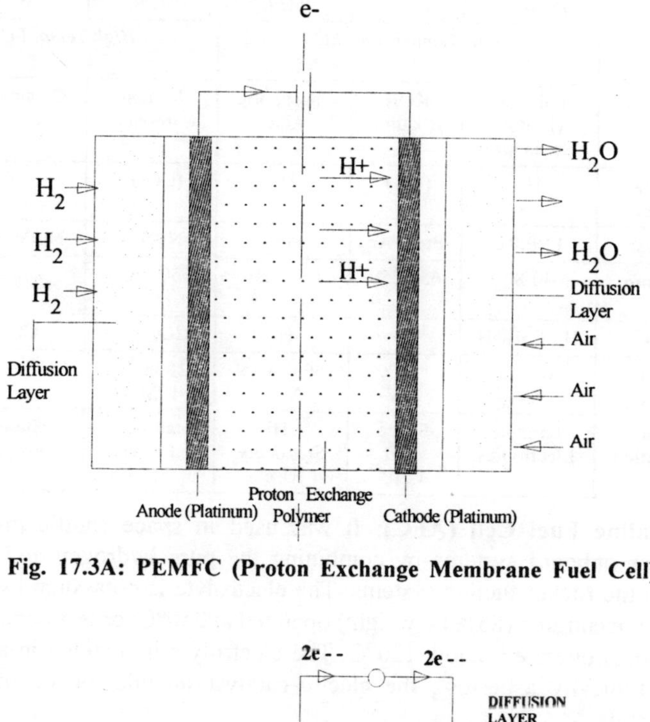

Fig. 17.3A: PEMFC (Proton Exchange Membrane Fuel Cell)

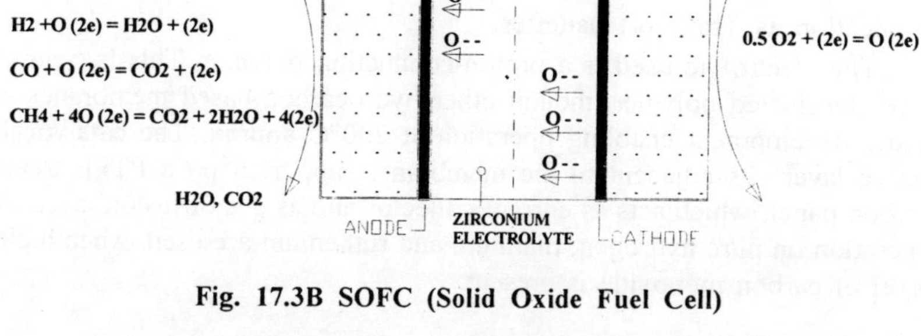

Fig. 17.3B SOFC (Solid Oxide Fuel Cell)

Fuel Cells are categorised into five main types on the basis of electrolytes as shown in Table 17.1. The electrolytes determine many important properties such as the operation temperature range and the appropriate catalyst materials used in the cells. Depending upon the operating temperature the FCs are divided into two categories – low temperature FC (PEMFC, AFC and PAFC) and high temperature FC (MC and SOFC). Of the five types, the PEMFC and SOFC are most studied and used from practical reasons.

Table 17.1: Fuel Cell Types and their Characteristics

Fuel Cell	*PEMFC*	*AFC*	*PAFC*	*MCFC*	*SOFC*
Temp. Range	*Low Temperature FC*			*High Temp. FC*	
Electrolytes	Polymer (Proton exchange)	KOH Alkaline	Phosphoric Acid	Molten Carbonate	Ceramics ZrO_2
Charge Carrier	H^+	$(OH)^-$	H^+	$(CO_3)^{2-}$	O^{2-}
Catalyst	Pt/PtRu	Pt/Ni/Ag	Pt	Ni/NiO	Ni/Perovskites
Op. Temp. C	40-80	65 - 250	150 - 250	550 - 650	600 -1000
Fuels	H_2, CH_3OH	H_2	H_2	CO, CH_4	CO, CH_4
Power Range	1 W – 100 KW	1 -100 KW	50 KW -5 MW	100KW-10MW	1KW-20MW
Main Application	Portable Electronics	Space craft CHP	CHP Stationary Use	Stationary Power Plant	Stationary Power Plant

Alkaline Fuel Cell (AFC): It was used in space shuttle producing power for onboard systems by combining the pure hydrogen and oxygen stored in the rocket fueling systems. The electrolyte is potassium hydroxide (KOH) concentrated (85% by weight) operated at 250°C, or less concentrated (35 to 50%) operated below 120°C. The electrolyte is retained in a porous matrix (typically asbestos), the electro-catalyst include nickel and other noble metals.

Polymer Electrolyte Membrane Fuel Cell (PEMC): Also called Proton Exchange Fuel Cell PEMFC. It has high power density, rapid start up, and low temperature operation (around 80 to 120°C) and is therefore ideal for application as Transport batteries.

The electrolyte used is a proton-conducting polymer. This is typically a perfluorinated polymer, though other hydrocarbon-based membranes are under development enabling operation at 200°C approx. The catalytically active layer sits adjacent to the membrane, supported on a PTFE treated carbon paper, which acts as current collector and as gas diffusion layer. For operation on pure hydrogen, platinum and ruthenium are used, when higher level of carbon monoxide is present.

The present cost of PEMFC system is high, but in future the price will come down with mass manufacturing programme. Demonstration plants of 250 KWe are being produced and used.

Phosphoric Acid Fuel Cell (PAFC): PAFC fuel cells have been successful as stationary co-generation plants with more than 220 commercial plants already delivered. Typical application is in hospital, where the waste heat can be used in the laundry etc.

PAFCs operate at 150-220°C, using 100 % phosphoric acid electrolyte retained in silicon carbide matrix. Generally, platinum electro-catalysts are used in both anode and cathode.

Molten Carbonate Fuel Cell (MCFC): The electrolyte in the MCFC is usually a combination of alkali carbonates, retained in a ceramic $LiAlO_2$ matrix. The operating temperature is 600-700°C, when the alkali carbonate forms a highly conducting molten salt with the carbonate ions providing the means for ionic conduction.

Solid Oxide Fuel Cell (SOFC): SOFC operates at elevated temperatures, generally above 800°C for all ceramic high temperature variants and 600°C for all metallic-ceramic intermediate temperature FC. The electrolyte is a dense ceramic, usually yttria-stabilised zirconium (YSZ), which is an oxide ion conductor at elevated temperature. The cathode is a perovskite material Strontium-doped Lanthanum magnetite, often mixed with YSZ in the form of a composite. The anode is a cernet of nickel and YSZ.

The main difference between HT-SOFC and IT-SOFC lies in (i) the thickness of the electrolyte, which is about 20 microns for IT-SOFC to reduce ionic resistance, (ii) the interconnect material; with stainless steel being used in lower temperature of IT-SOFC.

SOFCs lend themselves to application in which their high temperature can be used. Recently, advances in microelectronics have made possible a combined power plant of 250 KWe with an efficiency approaching 70%.

Fuel-cell powered Automobile: In a developed economy, nearly two-thirds of the fossil fuel is consumed by the automobile transportation, which is then accountable for the highest carbon dioxide pollution. The greatest promise of the hydrogen economy rests on FC-sourced electric drive of automobile, which is capable of eliminating automobile emission effectively. A number of agencies are working on FC car projects in USA. The present status is that compared with a gasoline-driven car with a cruising range of 300 miles and hence carrying 20 gallons of petrol net weight 52 kg, net volume 77.5 litres (gross weight 75 kg including containment), a FC car may presently require to carry 75 gallons, 20 kg net weight of liquid hydrogen, net volume 290 litres (at -253° K, gross weight comparable to gasoline but gross volume much higher for super-insulation and containment). Hence, the development process will need some more years of work to attain the desired technical and cost equivalence with the gasoline car.

Future Outlook: The prospect and potential of Hydrogen fuel for automobile application in particular are high. Only the FC cars can save the earth from the catastrophe of carbon invasion of the earth's atmosphere and can effectively mitigate the global warming. There are a few gaps in developments before the hydrogen economy can unfold.

1. Hydrogen must be produced economically from sea water or otherwise.
2. The problem of storage of hydrogen in a car, either in compressed cylinders, in liquid insulated tank or as solid or liquid hydride, must be solved economically and technologically.
3. The supply chain management of hydrogen must be solved at the same efficient level as the fossil fuel, oil, coal or natural gas.

C
HYDROGEN ECONOMY

A Short Outline

Before we close this chapter, it is as well that we briefly review the versatility of Hydrogen as a fuel of the future. As we know Hydrogen is one of the most abundant elements in the universe like carbon, oxygen and nitrogen. But like those other abundant elements, hydrogen does not occur in the free state on earth. The other superior property of hydrogen is that when burnt as fuel, emits only water vapour and heat and it is one of the cleanest source of energy available on earth.

Hydrogen is also safer to carry than gasoline or natural gas, for it diffuses very fast before it catches fire. On the other hand, hydrogen can catch fire even when its concentration is as low as 2 per cent or as high as 75 per cent compared to natural gas which is inflammable only when the concentration is about 5 – 15 per cent only.

Today only about 5 per cent of hydrogen consumed by industry is merchant hydrogen, produced mostly by steam reforming of natural gas. It is transported either in pipe line or in vessel in compressed condition or in tanks in rail or truck as a cryogenic liquid at a temperature of -253°C, only 20 degrees above the absolute zero. Even today, the airspace industry uses LH_2 for fuel cells aboard manned space station to produce electricity and potable water.

In 1999, Iceland announced a long term plan to become the world's first hydrogen economy by totally eliminating fossil fuel by 2050. Iceland motor vehicles and fishing vessels will run on hydrogen produced by electrolysis of water and electricity will be generated from hydropower and geothermal sources.

In the hydrogen economy there will be very large demand of electricity. The principal mode of generating hydrogen will be by electrolysis. In the electrolysis cell, the anode terminal will be connected to the positive electricity supply line, while the cathode terminal will be connected to the negative supply. The slightly acidified ionised water will be split into anion oxygen atoms collecting at the anode and cation hydrogen will collect at the cathode. However, electrolysis is very costly process, the cost may amount to four times that of hydrogen from steam reforming of natural gas. That is the reason presently only 4 per cent of merchant hydrogen is produced by electrolysis. In the future hydrogen economy, hydrogen is likely to be generated from high temperature nuclear reactor and supplemented by wind and solar.

Hydrogen is the lightest element and that always will pose logistic problem in its storage on automobile. But at the same time hydrogen-fuelled Fuel Cell has by far the highest energy efficiency compared to any fossil-fuelled thermal prime mover driven vehicles.

Perhaps in the ensuing hydrogen economy, it may require every large building to be fitted with a network of solar array and wind turbines operating an electrolysis plant coupled to a hydrogen storage unit and an array of appropriately sized Fuel Cell assembly.

References

Berry, G.D., Hydrogen Production 2004 (p. 253), *Encyclopedia of Energy*, Vol. 3, ed. C.J. Cleveland, Elsevier.

Berry, G.D., Hydrogen Storage and Transportation 2004 (p. 267), *Encyclopedia of Energy*, Vol. 3, ed. C.J. Cleveland, Elsevier.

Brandon, N. 2004, "Fuel Cells", *Encyclopedia of Energy*, Vol. 3, ed. Cleveland, Elsevier.

De, Rosa, A.V. 2009, Fundamentals of Renewable Energy (Ch 9, Fuel Cells, pp. 287-416), Elsevier.

Diccoco, J.M. 2004, Fuel Cell Vehicles, *Encyclopedia of Energy*, Vol. 3, ed. Cleveland, Elsevier.

Dunn, Seth, History of Hydrogen, 2004, *Encyclopedia of Energy*, Vol. 3, Ed. C.J. Cleveland, Elsevier.

Ghosh Roy, M.K., 2010, *Sustainable Development: Environment, Energy and Water Resources* Ane Books, New Delhi.

John, M.G. 2004, Hybrid Electric Car, *Encyclopedia of Energy*, ed. Cleveland, Elsevier.

Nersesian, R.L., 2007, *Energy for the 21st Century*, (Ch 10, *Hydrogen Economy*, p. 331), Pentagon Energy Press New Delhi/M.E. Sharpe.

APPENDIX

Indian Water Resources

"Water is food and fire is the eater of the food.
Fire is established in water and
Water is established in fire."
Taittiriya Upanishad 3.8

India is a land of many rivers and mountains. Its geographical area of about 329 million ha is criss-crossed by a large number of small and big rivers, some of them figuring amongst the mighty rivers of the world. The rivers and mountains have a greater significance in the history of Indian cultural development, religious and spiritual life. It may not be an exaggeration to say that the rivers are the heart and soul of Indian life.

Climate

The presence of the great mountain mass formed by the Himalayas in the North and of the ocean in the South is the two major influences operating on the climate of India. The first poses an impenetrable barrier to the influence of cold winds from central Asia, and gives the sub-continent the elements of tropical type of climate. The second, which is the source of cool moisture-laden winds reaching India, gives it the elements of the oceanic type of climate. India has a very great diversity and variety of climate and an even greater variety of weather conditions. The climate ranges from continental to oceanic, from extremes of heat to extremes of cold, from extreme aridity and negligible rainfall to excessive humidity and torrential rainfall. It is, therefore, necessary to avoid any genralisation as to the prevalence of any particular kind of climate, not only over the country as a whole but over major areas in it. The climatic condition influences to a great extent the water resources utilisation of the country.

Indian Rainfall

Rainfall in India is dependent in differing degrees on the South-West and North-East monsoons, on shallow cyclonic depressions and disturbances and on violent local storms which form in regions where cool humid winds of the sea meet hot dry winds from the land and occasionally reach cyclonic dimension. Most of the rainfall in India takes place under the influence of

South West monsoon between June to September except in Tamil Nadu where it is under the influence of North-East monsoon during October and November. The rainfall in India shows great variations, unequal seasonal distribution, still more unequal geographical distribution and the frequent departures from the normal. It generally exceeds 1000 mm in areas to the East of Longitude 78 degrees. It extends to 2500 mm along almost the entire West Coast and Western Ghats and in the East over most of Assam and Sub-Himalayan West Bengal. On the West of the line joining Porbandar to Delhi and thence to Ferozpur, the rainfall diminishes rapidly from 500 mm to less than 150 mm in the extreme west. The Peninsula has large areas of rainfall less than 600 mm with pockets of even 500 mm. The estimate of a real average rainfall is subjective depending on the method adopted. Therefore, estimates of local rainfall over the country obtained by employing other techniques may differ, especially in a vast country like India.

Monsoon Winds

The rains in India are mostly caused by South-West Monsoon winds. The Monsoon is believed to be an Arabic word meaning winds changing direction. Around end May and early June, when the water gets very warm over India, the South-Eastern Trade winds blowing over the Indian ocean gets deflected eastwards, while crossing over the equator due to the rotation of the earth. This trade wind now becomes South-West monsoon and blows over India carrying high moistures, as indicated in Fig. A.1. A part of the South-West monsoon coming from the Arabian sea sweeps over the west of South India bringing rains. The other part of the same southwest monsoon coming from the Bay of Bengal sweeps over the Eastern India causing rain. Gradually, the Monsoon covers the whole of India by about the second half of June. The S-W monsoon prevails over India June to September.

Indian Geography and patterns of Rainfall

The long Himalayan range extends 2500 km from the east to the west like a garland over the whole of India (see Fig. A.2). The range is believed to be geologically young, having been formed due to tectonic drift of the continent northward. In the South Peninsula, there are number of very old hill ranges:

- Western Ghat extending 1600 km along the west coast, ranging in height up to 2500 m and the hills are as old as 2500 million years;
- Eastern Ghats extending along the east coast from Orissa in the east to Nilgiri in the south, aged about 1300 million years, height ranging from 500 m to 2500 m;
- The Aravalli Range in Rajasthan in the north-western India, about 800 km long, height ranging to 1000 m;
- In the central India, the Deccan plateau consists of Vindhya Hills (1400 million years old) and the Satpura range (1000 million years old);
- In the northeast India, there are several minor ranges including Garo Hills, Jaintea Hills and the Khasi Hills.

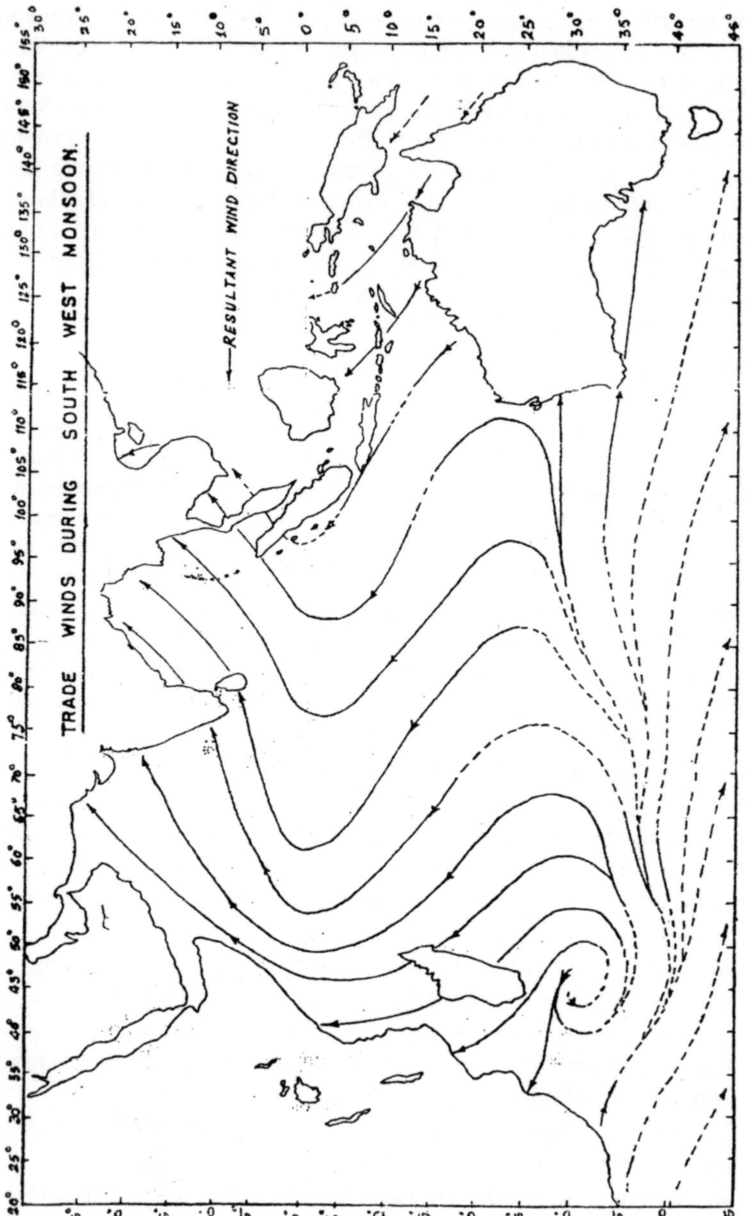

Fig. A.1: Trade Winds and South-West Monsoons.

The alignment of the above hill ranges are shown in Fig. A.2 and also shown in the Figure are the amounts of rainfall in millimeters per annum in both the Lee side and the downside of the hills. It will be seen how profoundly these hills interfere with the flow of the Monsoon winds, causing high rainfall in the lee side and scanty rainfall on the downside.

The highest rainfall of 11,400 mm occurs at Cherapunjee at an elevation of 1330 m in the northeast India. At the other places at Khasi and Jaintea Hills of the Northeast, the rainfall averages 10,000 mm.

* The discourse of this Chapter is based mainly on the seminal work of "*India's Water Wealth*" by Dr. K.L. Rao (1979).

Along the west coast of India on the windward side there occurs quite high rainfall, at places around 6000 mm but on the leeside, the monsoon cloud is obstructed and the rains become scanty, as low as 500 to 600 mm. Such conspicuous differences of rain, of course, do not occur on the plains of India.

Along the Himalaya foothills, the rainfall varies from 1000 mm to 2000 mm around the eastern India.

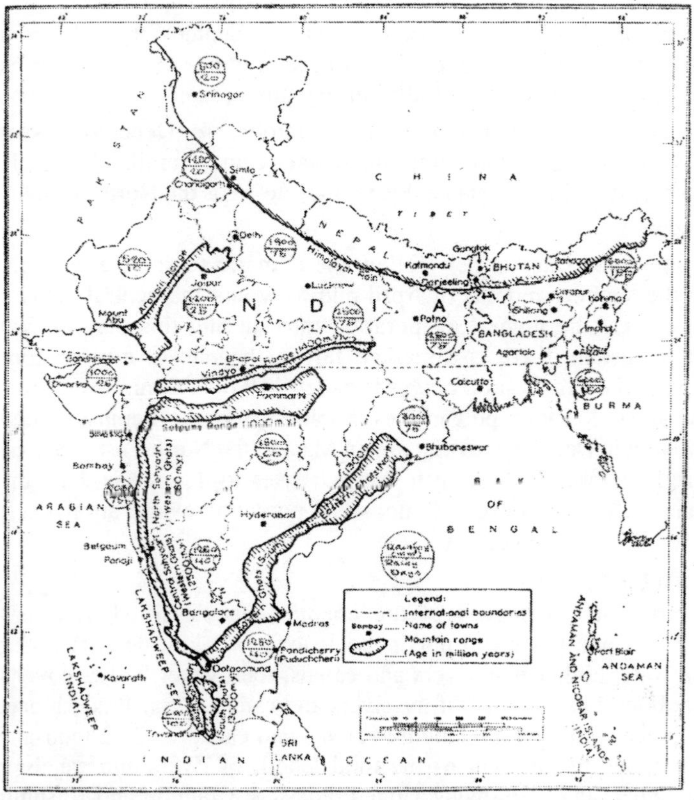

Fig. A.2: Alignment of Mountain Ranges of India and Rainfall Data.

In the eastern India, the S-W monsoon hits the coast hardest and wanes as the monsoon moves inland. In West Bengal and Orissa, the rain averages 1500 mm annually but in Rajasthan, deep in the West, the rain gradually subsides to 500 mm or even lower in the same region.

In Central India above the Vindhya Hill range, the rainfall hovers around 900 mm. The various rainfall figures over India are indicated encircled in Fig. A.2

The rain does not fall continuously even during the monsoon, but only intermittently. In Fig. A.2 are also shown the number of rainy days, again encircled (lower part) for identification. The number of actual wet days varies from 150 days in the North-East, 75 days in the plains of West Bengal to even 10 days in the North-Western deserts.

Seasonal Variation of Rains

During the winter (January-February), it rains in Kashmir and its neighbourhood to about 200 mm which helps the winter crops in the northern India and snowfall in the lower Himalayas.

From March to May, rainfall occurs in Kerala, Assam and also in Jammu and Kashmir.

The Southwest monsoon season, June to September is the principal rainy season in India when over 75 % of the rainfall occurs all over. Severe floods also occur in Eastern India during this season.

During the post-monsoon period, October to December, southern Peninsula including Andhra Pradesh, Rayalaseema, Tamilnadu and Kerala receive some rain. In Tamilnadu, this rain is known as the Northeast monsoon.

Evaporation

Evaporation rates closely follow the climatic seasons, and reach their peak in the summer months of April and May and the central areas of the country display the highest evaporation rates during this period. With the onset of monsoon, there is a marked fall in the rate of evaporation. The annual potential evaporation ranges between 150 to 250 mm over most parts of the country. Monthly potential evaporation over the Peninsula increases from 150 mm in December to 400 mm in May. In the North-East, it varies from 6 mm in December to 20 mm in May. It rises to 40 mm in June in West Rajasthan. After the onset of monsoon potential evaporation decreases generally all over the country.

Inland Water Systems

Inland Water resources of the country are classified as rivers and canals, reservoirs, tanks and ponds, beels, oxbow lakes, derelict water; and brackish water. Other than rivers and canals, total water bodies cover gross areas of about 7 million ha. Of the rivers and canals, Uttar Pradesh occupies the first place with the total length of rivers and canals as 31.2 thousand km, which is about 17 per cent of the total length of rivers and canals in the country. Other states following Uttar Pradesh are Jammu and Kashmir and Madhya Pradesh. Among the remaining forms of the inland water resources, tanks and ponds have maximum area (2.9 million ha. followed by reservoirs 2.1 million ha.).

Most of the area under tanks and ponds irrigation lies in Southern States of Andhra Pradesh, Karnataka and Tamil Nadu. These states along with West Bengal, Rajasthan and Uttar Pradesh, account for 62 per cent of total area under tanks and ponds irrigation in the country.

As far as reservoirs are concerned, major states like Andhra Pradesh, Gujarat, Karnataka, Madhya Pradesh, Maharashtra, Orissa, Rajasthan and Uttar Pradesh account for larger portion of area under reservoirs. More than 77 per cent of area under beels, oxbow, lakes and derelict water lies in the states of Orissa, Uttar Pradesh and Assam. Orissa ranks first as regards the total area of brackish water and is followed by Gujarat, Kerala and West Bengal. The total area of inland water resources is, thus, unevenly distributed over the country with five states namely Orissa, Andhra Pradesh, Gujarat, Karnataka and West Bengal accounting for more than half of the country's inland water bodies.

Indian River Systems

Major Rivers

There are fourteen major rivers in India. Of these, three rivers Indus (west-flowing to Pakistan to Arabian Sea), Ganga and Brahmaputra are of

Table A.1: Discharges of Major River Basins

Sl. No.	*Name of Rivers*	*Area (sq.km.)*	*Annual Discharge Million cu.m.*
1	The Indus	(468,068) 321,289	(207,800) 41,955
2	The Ganga	(1,050,000) 861,404	493,400
3	Brahmaputra	(580,000) 187,113+70,895	510,450
4	Sabarmati	21,674	3,200
5	Mahi	34,842	8,500
6	Narmada	98,796	40,705
7	Tapi	65,145	17,982
8	Subarnarekha	19,296	7,940
9	Brahmani	39,033	18,310
10	Mahanadi	141,589	66,640
11	Godavari	312,812	105,000
12	Krishna	258,948	67,675
13	Pennar	55,213	3,238
14	Cauvery	87,900	20,950
	Total	2,575,949	1,405,945

Note: The figures in brackets indicate total for the river basin including parts in neighbouring countries.

Himalayan origin and situated in North India above the Tropic of Cancer (23.5° latitude). Between the Tropic of Cancer and the 20° latitude, there are seven rivers including three east-flowing Subarnarekha, Brahmani and Mahanadi and four west-flowing Sabarmati, Mahi, Narmada and Tapi. In peninsular India below 20° latitude, there are again four east-flowing rivers Godavari, Krishna, Pennar and Cauvery. The geographical distribution of rivers is shown on the map of India (Fig. A.3). The catchment areas and the discharges of these rivers are also presented in Table A.1.

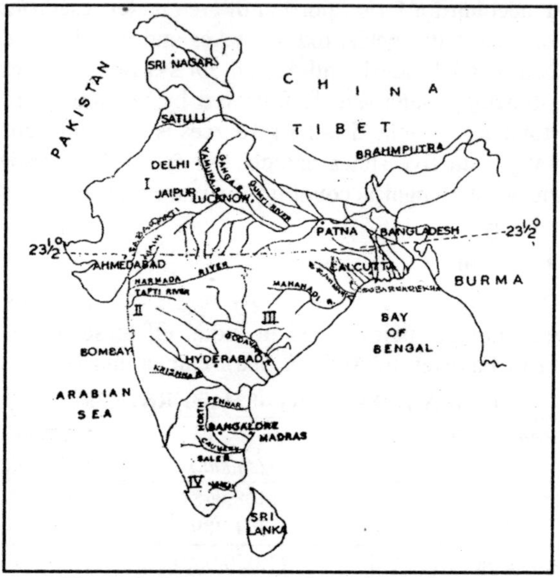

Fig. A.3. Fourteen Major Rivers of India

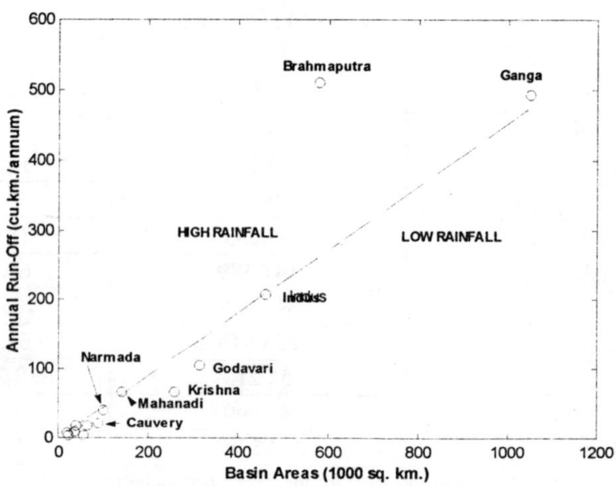

Fig. A.4. Catchment Areas and Flow of Major Rivers

From this Table A.1, it may noted that 14 major rivers have each catchment area of about 20,000 sq. m. or more; that is the reason of labeling them as major rivers. The other point to note is that rivers, which flow through areas of higher rainfall, have usually higher discharge, while the others have lower discharges. This situation is diagrammatically in Fig. A.4 to a base of the catchment areas.

Table A.2 gives total available water, consisting of both surface water and ground water for each major basin and also estimated per capita water availability in such basins.

Table A.2: Water Availability in different River Basins

River Basins	*Average Annual Surface Water (bcm)*	*Reple-nishable Ground Water (bcm)*	*Per Capita Surface Water (bcm)*	*Per capita Surface + ground Water (bcm)*
Indus	73.31	26.55	1749	2383.30
Brahmaputra-Barak	585.60	206.07	16,589	17,612.66
Ganga	525.02	171.00	1471	1950.73
Godavari	110.54	40.60	2048	2799.93
Krishna	78.12	26.40	1285	1719.64
Cauvery	21.36	12.30	728	1147.63
Subar narekha	12.37	1.8	1307	1497.89
Brahmani-Baitarini	28.48	4.05	2915	3329.58
Mahanadi	66.88	16.50	2513	3134.59
Pennar	6.32	4.39	651	1159.79
Mahi	11.02	-	1052	1052
Sabarmati	3.81	-	360	360
Narmada	45.64	10.80	3109	3869.46
Tapi	14.88	8.27	1007	1564.19
West-Flowing Rivers				
Tapi to Tadri	87.41	28.90	3383	3387.98
Tadri – KannyaKumari	113.53	-	3480	3480
Kutch,Saurastra (Luni)	15.10	-	683	683
East-Flowing Rivers				
Mahanadi to Godavari	22.52	-	953	953
Pennar- Kannyakumari	16.46	-	366	366
Iinland drainage in Rajasthan	0.0		0.0	
Minor Rivers drain to Bangladesh/Myanmar	31.00	-	14623	14623

Source: Indian Water Resource Society 1997.

Medium and Minor Rivers

Rivers, which have individual catchment area generally below 20,000 sq. m. but above 2,000 sq. m., are identified as medium rivers. In Table A. 3 are listed 44 such medium rivers flowing out to the east coast or the west coast, excepting the rivers in the North-East flowing into the neighbouring countries. The medium rivers generally flow within a coastal State; hence they are grouped shown by States. Their total discharges constitute less than 10 per cent of the total countrywide run-off.

Table A.3: Discharges of Medium Rivers

Sl. No	*Name of River*	*Source Indian State*	*Length*	*Area sq km*	*Annual Disch-arges (million cu m)*
West Flowing Rivers					
1	Shetrunji	Village Dalkania	182	5,5514	280
2	Bhadra	Rajkot Dist.	198	7,094	350
3	Dhadhar	Panchmaqhal Dist.	135	2,770	690
4	Burhabalang	Mayurbhanj Dist.	164	4,837	2,177
5	Baitarni	Keonjhar Dist.	365	12,789	5,755
6	Purna	Dhulia Dist.	142	2,431	1,093
7	Ambika	Dhangs District	142	2,715	1,247
8	Vaitarna	Nasik Dist.	126	2,572	4,350
9	Ulhas	Pune Dist.	122	4,637	3,014
10	Savitri	Kolaba Dist.	80	2,257	1,467
11	Mandavi.	Belgaum Dist.	87	2,0332	1,320
12	Kalindi	Belgaum Dist.	153	5,179	6,537
13	Gangavati/Bedti	Dharwar	152	3,902	4,925
14	Sharavati	Shimoga	122	2,209	4,545
15	Netravati	Canara	103	3,657	4,615
16	Ghalia/Beypore	Elamtalvi Hills	169	2,788	5,200
17	Bharatpuzha	Anamali Hills	251	5,397	8,800
18	Periyar	Sivsjini Hills	228	5,243	12,300
19	Pamba	Sivsjini Hills	177	1,961	6,300
East Flowing Rivers					
20	Rushikulya	Phulabani Dist.	146	7,753	1,800
21	Yamasadhara	Phulbani	221	10,830	3,500
22	Nagavali	Kalahandi	217	9,410	2,400
23	Sarda	Vishakhapat. Dist.	104	2,725	700
24	Yeleru	Vizag Dist.	125	3,809	9,000
25	Gundlakamma	Kurnool Dist.	220	8,494	1,000
26	Musi	Nellore Dist.	112	2,219	260
27	Paleru	Nellore Dist.	104	2,480	300
28	Munru	Nellore Dist.	122	3,734	450
29	Kunleru	Velikonda Dist.	73	3,534	420
30	Swarnamukhi	Palka	130	3,225	500
31	Kortalyaiar	Chingelpet	131	3,521	350

(contd...)

Sl. No	Name of River	Source Indian State	Length	Area sq km	Annual Disch-Arges(million cu m)
32	Palar	Kolar Dist.	348	17,871	1,780
33	Gingee	South Arcot	94	3,044	300
34	Ponnaiyar	Kolar Dist.	396	16,019	1,600
35	Vellar	Chithri Hills	193	8,558	850
36	Vaigai	Madurai District	258	7,741	770
37	Varshalli	Madurai Dist.	125	2,104	310
38	Cundar	Madurai Dist.	146	4,838	480
39	Vaippar	Tirunelveli Dist.	130	5,288	530
40	Tamarpani	Tirunelveli Dist.	130	5,482	1.370
Rivers flowing into other countries					
41	Karnaphuli	Mizoram	144	3,999	2,600
42	Kaldan	Mizoram	290	7,933	5,160
43	Imphal	Manipur	-	7,255	4,716
44	Tixu Nanitaluk	Nagaland	148	6,449	4,200
	Total			239,302	112, 211

Table A.4: Discharges of Minor Rivers

S. No.	Indian States	Catchment Area Sq. km.	Run-off Million cu m
1	Gujarat	4,634	1,160
2	Karnataka, Goa, Maharashtra	2,146	966
3	Kerala	19,489	40,073
4	West Bengal, Orissa	10,000	6,500
5	Andhra Pradesh	44,142	9,746
6	Tamil Nadu	16,162	1,625
	Total	96,537	60,088
	Other Minor Rivers	109,000	57,000
	Total	205,537	117,088
Desert rivers:			
	Luni & Other rivers	80,000	
	The Macha	2,515	
	The Rupen	4,000	
	The Saraswati	2,934	
	The Banas	8,674	
	The Ghaggar	1309	
	Others	1,309	
Total (Desert Rivers)		99,432	
Grand Total of Run-off for Minor and Desert rivers			127,000

A third category of rivers, identified as minor rivers have very limited catchment areas of 2000 sq m or below each flows relatively short distances from the coastal hills (Eastern Ghat or Western Ghat as the case may be) to the sea. Table A.4 shows the estimated Statewise total flows emanating from the minor rivers in each State.

There are some minor desert rivers (Rao, 1975), which are also listed along with the minor rivers in the Table A.4.

In Table A.5 are presented the totals of all discharges and catchment areas of all Indian rivers.

Table A.5: Details of Catchment Areas of 4 Groups of Rivers

<table>
<tr><th>River Basins</th><th>Catchment Million sq km</th><th>Total area %</th><th>Run-off Cu km</th><th>Run-off %</th><th>Popul-ation %</th></tr>
<tr><td>Group I: Major River Basins</td><td>North 1.44
Central 0.42
South 0.72
Total 2.58</td><td>83%</td><td>1,406</td><td>85%</td><td>80%</td></tr>
<tr><td>Group II: Medium</td><td>0.24</td><td>8%</td><td>112</td><td>7%</td><td rowspan="3">20%</td></tr>
<tr><td>Group III: Minor</td><td>0.20</td><td rowspan="2">9%</td><td rowspan="2">127</td><td rowspan="2">8%</td></tr>
<tr><td>Group IV: Desert Rivers</td><td>0.10</td></tr>
<tr><td>Total</td><td>3.12</td><td>100%</td><td>1,645</td><td>100%</td><td>100%</td></tr>
</table>

Groundwater Systems

Groundwater Resources of the Country

The ground water resources of the country have been estimated for freshwater as shown in Table A.6. The total annual replenishable ground water resources of the country have been estimated as 433 billion cubic meter (bcm). Keeping 34 bcm for natural discharge, the net annual ground water availability for the entire country is 399 bcm (Table A.6).

The annual ground water draft is 231 bcm out of which 213 bcm is for irrigation use and 18 bcm is for domestic and industrial use.

Indian Ground Water Exploration by CGWB

Ground water exploration aided by drilling is one of the major activities of the Central Ground Water Board (CGWB) with an objective to discover aquifers in different hydro-geological conditions and determination of hydraulic parameters. Large-scale sub-surface exploration programme for ground water was initiated during 1954. In the initial years, exploratory drilling activities were confined to alluvial tracts in major river basins and sub-mountainous bouldary tracts of Himalayan foothills. In mid eighties, CGWB added 26 new DTH drilling rigs in its fleet with which the exploratory drilling in hard rock regions gained momentum. The major thrust of exploratory drilling programme in nineties was in areas underlain by hard rock. Another important development in first half of nineties was introduction of open hole drilling technology in India. CGWB acquired seven percussion drilling rigs for exploratory drilling in bouldary/ semi-consolidated formations in Himalayan foothills from Jammu and Kashmir in northwest and to Arunachal Pradesh in north east. These exploration programmes formed the background of scientific evaluation of the water bearing properties of various rock formations.

About 29,500 wells have been drilled by Central Ground Water Board throughout the country. High yielding wells were drilled for underground water exploration programme in water deficient areas in the country, including tribal and drought prone areas. Most of these high yielding wells have been handed over to the respective State Governments for public water supply. Board had also come forward in disaster mitigation activities during Latur earthquake during 1993, Bhuj earthquake during 2001, and Super cyclone in Orissa during 2000 and Tsunami hit coastal belt of Tamil Nadu and Kerala and Andaman and Nicobar Islands during 2004 by way of construction of tube wells for water supply.

Status of Ground Water Development

The stage of ground water development in the country is 58%.The development of ground water in different areas of the country has not been uniform. Highly intensive development of ground water in certain areas in the country has resulted in over exploitation leading to decline in the levels of ground water and sea water intrusion in coastal areas. Out of 5723 numbers of assessment administrative units (Blocks/Taluks/ Mandals/ Watershed), 839 units are "overexploited", 226 units are "critical", 550 units are "semi-critical", 4078 units are "safe" and 30 units are "saline".

Artificial Recharge and Rainwater Harvesting

In order to augment the depleting ground water resources, it is essential that the surplus monsoon runoff that flows into the sea is conserved and recharged to augment ground water resources. Ground water storage that could be feasible has been estimated as 214 billion cubic meters (bcm) of which 160 bcm is considered retrievable. Central Ground Water Board has prepared a conceptual plan for artificial recharge of ground water for the country. Out of total geographical area of 3,287,263 sq. km. of the country, an area of 448,760 sq. km. has been identified suitable for artificial recharge. The total quantity of surplus monsoon runoff to be recharged works out as 36.4 bcm.

The Central Ground Water Board has prepared a Manual and subsequently a Guide on Artificial Recharge to Ground Water which provides guidelines on investigation techniques for selection of sites, planning and design of artificial recharge structures, economic evaluation and monitoring of recharge facility. These are of immense use to States/ U.T.s in planning

Table A.6: Statewise Groundwater Availability

Sl. No.	States/ Union Territories	Annual Replenishable Ground Water Resources					Natural Discharge during non-Monsoon	Net Annual Ground Water Availability	Stage of Ground Water Development
		Monsoon Season		Non-Monsoon Season		Total			
		Recharge from rainfall	Recharge from other sources	Recharge from rainfall	Recharge from other sources				
	Units	← Billion Cubic Metre →							%
	STATES								
	Andhra Pradesh	16.04	8.93	4.20	7.33	36.50	3.55	32.95	45
	Arunachal	1.57	0.0001	0.98	0.0002	2.56	0.26	2.30	0.04
	Assam	23.65	1.99	1.05	0.54	27.23	2.34	24.89	22
	Bihar	19.45	3.96	3.42	2.36	29.19	1.77	27.42	39
	Chattisgarh	12.08	0.43	1.30	1.13	14.931.	1.25	13.68	20
	Delhi	0.13	0.06	0.02	0.09	0.30	0.02	0.28	170
	Goa	0.22	0.01	0.01	0.04	0.28	0.02	0.27	27
	Gujarat	10.59	2.08	0.00	3.15	15.81	0.79	15.02	76
	Haryana	3.52	2.15	0.92	2.72	9.31	0.68	8.63	9.10
	Himachal	0.33	0.01	0.08	0.02	0.43	0.04	0.39	30

(contd...)

(contd...)

Sl. No.	States/ Union Territories	Annual Replenishable Ground Water Resources					Natural Discharge during non-Monsoon	Net Annual Ground Water Availability	Stage of Ground Water Development
		Monsoon Season		Non-Monsoon Season		Total			
		Recharge from rainfall	Recharge from other sources	Recharge from rainfall	Recharge from other sources				
	Units	← Billion Cubic Metre →							%
	Jammu Kashmir	0.61	0.77	1.00	0.32	2.70	0.27	2.43	14
	Jharkhand	4.26	0.14	1.00	0.18	5.58	0.33	5.25	21
	Karnataka	8.17	4.01	1.50	2.25	15.93	0.63	15.30	70
	Kerala	3.79	0.01	1.93	1.11	6.84	0.61	6.23	47
	Madhya Pradesh	30.59	0.96	0.05	5.59	37.19	1.86	35.33	48
	Maharastra	30.59	2.51	1.94	8.36	32.96	1.75	31.21	48
	Manipur	0.20	0.005	0.16	0.01	0.38	0.04	0.34	0.65
	Meghalaya	0.79	0.03	0.33	0.005	1.15	0.12	1.04	0.18
	Mizoram	0.03	0.00	0.02	0.00	0.04	0.004	0.04	0.90
	Nagaland	0.28	0.00	0.08	0.000	0.36	0.04	0.32	3
	Orissa	12.81	3.56	3.58	3.14	23.09	2.08	21.01	18
	Punjab	5.98	10.91	1.36	5.54	23.78	2.33	21.44	145
	Rajasthan	8.76	0.62	0.26	1.92	11.56	1.18	10.38	125

(contd...)

(contd...)

Sl. No.	States/ Union Territories	Annual Replenishable Ground Water Resources					Natural Discharge during non-Monsoon	Net Annual Ground Water Availability	Stage of Ground Water Development
		Monsoon Season		Non-Monsoon Season		Total			
		Recharge from rainfall	Recharge from other sources	Recharge from rainfall	Recharge from other sources				
	Units	← Billion Cubic Metre →							%
	Punjab	5.98	10.91	1.36	5.54	23.78	2.33	21.44	145
	Rajasthan	8.76	0.62	0.26	1.92	11.56	1.18	10.38	125
	Sikkim	-	-	-	-	0.08	0.00	0.08	16
	Tamilnadu	4.91	11.96	4.53	1.67	23.07	2.31	20.76	85
	Tripura	1.10	0.00	0.92	0.17	2.19	0.22	1.97	9
	Uttar Pradesh	38.63	11.95	5.64	20.14	76.35	6.17	70.18	70
	Uttaranchal	1.37	0.27	5.44	4.86	30.36	2.90	27.46	66
	West Bengal	17.87	2.19	5.44	4.86	30.36	2.90	27.46	42
	TOTAL STATES	**247.86**	**69.51**	**41.83**	**73.15**	**432.42**	**33.73**	**398.70**	**58**
	Union Territories								
	Andaman Nicobar	-	-	-	-	0.33	0.005	0.32	4
	Chandigarh	0.016	0.001	0.005	0.001	0.023	0.002	0.02	0
	DadraNagarhaveli	0.059	0.005	-	-	0.063	0.0003	0.008	14

(contd...)

(contd...)

Sl. No.	States/ Union Territories	Annual Replenishable Ground Water Resources					Natural Discharge during non-Monsoon	Net Annual Ground Water Availability	Stage of Ground Water Development
		Monsoon Season		Non-Monsoon Season		Total			
		Recharge from rainfall	Recharge from other sources	Recharge from rainfall	Recharge from other sources				
	Units	←			Billion Cubic Metre			→	%
	Chandigarh	0.016	0.001	0.005	0.001	0.023	0.002	0.02	0
	DadraNagarhaveli	0.059	0.005	-	-	0.063	0.0003	0.008	14
	Daman Diu	0.006	0.002	0.00	0.001	0.009	0.0004	0.008	107
	Lakshadweep	-	-	-	-	0.012	0.009	0.004	63
	Pondicherry	0.057	0.067	0.007	0.029	0.16	0.016	0.144	105
	TOTAL UTS	0.138	0.075	0.012	0.031	0.597	0.036	0.556	33
	GRAND TOTAL	**248.0**	**69.59**	**41.85**	**73.19**	**433.02**	**33.77**	**399.25**	**58**

Source: Ministry of Water Resources, Government of India.

and implementation of recharge schemes for augmentation of ground water in various parts of the country.

During the Ninth Five Year Plan, a Central Sector Scheme on "Studies on Recharge of Ground Water" was undertaken, under which 165 recharge projects were implemented in 27 States/UTs. The impact assessment of completed recharge projects before monsoon of 2003 had indicated rise in water levels and sustainability of dug wells/ tube wells, decrease in soil erosion and improvement in social-economic status of farmers of benefited zone due to increase in crop production.

During the X Plan demonstrative studies on Artificial Recharge to Ground Water and Rain Water Harvesting have been taken up during 2006-07 by CGWB in 8 districts in Andhra Pradesh, Madhya Pradesh, Karnataka and Tamil Nadu at a total cost of Rs. 5.95 Crores. The construction of 189 artificial recharge structures out of 200 has been completed and remaining is under construction. The impact assessment studies of the completed recharge projects are being conducted by CGWB.

During the XI Plan, for undertaking demonstrative recharge project, CGWB and state agencies in coordination are coming out with studies for feasible areas, structures and other modalities. The detailed project reports for Artificial Recharge projects received from Tamil Nadu, Punjab, Kerala, Andhra Pradesh and Arunachal Pradesh are under consideration.

Inter-Basin Transfer of Water – National Water Grid

National Water Development Agency (NWDA)

It was set up in July, 1982 as an Autonomous Society under the Societies Registration Act, 1860, to carry out the water balance and other studies on a scientific and realistic basis for optimum utilisation of Water Resources of the Peninsular rivers system for preparation of feasibility reports and thus to give concrete shape to Peninsular Rivers Development Component of National Perspectives. In 1990, NWDA was also entrusted with the task of Himalayan Rivers Development Component of National Perspectives. Recently, the functions of NWDA have been further enlarged and the work of preparation of detailed Project Reports (DPR) of various link proposals and Pre-feasibility Reports and feasibility reports of intra-State links as proposed by the States have been included in the functions of NWDA.

Criticality of Water Availability and Need of Inter-basin Water Transfer

The rainfall over the country is primarily associated with tropical depressions originating in the Arabian Sea and the Bay of Bengal. The summer monsoon accounts for more than 85 per cent of the precipitation. The uncertainty of occurrence of rainfall marked by prolonged dry spells and fluctuations in seasonal and annual rainfall is a serious problem for the country. Large parts of Haryana, Maharashtra, Andhra Pradesh, Rajasthan, Gujarat, Madhya Pradesh, Karnataka and Tamil Nadu are not only deficit in rainfall but also subject to large variations, resulting in frequent droughts and causing immense hardship to the population and enormous loss to the nation. The water availability even for drinking purposes becomes critical, particularly in the summer months as the rivers dry up and the ground water recedes. Regional variations in the rainfall lead to situations when some parts of the country do not have enough water even for raising a single crop. On the other hand, excess rainfall occurring in some parts of the country creates havoc due to floods. Irrigation using river water and ground water has been the prime factor for raising the food grain production in our country from a mere 50 million tonnes in the 1950s to more than 200 million tonnes at present, leading us to attain self sufficiency in food. Irrigated area has increased from 22 million hectares to 95 million hectares during this period. The population of India, which is about 1200 million at present, is expected to increase to about1500 million in the year 2050 and that would require about 400 million tonnes of food grains. For meeting this requirement, it would be necessary to increase irrigation potential to 160 million hectares for all crops by 2050. India's maximum irrigation potential that could be created through conventional sources has been assessed to be about 140 million hectares. For attaining a potential of 160 million hectares, other strategies shall have to be evolved. Floods are a recurring feature, particularly in Brahmaputra and Ganga rivers, in which almost 60 per cent of the river flows of our country occur. Flood damages, which were Rs. 52 crores in 1953, have gone up to Rs. 5,846 crores in 1998 with annual average being Rs. 1,343 crores affecting the States of Assam, Bihar, West Bengal and Uttar Pradesh along with untold human sufferings. On the other hand, large areas in the States of Rajasthan, Gujarat, Andhra Pradesh, Karnataka and Tamil Nadu face recurring droughts. As much as 85 percentage of drought prone area falls in these States. One of the most effective ways to increase the irrigation potential for increasing the food grain production, mitigate floods and droughts and reduce regional imbalance in the availability of water is the Inter Basin Water Transfer (IBWT) from the surplus rivers to deficit areas. Brahmaputra and Ganga particularly their northern tributaries, Mahanadi, Godavari and West Flowing Rivers originating from the Western Ghats are found to be surplus in water resources. If we can build storage reservoirs on these rivers and connect them to other parts of the country, regional imbalances could be reduced significantly and lot of benefits by way of additional irrigation, domestic and industrial water supply, hydropower generation, navigational facilities etc. would accrue.

Earlier Proposals

Suggestions for a National Water Grid for transferring surplus water available in some regions to water deficit areas have been made from time to time. The two such proposals put forth earlier in the seventies, which attracted considerable attention, were:

- National Water Grid by Dr. K.L. Rao (1972)

- Garland Canal by Capt. Dastur (1977)

Dr. K.L. Rao's Plan (1972), which had 2640 km. long Ganga - Cauvery link as its main component involved large scale pumping over a head of 550 m. The power requirement for lifting the water was huge, estimated to be 5000 to 7000 MW, for irrigating an additional area of 4 million hectares only. The scheme was also not having any flood control benefit. Dr. Rao had estimated this proposal to cost about Rs. 12,500 crores, which at 2002 price level comes to about Rs.150,000 crores. The Central Water Commission, which examined the proposal, found it to be grossly under estimated and economically prohibitive.

Capt. Dastur's Proposal (1977) envisaged construction of two canals – the first 4200 km. Himalayan Canal at the foot of Himalayan slopes running from the Ravi in the West to the Brahmaputra and beyond in the east; and the second 9300 km Garland Canal covering the central and southern parts, with both the canals integrated with numerous lakes and interconnected with pipelines at two points, Delhi and Patna. The cost estimated by Capt. Dastur was Rs. 24,095 crores. The proposal was examined by two committees of experts comprising Senior Engineers from CWC, State Governments, Professors from the IIT, Delhi and Roorkee University and Scientists from Geological Survey of India and Indian Meteorological Department who opined that the proposal was technically infeasible. The cost estimated by the experts in 1979 was about Rs. 12 million crores. The realistic cost at 2002 price level comes to about Rs. 70 million crores.

Surface Water Availability – All India Scenarios

From Table A.2, we see that the per capita availability of surface water is excessively high in the Brahmaputra basin; in many other basins, it is quite sufficient, being about 2000 cu.m. per capita per annum, and water is scarce or at stress level in many basins, e.g. Cauvery, Pennar, etc. Again we observe from Fig. A.2 that rainfall is highly uneven on all India basins. The task is therefore to identify the scarcity areas on regional bases and to examine if it is technically feasible to transfer water from the nearest surplus basin rivers to balance out the scarcity over the geographical spread of India. The second task is to store water in sufficient quantities from existing low percentage (see Table A.7) to minimum twenty five per cent of the annual run-off. These measures will demand heavy investments in fund, organisation and time of few decades, but once implemented can relieve India of potential scarcity for much of twenty first century.

We will now briefly outline the National Water Grid system, which was conceived by the National Water Power Commission several decades ago but as yet remains unexecuted because of its gigantic scale and lack of political consensus; but also to a lesser extent on issues of availability of fund and doubt of technical feasibility.

Table A.7: Storages existing or under construction in various Basins

<table>
<tr><th>S. No.</th><th>Basins</th><th colspan="2">Storage Million cu m</th><th>Percentage of Basin Run-off</th></tr>
<tr><td>1</td><td>Indus</td><td colspan="2">14,419</td><td>34</td></tr>
<tr><td>2</td><td>Ganga</td><td colspan="2">33,476</td><td>7</td></tr>
<tr><td>3</td><td>Bramhaputra</td><td colspan="2">142</td><td>-</td></tr>
<tr><td>4</td><td>Sabarmati</td><td colspan="2">1,017</td><td>33</td></tr>
<tr><td>5</td><td>Mahi</td><td colspan="2">4,286</td><td>56</td></tr>
<tr><td>6</td><td>Narmada</td><td colspan="2">2,550</td><td>6</td></tr>
<tr><td>7</td><td>Tapi</td><td colspan="2">8,148</td><td>45</td></tr>
<tr><td>8</td><td>Subarnarekha</td><td colspan="2">283</td><td>4</td></tr>
<tr><td>9</td><td>Brahmani, Baitarini, Rishikulya etc</td><td colspan="2">3,953</td><td>22</td></tr>
<tr><td>10</td><td>Mahanadi</td><td colspan="2">and,926</td><td>12</td></tr>
<tr><td>11</td><td>Godavari</td><td colspan="2">14,858</td><td>15</td></tr>
<tr><td>12</td><td>Krishna</td><td colspan="2">29,860</td><td>44</td></tr>
<tr><td>13</td><td>Pennar</td><td colspan="2">1,9and8</td><td>61</td></tr>
<tr><td>14</td><td>Cauvary</td><td colspan="2">5,4and8</td><td>26</td></tr>
<tr><td>15</td><td>West Flowing Rivers (Kanyakumari to Tapi, Medium & Minor rivers)</td><td>14,121</td><td rowspan="3">18,452</td><td rowspan="3">8</td></tr>
<tr><td>16</td><td>East Flowing Rivers of AP. & Tamilnadu (medium & Minor Rivers)</td><td></td></tr>
<tr><td>1and</td><td>Rivers of Saurashtra & Kutch (Luni & other minor rivers)</td><td>1,810</td></tr>
<tr><td colspan="2">Total</td><td colspan="2">146,826</td><td></td></tr>
<tr><td></td><td>Storage in Tanks and small ponds</td><td colspan="2">15,400</td><td></td></tr>
<tr><td colspan="2">Total</td><td colspan="2">162,226</td><td></td></tr>
</table>

Proposed Network of National Water Grid

Considering the locations of the surplus rivers systems in the North to the deficient river systems in the South, it could be inferred that the general direction of the main arteries of the Grid should be north to south. Secondly, most of the rivers in India either flow to the East or the West; the North-south alignment of the major grid artery will be thus able to connect a number of rivers. With the above broad concepts of the grid, the inter-river linkages can be visualised as follows:

1. Ganga-Cauvery Link connecting the Ganga in the North with the Cauvery in the South and passing through the basins of the Son, Narmada, Tapi, Godavari, Krishna and Pennar.
2. To further boost the flow in Ganga, a link from Brahmaputra to Ganga

through Bangladesh.

3. Canal from Narmada to Gujarat and West Rajasthan, inter-connecting rivers Mahi, Sabarmati and Luni and which are already under construction.
4. Canal from the Chambal river to Central Rajasthan, which has been also implemented in some form;
5. A number of network link canals from the surging rivers of the western Ghat rivers to the East.

The above schemes of inter-linking canal network have been superimposed in Fig. A.5. Another important aspect of inter-basin transfer will be to build a series of storages and reservoirs along the arterial routes to feed the southern basins during the dry seasons. In any case, such reservoirs may be necessary to install major pumping stations to over geodetic elevations.

Proposed Ganga-Cauvery Link

Ganga has good perennial flow near Patna and downstream of the two tributaries Ghagra and Gandak. Taking the average annual discharge of Ganga from Table A.1 at 493,400 million cu m, at an equivalent flow of 15,646 cumecs (cubic metre per second), the National Grid of the CWPC planned to undertake the following inter-connections:

(i) To draw 1700 cumecs of water from Ganges downtown of Ghagra and Gandak for 150 days;

(ii) This drawal of 1700 cumecs (cubic metre per second) amounted to 10.9 % of the Ganga average flow of 15,646 cumecs. Further, this drawal of 1700 cumecs for 150 days amounted to a diversion of 22,032 million cu m annually, which amounts to 4.47 % of the Ganga discharges of 493,400 million cu m.

(iii) Out of this withdrawal of1700 cumecs from Ganga, 290 cumecs will be pumped to draught-prone areas of South Uttar Pradesh and south Bihar, which are, of course, situated within the Ganga basin itself.

(iv) There is a further parallel proposal of diverting 1150 cumecs from Brahmaputra near Dhubri through Bangladesh to upstream off Farakka barrage on the Ganga. That means the Ganga basin will lose very little water by diversion to the southern grid.

(v) The diversion route of the grid will be such that water from Ganga will be lifted along the Son river to the proposed Bagri reservoir on Narmada to an elevation of + 423 m. With a total lift of +400 m to overcome the Vindhya mountain range, the grid canal will flow along Son, Narmada, Wainganga and Godavari over a distance of 960 km. From here the canal has to traverse another 960 to the Cauvery River. The last leg of the canal can be aligned after detailed survey.

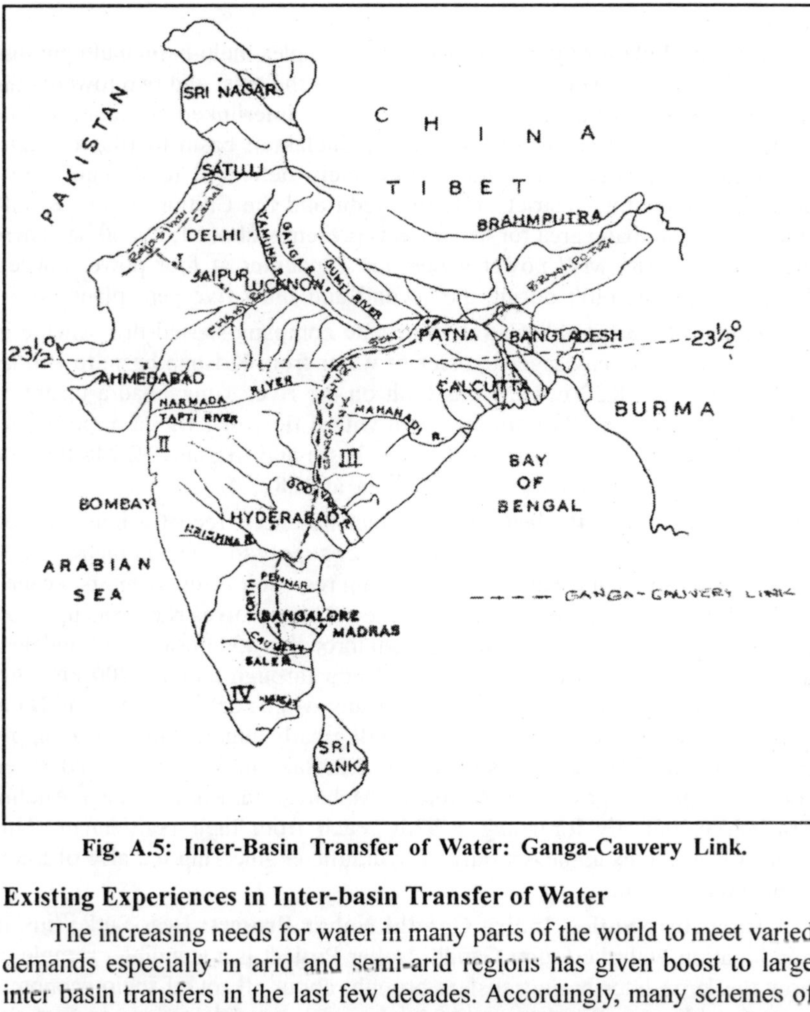

Fig. A.5: Inter-Basin Transfer of Water: Ganga-Cauvery Link.

Existing Experiences in Inter-basin Transfer of Water

The increasing needs for water in many parts of the world to meet varied demands especially in arid and semi-arid regions has given boost to large inter basin transfers in the last few decades. Accordingly, many schemes of large scale water transfer projects have been planned and some of them implemented and are standing as a landmark for the overall development of the respective regions. A few of them are briefly described here.

Periyar Project: The project is the most notable endeavor of the 19th century in trans-basin diversion. The project envisages transfer of water from Periyar basin to Vaigai basin. A masonry gravity dam of 47.28 m high has been constructed across a gorge on west flowing Periyar River. A 1,740 m long tunnel with a discharging capacity of 40.75 cumecs has been driven across the mountain barrier to convey the water eastwards to Vaigai basin. The project was commissioned in 1895 and provided irrigation to 57,923 ha initially, which has since been extended to 81,069 ha. There is also a power station of capacity 140 MW.

(contd...)

Parambikulam Aliyar: The project is a complex multi-basin multi-purpose project of seven streams, five flowing towards the west and two towards the east, have been dammed and their reservoirs interlinked by tunnels. The project envisages transfer of water from Chelakudi basin to Bharatapuzha and Cauvery basins. The water is ultimately delivered to drought prone areas in Coimbatore district of Tamil Nadu and the Chittur area of Kerala states. The command area for irrigation is presently about 1,62,000 ha. There are a total of 185 MW power generation capacities at four power houses. This project was built during the second and third five year plans.

Kurnool Cudappah Canal: A private company started this scheme in 1863. The project envisages transfer of water from Krishna basin to Pennar basin. An anicut 8.23 m high was built on the river Tungabhadra upstream of Kurnool town. A 304 km long canal with a flow of 84.9 cu m per sec at its head extends from Krishna to Pennar basin and irrigates 52,746 ha. The scheme was taken over by Govt. of India in 1882.

Telugu Ganga Project: This project has been recently implemented primarily to meet the pressing need of water supply to Chennai metropolitan area. It brings Krishna water from Srisailam reservoir through an open canal, first to Somasila reservoir in Pennar valley. This involves rock cuts up to 35 m deep. From Somasila, the water is taken through a 45 km canal to Kandaleru and then to Poondi reservoir in Tamil Nadu through another 200 km long canal. By mutual agreement, 12 TMC of water will be delivered to Tamil Nadu at the border from Krishna basin. This will greatly augment the water supply to Chennai city. The canal also irrigates 2.33 lakh ha in Andhra Pradesh en route. The project was made possible by Maharashtra, Karnataka and Andhra Pradesh voluntarily foregoing 5 TMC each from their entitlement. This project is a fine example not only of hydraulic engineering but also of Inter-State cooperation.

Ravi-Beas-Sutlej- Indira Gandhi Nahar Project: Beas-Sutlej link in combination with the Indira Gandhi Nahar Project is a standing example of how the large inter basin transfers brought about all round socio-economic growth with overall enhancement in the ecology and environment of the region. Under the Indus Water Treaty, the water of three eastern rivers viz. Sutlej, Beas and Ravi were allocated to India. As the land to be benefited in India, lies mostly to the east and south of these rivers, the rivers had to be interlinked and the water conveyed to canal systems for serving vast tracts in India. The main storage on Sutlej is at Bhakra, while that on Beas is at Pong. Bhakra system provides irrigation to 26.3 lakh ha. of new area besides stabilisation of existing irrigation of 9 lakh ha. The aggregate generation capacity of power on Bhakra Nangal Project is 1,354 MW. A diversion dam, Pondoh, 140 km upstream of Pong on Beas, enables diversion of water from Beas to Bhakra reservoir and generates 165 MW of power. The Beas-Sutlej link is 37.25 km long of which 25.45 km is tunnel through difficult rock formations. The capacity of the tunnel is 254.70 cumecs(cu.m. per sec).

Another dam on Ravi namely, Ranjit Sagar dam will provide additional water to Beas and also generate a large block of power. Subsequently, it was decided to link the Indira Gandhi Nahar Project with the river systems to provide 9.36 bcm of water to Rajasthan Canal for irrigating the areas of Thar Desert.

It is no exaggeration to say that the transfer of surplus waters of Ravi, Beas and Sutlej to Rajasthan right up to Jaisalmer and Barmer through Indira Gandhi Nahar Pariyojana has eliminated drought conditions and provided power benefits to the entire North grid, transformed desert waste land into an agriculturally productive area by bringing irrigation and vegetation to about 2 million hectare area. Contribution in agricultural production due to implementation of the project is worth Rs. 1,750 crores annually. Canal water is also available for meeting domestic needs. The Indian military at western boundary receive water from this canal. The project has miraculously changed the living standard and socio-economic conditions of the people in the area.

Inter Basin Water Transfers in Other Countries: Many large-scale water transfer schemes have been planned and implemented in other countries also. In Canada, sixteen inter basin water transfer schemes have been implemented for hydropower development.

In United States, the California's State Water Project, first phase of which was completed in 1973, provides for the diversion of 4 cubic km of flow from better-watered northern California to the drier central and southern parts of the State. The conveyance system comprises of 715 km long California Aqueduct, a complex system of lined and unlined canals, pumping stations, siphons and tunnels. The lift involved is nearly 1,000 m.

The Texas Water Plan envisages redistribution of water in Texas and New Mexico to meet the needs of the year 2020.

Similarly, the waters of the Colorado River (an international river between USA and Mexico) are being supplied outside the basin to the Imperial valley in California.

Major existing and under construction inter basin transfers in Canada include Kemano, Churchill Diversion, Well and Canal, James Bay, Churchill Falls, Bay d' Espoir etc. Proposed inter basin transfers in Canada include Ogoki, Long Lake (for transfer within Canada) and North American Water and Power Alliance (NAWAPA), Grand Canal Concept, Canadian Water, Magnum Plan, Central North American Water Project (CNAWP), Smith Plan etc. for transfer from Canada to USA.

In Mexico, for the Mexico City water supply, transfer of ground waters from the Lerma basin was completed in 1958. The Water Plan for the North Western Region (PLHINO) conceived a set of inter basin transfers. .

Mahaveli-Ganga Project of Srilanka includes several inter basin transfer links.

Inter basin transfer projects have also been planned and implemented in China and former USSR. A notable scheme executed in the USSR is Irtysh

Karganda scheme in the central Kazakhstan. The link canal is about 450 km long with a maximum capacity of 75 cumec. Lift involved is 14 to 22 m. There is another plan to transfer 90,000 million cu.m. of water from the north flowing river to the area in south. Other proposals include partial redistribution of water resources of northern rivers and lakes of European part to the Caspian Sea basin involving 2 M. Ha. m of water.

The Lingua Canal was completed in China in 214 BC and the Grand Canal was completed in 605 AD. Recently completed projects in China include Biliuha - Dalian inter basin water supply system, Trans basin transfer of Luhana River to Tiajian and Tengshan, Inter basin diversion of Guanglong province and Inter basin diversions of Fujian province. Diversion of Quiantang river water, diversion of Yellow river surpluses and South to North transfer projects with the West route, Middle route and East route are other proposed projects.

References

Ground Water; http://wrmin.nic.in/index3.asp?subsublinkid=8318&langid1 &sslid=419

Inter-Basin Transfer of Water; http://www.gov.in/outerwin, php2id=http:/nwda.gov.in

Rao, K.L., 1979, *India's Water Wealth*, Orient Longman, New Delhi.

Surface Water; http://wrmin.nic.in/index3.asp?subsublinkid=8208&langid =l&sslid=331

Index

A

ACC Ltd, 325
Acceler Mittal India, 112
Agency for Non-Conventional Energy and Rural Technology, 325
Agenda-21, 37, 111, 184
Agro-Forestry, 63
Albedo, 170-171
Alexander F. Yuan, 26
Anand, 162
Anil Kakodkar, Dr. 280-281
Anthropogenic Warming, 174
Agriculture and Processed Food Products Expert Development Agency, 74-76
Aquaculture,
  Developments, 91
  Freshwater, 82
  Global, 88
  Integrated Coastal, 82, 84
  Production, 92
  Systems, 91
  World, 87
Aravali Range, 345
Atomic Energy Commission
  Al Gore, 180

B

Bali Action Plan, 103, 183-185
Ban Ki Moon, 61, 184
Banerjee, 38
Basin,
  Amazon, 143, 150
  Amu Darya, 150
  Bharatapuzha, 366
  Brahmputra-Barak, 142, 153, 157-158
  Cauvery, 157, 366
  Chelakudi, 366
  Ganga, 157-158
  Godavari, 158
  Indus River, 150
  Krishna, 366
  Krishna-Godavari, 245-246
  Luni, 158
  Mahanadi, 158, 245
  Pennar, 157-158, 366
  Periyar, 365
  Sabarmati, 158
  Salt River, 142
  Syr Darya, 150
  Tisza River, 142
  Vaigai, 365
  Yellow River, 150
Baumgartner, 133
Beas-Satluj Link Project, 258
Bell-shaped relationship, 92
Ben Jerry's Ice-cream, 123
Berlin Mandate, 182, 190
Bhakra-Beas Management Board, 161
Bio-diesel, 177, 229, 291-294, 300, 303-305
Bio-diversity, 33-34, 63, 149

Bio-Fertilizers, 71, 74
Bio-mass, 95, 289-291, 294, 297, 330
Bio-metric identification-
  Universal Identification Device, 50
Bio-pesticides, 71
Bio-process, 41
Bio-resources, 40
Bio-technology, 41-44, 79
Biswas, A.K., 148, 151
Bonem, 138
Botanical Survey of India, 98
Brahmaputra Flood Control Board, 161
Brita Filters, 123
  Brundtland-
  Gro Harlem, 10, 36
  Report, 10, 36-37, 58, 110-111, 184
Budyko, 128

C

Caltech, 25
Canal,
  Grand, 367
  Garland, 362
  Kurnool Cudappah, 366
  Lingua, 368
Cancun Agreements, 200
Cape Kennedy, 335
Capt. Dastur's Proposal, 362
Caspian Sea, 253, 368
Cattle Development Centre, 117
Cauvery Delta, 157
Cellulose, 298, 300
Central Electricity Act, 2003, 220
Central Ground Water Board, 161, 354-355
Central Pollution Control Board, 98
Central Statistical Organisation, 64
Central Water Commission, 161
Central Zoo Authority, 98
Centre for Forestry Research and Human Resource Development, 98
Centre for Social Forestry and Eco-habitation, 98
Chandi Prasad Bhatt, 100
Chicago Centre for Green Technology, 23
Child Hunger, 54
China Light Power, 324
Chipko Movement, 100
Churchill-
  Diversion, 367
  Falls, 367
Civil society, 34
Civilion Nuclear Treaty, 280
Clean coal Technology, 233
Clean Development
  Mechanism, 63, 102, 104-105, 112, 119, 183, 185, 191
  Management Credit, 63
Coal Bed Methane, 247-248
Coal India Ltd., 220
Copenhagen Accord, 183, 186, 188, 196
Cosgrove, 145

D

Da Rosa, 302
Dabhol, 246
Dahej Terminal, 246
Dam,
  Aswan High, 149
  Bhakra, 258
  Koyna, 148
  Pandoh, 366
  Kremastha, 148
  Ranjit Sagar, 367
Damodar Valley Corporation, 161
Daniel Davies, 25
David King, Sir, 205
Deccan Plateau, 345
Dhruv, 283

Diesel, Prof., 295
Directorate of Forest Education, 98
Diurnal, 310
Donthi N. Reddy, 75

E

Eastern Ghats, 345, 354
Echoworld, 21, 28-29
Enercon, 323
Energy Poverty, 217
Energy,
  Crisis, 210
  Demand, 210, 218
  policy, 204, 207
  Scenario, 209-211, 216-217, 229
  Security, 215, 219
Ennore LNG , 246

F

FAO Codex Standards, 75, 81
Farming,
  Dry, 147
  Organic, 20, 11
  Rain-fed, 147
Fish Farmers Development Agencies, 82
Fishery Survey of India, 83, 98,
Fishing Harbour, 82
Food Security Act, 49-50
Forest Advisory Committee, 107
Forest Conservation Act, 98, 100, 106-108
Forest Research Institute, 98
Forest Rights Act, 2006, 100, 113
Forest Stewardship Council, 34
Forest, Hinglajgarh, 93
Forestry,
  Agro, 96
  Community (Rural), 96
  Extension (Urban), 96
  Farm, 95-96, 107, 110, 119
  Global, 102
Fossil-
  Energy, 229
  Fuel, 240, 332-333
Francis Check, 42
Fujitsu,-
  Environment care, 28
  Green Procurement Policy, 28
Fukushima Daichi, 286

G

G.B. Pant Institute of Himalayan Environment and Development, 98
Gandhian dictum, 19
Ganga Flood Control Commission, 161
Garibay, S.V. 72, 75
Gartner, 125
Gasohine driven car, 341
Geiger, 288
Geological Survey Report, 219
Global
  Forest Policy, 101
  Hunger Index, 49-50, 54
  Hunger, 54
  Nuclear Fuel Market, 265
  Warming, 102,165-167, 169, 175-176, 178
Goals, Millennium Development, 58-61, 146, 155, 184
Goddard Institute for Space Sciences, 178
Green
  Company, ITC, 27
  Agenda, 111
  Agriculture, 62, 74-75, 76
  Agriculture, 76
  Building, 21, 23, 27
  Business, 122-123
  Corporates,
    Tata, ITC and Fujitsu, 26
    Miljo Greenland As, 26
  corporates, 110, 114

energy, 12-13, 23-24, 26, 204, 221, 307, 319, 333
Governance, 122
Outputs Market, 71
Startegies, 124
Technology, 11-12, 20-21, 31
thinking,
Corporates changing to, 111
the philosophy of, 19
The principle of sustainable, 19
The principles of, 20
Procurement Policy, 28
World, 12-13
Gujarat State Petroleum Corp., 245
Gulf of Mexico, 253, 329

H

Haile, Professor Sessina,
of the Califormia Institute of Technology (Caltech), 25
Hazira LNG , 246
Hebei, 25-26
Heliostat, 309
Hemicellulase, 298
Henry Hub, 242
Hills,
Garo, 345
Jaintea, 345, 347
Khasi, 345, 347
Nagda, 325
Vindhya, 345, 348
Himalayas, 345, 347
Holistic Philosophy, 44
Homi Bhabha, Dr., 279
Hydel Scheme,
Yamuna, 258
Sharavath, 258
Hydraulic Fracturing, 250
Hydrologic Cycle, 127-128

I

International-
Food Policy Research Institute, 49
Iodine deficiency disorders, 57-58
Energy Agency, 207, 279
Federation for organic Agricultural Movement, 74
Nuclear Agency, 213
Indira Gandhi National Forest Academy, 98
Indian Council of Forestry Research and Education, 98
Institute of Forest Genetics and Tree Breeding, 98
Infosys Technology, 112
ITC-Welcome Group, 112, 114, 118-121
Green Centre at Gurgaon, 121
Green Corporate Giant, 114
INOX Group, 325
Indian
Atomic Energy Establishment, 337
Renewable Energy Development Agency, 326
Water Resource Society, 351
Water Treaty, 366

J

James Bay, 367
James Mason, 314
James Watson, 42
Jan Martel, 44
John Young, 44
Joule-Thompson effect, 334
Jules Pretty, 63
Jules Verne, 333
Jyoti, K., 72, 75

K

Ken Zweibel, 314
Kiev, 285
Klass, 292
Kutch, 324
Kyoto Protocol, 102-103, 105, 110, 180-186, 190, 192, 319,

L

Lake Nasser, 148
Land Acquisition Act, 106
Leonardo da vinci, 125
Libowitz, 335
Lungs of the earth, 32
Luvovich, 133
Law, Stepham-Boltzmann, 169

M

Mackerron, 204
Malnutrition,
 Protein-energy, 53
 Micro-nutrients, 57
Mariculture, 84
Mariner, 312
Mark A. Delucchi, 330-331
Mark Z. Jocobson, 330-331
Marrakech Accord, 191
Maru-ship System, 91
Mercaptan, 242
Metabolic Products, 41
Mike Valenti, 21
Mohina Rowland, 175
Mojave desert, 308
Mongilo, 75
Monju, 275
Montreal Protocol, 175, 177, 192
Mordel Plata, 126, 131
Mukti fields, 245
Muller-Steinhagen, 309
Mumbai High Complex, 245-246
Murray, R.L., 267, 270

N

NABARD, 116
Nano electric Vehicle, 26
 Bacteria, 44
 Scale, 43-44
 Technology, 43
National
 Commission on Agriculture, 94
 Fisheries Development Board, 83-84
 Food Security Mission, 65
 Forest Act, 1980, 99
 Forest Policy, 98-99
 Horticultural Mission, 66
 Institute of Animal Welfare, 98
 Mission for Enhanced Energy Efficiency, 222-223
 Mission on Sustainable Agriculture, 67, 73
 Museum of Natural History, 98
 Organic Standards, 74
 Research Council on Climate Change, 178
 Thermal Power Corp, 220
 Water Develolpment Agency, 161
 Water Policy, 155, 159, 162
 Water Power Commission, 362
 Zoological Park, 98
Nature at Service, forests, 31, 33
Neil Bowdler, 24
Nevada, 316
New York Times, 184
Niko Resources, 246
Nokia India, 111
Nuclear Power Corp. of India, 282
Nuclear Power Parks, 280

O

Oak Ridge National Laboratory, 317

P

Panchayat (Extension to the Scheduled Areas) Act, 108,

Panna, 245
Paul Polman, 112-113
Pawan Wadhwa, 70, 76
Petersen, 321
Petronet, LNG, 246
Plan,
  Magnum, 367
  Smith, 367
  Zweibel, 314-315
POSCO, Korean Steel Company, 108
  Carbon, 120
  Water, 120

R

Raheman and Mondol, 304
Rain Water Harvesting, 360
Rainforest Alliance, 22
Rao, K.L., Dr., 361-362
Rare Earth Metals, 25
Rashtriya Krishi Vikas Yojna, 65-66
Reliance, 246, 248
Renewable Resources, 40
Reservoir,
  Bhakra, 366
  Koyana, 324
  Poondi, 366
  Somasila, 366
Rio de Jenerio, 37, 111, 180, 184
Rio Declaration, 37
Rio Meet, 111
River,
  Basin, 349, 351
  Beas, 366-367
  Brahmani, 350, 363
  Brahmaputra, 140, 349, 361-364
  Cauvery, 350, 362-365
  Chambal, 364
  Colorado, 143, 367
  Denube, 140
  Elbe, 140
  Ganga, 140, 153, 349, 361-364
  Ghagra, 364
  Godavari, 350, 361, 363-364
  Indus, 143, 349, 363
  Kemano, 367
  Krishna, 350, 363
  Luhana, 368
  Luni, 364
  Mahanadi, 350, 361, 363
  Mahi, 350, 363-364
  Mississippi, 140
  Narmada, 350, 363-364
  Ohio and Rhine, 132, 140
  Pennar, 350, 362-363
  Periyar, 365
  Ravi, 362, 366-367
  Rijsberman, 145
  Sabarmati, 350, 363-364
  Seine, 140
  Son, 364
  Subernarekha, 350, 363
  Sutlej, 366-367
  Tapi, 350, 363
  Tapti, 245
  Tengsan, 368
  Tiajian, 368
  Volga-Don, 140
  Wainganga, 364
  Yellow, 143, 150, 368

S

Sandler, 168
Satpura range, 345
SBSTA-24, 104-105
SBSTA-25, 104-105
Scheduled Tribe and other Traditional Forest Dwellers (Recognition of Forest Right) Act, 108
Schneiker, 45
Science Reporter, 24
Scientific American India, 330
Scrase, 204
Seed Village, 67
Seeds Act, 1966, 67
Sen, A.K., 219
Shale, and total, 246

Barnet, 250
Fayetteville, 250
Haynesville, 250
Marcellus, 250
Mithene Gas, 248-255
gas, 248-251
Utica, 250
Shell, 209
Shriram, 323
Skylab, 312
Solar Century,
British Photovoltaic Company, 25
Strategy of Management,
Tripple Bottom Line, 27, 114, 118-119
Stern Report, 205
Sunderlal Bahuguna, 100
Supercritical Evaporation Technology, 13
Sustainable Development, 36, 44
Suzlon, 322-324, 326,
Syngas, 25

T

Tata Power, 324
Technological Evolution, 40
Thales Miletus, 125
The Central Institute of-
Coastal engineering for Fisheries, Bengaluru, 83
Fisheries, National and Engineering Training, Kochi, 83
Three Mile Island, 278-279, 319, 329
Town Gas, 333
Toyota,
World Auto Giant, 123
Turbine,
Francis, 259
Kaplan, 259-260
Pelton, 259

U

UN Commission on Environment and Development, 184
Environment Programme, 31, 179
Framework Convention on Climate Change 63, 102-103, 105, 112, 118-119, 180-186, 189-190, 192, 200, 218, 240, 319
Millennium Developmental Goals, 137, 158, 146, 155
Reducing Emissions from Deforestation and Forest Degradation, 34
Standing Committee on Nutrition, 55
World Food Programme, 50-53
Upper River Yamuna Board, 161
US Environmental Protection Agency, 44
US Green Building Council, LEED, 22
US High Commissioner for Refugees, 56

V

Vasites Fthenakis, 314
Valley,
Bhakra Nangal, 142
Damodar, 142
Euphrates, 152
Hirakund, 142
Hwang HO River, 152
Indus, 152
Narmada, 152
Nile, 152
Tigris, 152
Vankusawada Wind Park, 324
Vindhya Mountain Range, 364
Von Weizsacker, 122
Vision-2025, 141, 144-145
Vienna convention, 175

W

Water-
Crisis, 146

Demand, 134, 139, 158
Development, 142, 161
Dispute Act, 160
Harveshing, 47, 63, 116, 121
Institutional Frames Work, 160
Quality, 127, 134, 147
Resources, 125, 130, 134, 142, 145, 149, 152-153, 160, 162-164
Rights, 162
Security, 125, 141, 152
Stocks, 127, 129
Stress, 145-146, 158
Supplies, 133, 137, 156, 161
Vision, 144
wealth, 347
Western Ghat, 345, 354, 361
Wien's Law, 169
Wild Life Institute of India, 98
William Dillon, Dr., 251
Wind-
Energy turbine, 26
energy, 25, 214
farm, 325, 329
Smile company of Japan, 25
smile generators, 25
Wipro Technologies, 112
Wollman, 138
World Commission on environment and development, 36
Energy Outlook, 206-208
Energy Policy, 205
Food Summit, 57, 77
Hunger, 49, 53
Nuclear Association, 263, 265
Nuclear Energy, 275
Resource Institute, 48, 158
Wind Energy Association, 327
Wuppertal Institute, 122
Wyman, 293

Y

Year of Forests (2011), 31, 33,

Z

Zoological Survey of India, 98
Zero-grasing, 63